Spring（第2版）
源码深度解析

郝佳 编著

人民邮电出版社

北京

图书在版编目（CIP）数据

Spring源码深度解析：第2版 / 郝佳编著. -- 北京：人民邮电出版社，2019.1
ISBN 978-7-115-49914-1

Ⅰ. ①S… Ⅱ. ①郝… Ⅲ. ①JAVA语言－程序设计 Ⅳ. ①TP312.8

中国版本图书馆CIP数据核字(2018)第244988号

内 容 提 要

本书从核心实现、企业应用和 Spring Boot 这 3 个方面，由浅入深、由易到难地对 Spring 源码展开了系统的讲解，包括 Spring 整体架构和环境搭建、容器的基本实现、默认标签的解析、自定义标签的解析、bean 的加载、容器的功能扩展、AOP、数据库连接 JDBC、整合 MyBatis、事务、SpringMVC、远程服务、Spring 消息、Spring Boot 体系原理等内容。

本书不仅介绍了使用 Spring 框架开发项目必须掌握的核心概念，还指导读者使用 Spring 框架编写企业级应用，并针对在编写代码的过程中如何优化代码、如何使得代码高效给出了切实可行的建议，从而帮助读者全面提升实战能力。

本书语言简洁，示例丰富，可帮助读者迅速掌握使用 Spring 进行开发所需的各种技能。本书适合于已具有一定 Java 编程基础的读者，以及在 Java 平台下进行各类软件开发的开发人员、测试人员等。

◆ 编　著　郝　佳
　　责任编辑　傅道坤
　　责任印制　焦志炜

◆ 人民邮电出版社出版发行　北京市丰台区成寿寺路 11 号
　邮编　100164　电子邮件　315@ptpress.com.cn
　网址　http://www.ptpress.com.cn
　北京七彩京通数码快印有限公司印刷

◆ 开本：800×1000　1/16
　印张：28.25　　　　　2019 年 1 月第 1 版
　字数：615 千字　　　2024 年 12 月北京第 20 次印刷

定价：99.00 元

读者服务热线：**(010)81055410** 印装质量热线：**(010)81055316**
反盗版热线：**(010)81055315**
广告经营许可证：京东市监广登字20170147号

作者简介

郝佳，计算机专业硕士学位，曾发表过多篇论文并先后被 EI、SCI 收录；2008 年辽宁省教育厅科技计划项目研究人之一；长期奋斗于 J2EE 领域，2013 年入职阿里巴巴，目前担任业务中间件软件架构师；一直专注于中间件领域，拥有 6 项技术专利，擅长系统的性能优化；热衷于研究各种优秀的开源代码并从中进行总结，从而实现个人技能的提高，尤其对 Spring、Hibernate、MyBatis、JMS、Tomcat 等源码有着深刻的理解和认识。

前言

源代码的重要性

Java 开发人员都知道，阅读源码是非常好的学习方式，在我们日常工作中或多或少都会接触一些开源代码，比如说最常用的 Struts、Hibernate、Spring，这些源码的普及与应用程度远远超过我们的想象，正因为很多人使用，也在推动着源码不断地完善。这些优秀的源码中有着多年积淀下来的精华，这些精华是非常值得我们学习的，不管我们当前是什么水平，通过反复阅读源码，能力都会有所提升，小到对源码所提供的功能上的使用更加熟练，大到使我们的程序设计更加完美优秀。但是，纵观我们身边的人，能够做到通读源码的真的是少之又少，究其原因，不外乎以下几点。

- 阅读源码绝对算得上是一件费时费力的工作，需要读者耗费大量的时间去完成。而作为开发人员，毕竟精力有限，实在没办法拿出太多的时间放在源码的阅读上。
- 源码的复杂性。任何一款源码经历了多年的发展与提炼，其复杂程度可想而知。当我们阅读源码的时候，大家都知道需要通过工具来跟踪代码的运行，进而去分析程序。但是，当代码过于复杂，环环相扣绕来绕去的时候，跟进了几十个甚至几百个函数后，这时我们已经不知道自己所处的位置了，不得不再重来，但是一次又一次地，最终发现自己根本无法驾驭它，不得不放弃。
- 有些源码发展多年，会遇到各种各样的问题，并对问题进行了解决，而其中有些问题对于我们来说甚至可以用莫名其妙来修饰，有时候根本想不出会在什么情况下发生。我们查阅各种资料，查询无果后，会失去耐心，最终放弃。

无论基于什么样的原因，放弃阅读源码始终不是一个明智的选择，因为你失去了一个跟大师学习的机会。而且，当你读过几个源码之后就会发现，它们的思想以及实现方式是相通的。这就是开源的好处。随着各种开源软件的发展，各家都会融合别家优秀之处来不断完善自己，这样，到最后的结果就是所有的开源软件从设计上或者实现上都会变得越来越相似，也就是说当你读完某个优秀源码后再去读另一个源代码，阅读速度会有很大提升。

以我为例，Spring 是我阅读的第一个源码，几乎花费了近半年的时间，其中各种煎熬可想而知，但是当我读完 Spring 后再去读 MyBatis，只用了两周时间。当然，暂且不论它们的复杂程度不同，至少我在阅读的时候发现了很多相通的东西。当你第一次阅读的时候，重点一定是在源码的理解上，但是，当读完第一个源码再去读下一个的时候，你自然而然地会带着批判或者说挑剔的眼光去阅读：为什么这个功能在我之前看的源码中是那样实现的，而在这里会是这样实现的？这其中的道理在哪里，哪种实现方式更优秀呢？而通过这样的对比及探索，你会发现，自己的进步快得难以想象。

我们已经有些纠结了，既然阅读源码有那么多的好处，但是很多读者却因为时间或者能力的问题而不得不放弃，岂不是太可惜？为了解决这个问题，我撰写了本书，总结了自己的研究心得和实际项目经验，希望能对正在 Spring 道路上摸索的同仁提供一些帮助。

本书特点

本书完全从开发者的角度去剖析源码，每一章都会提供具有代表性的实例，并以此为基础进行功能实现的分析，而不是采取开篇就讲解容器怎么实现、AOP 怎么实现之类的写法。在描述的过程中，本书尽可能地把问题分解，使用剥洋葱的方式一层一层地将逻辑描述清楚，帮助读者由浅入深地进行学习，并把其中的难点和问题各个击破，而不是企图一下让读者理解一个复杂的逻辑。

在阅读源码的过程中，难免会遇到各种各样的生僻功能，这些功能在特定的场合会非常有用，但是可能多数情况下并不是很常用，甚至都查阅不到相关的使用资料。本书中重点针对这种情况提供了相应的实用示例，让读者更加全面地了解 Spring 所提供的功能，使读者对代码能知其然还知其所以然。

本书按照每章所提供的示例跟踪 Spring 源码的流程，尽可能保证代码的连续性，确保读者的思维不被打乱，让读者看到 Spring 的执行流程，尽量使读者在阅读完本书后，即使在不阅读 Spring 源码的情况下也可以对 Spring 源码进行优化，甚至通过扩展源码来满足业务需求（这对开发人员来说是一个很高的要求）。本书希望能帮助读者全面提升实战能力。

本书结构

本书分为3部分：核心实现、企业应用和 Spring Boot。

- 第1部分，核心实现（第1~7章）：是 Spring 功能的基础，也是企业应用部分的基础，主要对容器以及 AOP 功能实现做了具体的分析。如果读者之前没有接触过 Spring 源代码，建议认真阅读这个部分，否则阅读企业应用部分时会比较吃力。
- 第2部分，企业应用（第8~13章）：在核心实现部分的基础上围绕企业应用常用的模块进行讨论，这些模块包括 Spring 整合 JDBC、Spring 整合 MyBatis、事务、SpringMVC、远程服务、Spring 消息服务等，旨在帮助读者在日常开发中更加高效地使用 Spring。
- 第3部分，Spring Boot（第14章）：对近期流行的 Spring Boot 的体系原理进行分析，剥离其神秘的面纱。Spring Boot 作为 Spring 外的一个独立分支，可以说将 Spring 的扩展能力应用得出神入化，仔细研读后一定会受益匪浅。

本书适用的 Spring 版本

截至完稿，Spring 已经发布了 5.x 版本。本书所讨论的内容属于 Spring 的基础和常用的功能，这些功能都经过长时间、大量用户的验证，已经非常成熟，改动的可能性相对较小，即使 Spring 后续更新到 10.x，相信这些内容也不会过时，因此值得读者去阅读。而且从目前 Spring 的功能规划来看，本书所涉及的内容并不在 Spring 未来改动的范围内，因此在未来的很长一段时间内本书都不会过时。

感谢

创作本书的过程是痛苦的，持续时间也远远超乎了我的想象，而且本以为自己对 Spring 已经非常熟悉，没想到在写作的过程中还是会遇到各种各样的问题，但是我很幸运我能坚持下来。在这里我首先应该感谢爸爸妈妈，虽然他们不知道儿子在忙忙碌碌地写些什么，但是他们对我始终如一的支持与鼓励使我更加坚定信心，在这里祝他们身体健康！同时还要感谢我的妻子，在刚刚生下宝宝后没有过哺乳期的情况下也一直默默支持我，没有因为缺少我的陪伴而埋怨。同时也感谢妹子王晶对稿件提出的建议与意见。最后感谢郭维云、郝云勃、郝俊、李兴全、梁晓颖、陈淼、孙伟超、王璐、刘瑞、单明、姚佳林、闫微微、李娇、时宇、李平、唐广亮、刘阳、黄思文、金施源等在本书编写过程中给予的支持与帮助。

联系作者

本书在编写过程中,以"够用就好"为原则,尽量覆盖到 Spring 开发的常用功能。所有观点都出自作者的个人见解,疏漏、错误之处在所难免,欢迎大家指正。读者如果有好的建议或者在学习本书的过程中遇到问题,请发送邮件到 haojia_007@163.com,希望能够与大家一起交流和进步。

在看得见的地方学习知识,在看不到的地方学习智慧。祝愿大家在 Spring 的学习道路上顺风顺水。

资源与支持

本书由异步社区出品,社区(https://www.epubit.com/)为您提供相关资源和后续服务。

提交勘误

作者和编辑尽最大努力来确保书中内容的准确性,但难免会存在疏漏。欢迎您将发现的问题反馈给我们,帮助我们提升图书的质量。

当您发现错误时,请登录异步社区,按书名搜索,进入本书页面,点击"提交勘误",输入勘误信息,点击"提交"按钮即可。本书的作者和编辑会对您提交的勘误进行审核,确认并接受后,您将获赠异步社区的 100 积分。积分可用于在异步社区兑换优惠券、样书或奖品。

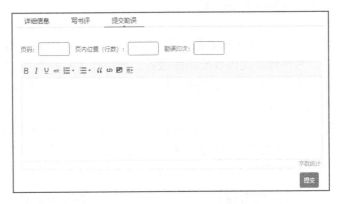

扫码关注本书

扫描下方二维码,您将会在异步社区微信服务号中看到本书信息及相关的服务提示。

与我们联系

我们的联系邮箱是 contact@epubit.com.cn。

如果您对本书有任何疑问或建议,请您发邮件给我们,并请在邮件标题中注明本书书名,以便我们更高效地做出反馈。

如果您有兴趣出版图书、录制教学视频,或者参与图书翻译、技术审校等工作,可以发邮件给我们;有意出版图书的作者也可以到异步社区在线提交投稿(直接访问 www.epubit.com/selfpublish/submission 即可)。

如果您是学校、培训机构或企业,想批量购买本书或异步社区出版的其他图书,也可以发邮件给我们。

如果您在网上发现有针对异步社区出品图书的各种形式的盗版行为,包括对图书全部或部分内容的非授权传播,请您将怀疑有侵权行为的链接发邮件给我们。您的这一举动是对作者权益的保护,也是我们持续为您提供有价值的内容的动力之源。

关于异步社区和异步图书

"**异步社区**"是人民邮电出版社旗下 IT 专业图书社区,致力于出版精品 IT 技术图书和相关学习产品,为作译者提供优质出版服务。异步社区创办于 2015 年 8 月,提供大量精品 IT 技术图书和电子书,以及高品质技术文章和视频课程。更多详情请访问异步社区官网 https://www.epubit.com。

"**异步图书**"是由异步社区编辑团队策划出版的精品 IT 专业图书的品牌,依托于人民邮电出版社近 30 年的计算机图书出版积累和专业编辑团队,相关图书在封面上印有异步图书的 LOGO。异步图书的出版领域包括软件开发、大数据、AI、测试、前端、网络技术等。

异步社区

微信服务号

目录

第 1 部分 核心实现

第 1 章 Spring 整体架构和环境搭建 ········· 2
- 1.1 Spring 的整体架构 ········· 2
- 1.2 环境搭建 ········· 4
  - 1.2.1 源码链接获取 ········· 5
  - 1.2.2 源码下载及 IDEA 导入 ········· 6
- 1.3 cglib 和 objenesis 的编译错误解决 ········· 9
  - 1.3.1 问题发现及原因 ········· 9
  - 1.3.2 问题解决 ········· 9
- 1.4 AspectJ 编译问题解决 ········· 10
  - 1.4.1 问题发现 ········· 10
  - 1.4.2 问题原因 ········· 12
  - 1.4.3 问题解决 ········· 13

第 2 章 容器的基本实现 ········· 19
- 2.1 容器基本用法 ········· 19
- 2.2 功能分析 ········· 20
- 2.3 工程搭建 ········· 21
- 2.4 Spring 的结构组成 ········· 22
  - 2.4.1 beans 包的层级结构 ········· 22
  - 2.4.2 核心类介绍 ········· 23
- 2.5 容器的基础 XmlBeanFactory ········· 26
  - 2.5.1 配置文件封装 ········· 27
  - 2.5.2 加载 Bean ········· 30

目录

2.6 获取 XML 的验证模式 ·· 33
   2.6.1 DTD 与 XSD 区别 ··· 33
   2.6.2 验证模式的读取 ··· 35
2.7 获取 Document ··· 37
2.8 解析及注册 BeanDefinitions ··· 40
   2.8.1 profile 属性的使用 ·· 42
   2.8.2 解析并注册 BeanDefinition ·· 42

第 3 章 默认标签的解析 ··· 44
3.1 bean 标签的解析及注册 ·· 44
   3.1.1 解析 BeanDefinition ·· 46
   3.1.2 AbstractBeanDefinition 属性 ·· 64
   3.1.3 解析默认标签中的自定义标签元素 ·· 67
   3.1.4 注册解析的 BeanDefinition ·· 69
   3.1.5 通知监听器解析及注册完成 ·· 72
3.2 alias 标签的解析 ·· 72
3.3 import 标签的解析 ·· 73
3.4 嵌入式 beans 标签的解析 ·· 76

第 4 章 自定义标签的解析 ·· 77
4.1 自定义标签使用 ·· 78
4.2 自定义标签解析 ·· 80
   4.2.1 获取标签的命名空间 ·· 81
   4.2.2 提取自定义标签处理器 ··· 81
   4.2.3 标签解析 ··· 83

第 5 章 bean 的加载 ·· 86
5.1 FactoryBean 的使用 ·· 92
5.2 缓存中获取单例 bean ··· 93
5.3 从 bean 的实例中获取对象 ··· 94
5.4 获取单例 ·· 98
5.5 准备创建 bean ··· 100
   5.5.1 处理 override 属性 ··· 101
   5.5.2 实例化的前置处理 ·· 102
5.6 循环依赖 ·· 104
   5.6.1 什么是循环依赖 ··· 104
   5.6.2 Spring 如何解决循环依赖 ··· 104

5.7 创建 bean ··· 108
5.7.1 创建 bean 的实例 ··· 111
5.7.2 记录创建 bean 的 ObjectFactory ······································ 120
5.7.3 属性注入 ··· 123
5.7.4 初始化 bean ·· 132
5.7.5 注册 DisposableBean ··· 136

第 6 章 容器的功能扩展 ·· 137
6.1 设置配置路径 ··· 138
6.2 扩展功能 ·· 138
6.3 环境准备 ·· 140
6.4 加载 BeanFactory ·· 141
6.4.1 定制 BeanFactory ··· 143
6.4.2 加载 BeanDefinition ·· 144
6.5 功能扩展 ·· 145
6.5.1 增加 SpEL 语言的支持 ·· 146
6.5.2 增加属性注册编辑器 ··· 147
6.5.3 添加 ApplicationContextAwareProcessor 处理器 ······················· 152
6.5.4 设置忽略依赖 ··· 154
6.5.5 注册依赖 ·· 154
6.6 BeanFactory 的后处理 ··· 154
6.6.1 激活注册的 BeanFactoryPostProcessor ································· 154
6.6.2 注册 BeanPostProcessor ··· 160
6.6.3 初始化消息资源 ·· 163
6.6.4 初始化 ApplicationEventMulticaster ···································· 167
6.6.5 注册监听器 ·· 169
6.7 初始化非延迟加载单例 ·· 169
6.8 finishRefresh ·· 172

第 7 章 AOP ··· 175
7.1 动态 AOP 使用示例 ··· 175
7.2 动态 AOP 自定义标签 ·· 177
7.3 创建 AOP 代理 ·· 181
7.3.1 获取增强器 ·· 184
7.3.2 寻找匹配的增强器 ·· 193
7.3.3 创建代理 ·· 195
7.4 静态 AOP 使用示例 ··· 209
7.5 创建 AOP 静态代理 ··· 211

目录

- 7.5.1 Instrumentation 使用 ····· 211
- 7.5.2 自定义标签 ····· 215
- 7.5.3 织入 ····· 217

第 2 部分　企业应用

第 8 章　数据库连接 JDBC ····· 222
- 8.1 Spring 连接数据库程序实现（JDBC）····· 223
- 8.2 save/update 功能的实现 ····· 225
    - 8.2.1 基础方法 execute ····· 227
    - 8.2.2 Update 中的回调函数 ····· 231
- 8.3 query 功能的实现 ····· 233
- 8.4 queryForObject ····· 237

第 9 章　整合 MyBatis ····· 239
- 9.1 MyBatis 独立使用 ····· 239
- 9.2 Spring 整合 MyBatis ····· 243
- 9.3 源码分析 ····· 245
    - 9.3.1 sqlSessionFactory 创建 ····· 245
    - 9.3.2 MapperFactoryBean 的创建 ····· 249
    - 9.3.3 MapperScannerConfigurer ····· 252

第 10 章　事务 ····· 262
- 10.1 JDBC 方式下的事务使用示例 ····· 262
- 10.2 事务自定义标签 ····· 265
    - 10.2.1 注册 InfrastructureAdvisorAutoProxyCreator ····· 265
    - 10.2.2 获取对应 class/method 的增强器 ····· 269
- 10.3 事务增强器 ····· 276
    - 10.3.1 创建事务 ····· 279
    - 10.3.2 回滚处理 ····· 289
    - 10.3.3 事务提交 ····· 295

第 11 章　SpringMVC ····· 298
- 11.1 SpringMVC 快速体验 ····· 298
- 11.2 ContextLoaderListener ····· 302
    - 11.2.1 ServletContextListener 的使用 ····· 302
    - 11.2.2 Spring 中的 ContextLoaderListener ····· 303

11.3 DispatcherServlet ... 306
11.3.1 servlet 的使用 ... 307
11.3.2 DispatcherServlet 的初始化 ... 308
11.3.3 WebApplicationContext 的初始化 ... 311
11.4 DispatcherServlet 的逻辑处理 ... 327
11.4.1 MultipartContent 类型的 request 处理 ... 333
11.4.2 根据 request 信息寻找对应的 Handler ... 333
11.4.3 没找到对应的 Handler 的错误处理 ... 337
11.4.4 根据当前 Handler 寻找对应的 HandlerAdapter ... 338
11.4.5 缓存处理 ... 338
11.4.6 HandlerInterceptor 的处理 ... 339
11.4.7 逻辑处理 ... 340
11.4.8 异常视图的处理 ... 341
11.4.9 根据视图跳转页面 ... 341

第 12 章 远程服务 ... 347
12.1 RMI ... 347
12.1.1 使用示例 ... 347
12.1.2 服务端实现 ... 349
12.1.3 客户端实现 ... 357
12.2 HttpInvoker ... 362
12.2.1 使用示例 ... 362
12.2.2 服务端实现 ... 364
12.2.3 客户端实现 ... 368

第 13 章 Spring 消息 ... 374
13.1 JMS 的独立使用 ... 374
13.2 Spring 整合 ActiveMQ ... 376
13.3 源码分析 ... 378
13.3.1 JmsTemplate ... 379
13.3.2 监听器容器 ... 383

第 3 部分 Spring Boot

第 14 章 Spring Boot 体系原理 ... 394
14.1 Spring Boot 源码安装 ... 396
14.2 第一个 Starter ... 397
14.3 探索 SpringApplication 启动 Spring ... 400

14.4 Starter 自动化配置原理

- 14.3.1 SpringContext 创建 ··· 401
- 14.3.2 bean 的加载 ··· 402
- 14.3.3 Spring 扩展属性的加载 ··· 403
- 14.3.4 总结 ··· 403

14.4 Starter 自动化配置原理 ··· 403
- 14.4.1 spring.factories 的加载 ··· 404
- 14.4.2 factories 调用时序图 ··· 405
- 14.4.3 配置类的解析 ··· 407
- 14.4.4 Componentscan 的切入点 ··· 410

14.5 Conditional 机制实现 ··· 413
- 14.5.1 Conditional 使用 ··· 413
- 14.5.2 Conditional 原理 ··· 415
- 14.5.3 调用切入点 ··· 418

14.6 属性自动化配置实现 ··· 420
- 14.6.1 示例 ··· 420
- 14.6.2 原理 ··· 422

14.7 Tomcat 启动 ··· 428

第 1 部分　核心实现

第 1 章　Spring 整体架构和环境搭建
第 2 章　容器的基本实现
第 3 章　默认标签的解析
第 4 章　自定义标签的解析
第 5 章　bean 的加载
第 6 章　容器的功能扩展
第 7 章　AOP

第 1 章　Spring 整体架构和环境搭建

Spring 是于 2003 年兴起的一个轻量级 Java 开源框架，由 Rod Johnson 在其著作 *Expert One-On-One J2EE Design and Development* 中阐述的部分理念和原型衍生而来。Spring 是为了解决企业应用开发的复杂性而创建的，它使用基本的 JavaBean 来完成以前只可能由 EJB 完成的事情。然而，Spring 的用途不仅限于服务器端的开发，从简单性、可测试性和松耦合的角度而言，任何 Java 应用都可以从 Spring 中受益。

1.1　Spring 的整体架构

Spring 框架是一个分层架构，它包含一系列的功能要素，并被分为大约 20 个模块，如图 1-1 所示。

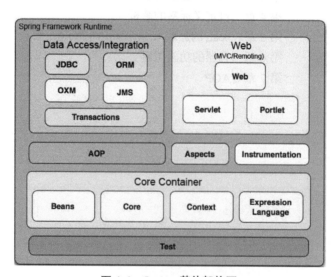

图 1-1　Spring 整体架构图

这些模块被总结为以下几部分。

1. Core Container

Core Container（核心容器）包含有 Core、Beans、Context 和 Expression Language 模块。

Core 和 Beans 模块是框架的基础部分，提供 IoC（转控制）和依赖注入特性。这里的基础概念是 BeanFactory，它提供对 Factory 模式的经典实现来消除对程序性单例模式的需要，并真正地允许你从程序逻辑中分离出依赖关系和配置。

- Core 模块主要包含 Spring 框架基本的核心工具类，Spring 的其他组件都要用到这个包里的类，Core 模块是其他组件的基本核心。当然你也可以在自己的应用系统中使用这些工具类。
- Beans 模块是所有应用都要用到的，它包含访问配置文件、创建和管理 bean 以及进行 Inversion of Control / Dependency Injection（IoC/DI）操作相关的所有类。
- Context 模块构建于 Core 和 Beans 模块基础之上，提供了一种类似于 JNDI 注册器的框架式的对象访问方法。Context 模块继承了 Beans 的特性，为 Spring 核心提供了大量扩展，添加了对国际化（例如资源绑定）、事件传播、资源加载和对 Context 的透明创建的支持。Context 模块同时也支持 J2EE 的一些特性，例如 EJB、JMX 和基础的远程处理。ApplicationContext 接口是 Context 模块的关键。
- Expression Language 模块提供了强大的表达式语言，用于在运行时查询和操纵对象。它是 JSP 2.1 规范中定义的 unifed expression language 的扩展。该语言支持设置/获取属性的值，属性的分配，方法的调用，访问数组上下文（accessing the context of arrays）、容器和索引器、逻辑和算术运算符、命名变量以及从 Spring 的 IoC 容器中根据名称检索对象。它也支持 list 投影、选择和一般的 list 聚合。

2. Data Access/Integration

Data Access/Integration 层包含 JDBC、ORM、OXM、JMS 和 Transaction 模块。

- JDBC 模块提供了一个 JDBC 抽象层，它可以消除冗长的 JDBC 编码和解析数据库厂商特有的错误代码。这个模块包含了 Spring 对 JDBC 数据访问进行封装的所有类。
- ORM 模块为流行的对象-关系映射 API，如 JPA、JDO、Hibernate、iBatis 等，提供了一个交互层。利用 ORM 封装包，可以混合使用所有 Spring 提供的特性进行 O/R 映射，如前边提到的简单声明性事务管理。

Spring 框架插入了若干个 ORM 框架，从而提供了 ORM 的对象关系工具，其中包括 JDO、Hibernate 和 iBatisSQL Map。所有这些都遵从 Spring 的通用事务和 DAO 异常层次结构。

- OXM 模块提供了一个对 Object/XML 映射实现的抽象层，Object/XML 映射实现包括 JAXB、Castor、XMLBeans、JiBX 和 XStream。
- JMS（Java Messaging Service）模块主要包含了一些制造和消费消息的特性。

- Transaction 模块支持编程和声明性的事务管理，这些事务类必须实现特定的接口，并且对所有的 POJO 都适用。

3. Web

Web 上下文模块建立在应用程序上下文模块之上，为基于 Web 的应用程序提供了上下文。所以，Spring 框架支持与 Jakarta Struts 的集成。Web 模块还简化了处理大部分请求以及将请求参数绑定到域对象的工作。Web 层包含了 Web、Web-Servlet、Web-Struts 和 Web-Porlet 模块，具体说明如下。

- Web 模块：提供了基础的面向 Web 的集成特性。例如，多文件上传、使用 servlet listeners 初始化 IoC 容器以及一个面向 Web 的应用上下文。它还包含 Spring 远程支持中 Web 的相关部分。
- Web-Servlet 模块 web.servlet.jar：该模块包含 Spring 的 model-view-controller（MVC）实现。Spring 的 MVC 框架使得模型范围内的代码和 web forms 之间能够清楚地分离开来，并与 Spring 框架的其他特性集成在一起。
- Web-Struts 模块：该模块提供了对 Struts 的支持，使得类在 Spring 应用中能够与一个典型的 Struts Web 层集成在一起。注意，该支持在 Spring 3.0 中已被弃用。
- Web-Porlet 模块：提供了用于 Portlet 环境和 Web-Servlet 模块的 MVC 的实现。

4. AOP

AOP 模块提供了一个符合 AOP 联盟标准的面向切面编程的实现，它让你可以定义例如方法拦截器和切点，从而将逻辑代码分开，降低它们之间的耦合性。利用 source-level 的元数据功能，还可以将各种行为信息合并到你的代码中，这有点像.Net 技术中的 attribute 概念。

通过配置管理特性，Spring AOP 模块直接将面向切面的编程功能集成到了 Spring 框架中，所以可以很容易地使 Spring 框架管理的任何对象支持 AOP。Spring AOP 模块为基于 Spring 的应用程序中的对象提供了事务管理服务。通过使用 Spring AOP，不用依赖 EJB 组件，就可以将声明性事务管理集成到应用程序中。

- Aspects 模块提供了对 AspectJ 的集成支持。
- Instrumentation 模块提供了 class instrumentation 支持和 classloader 实现，使得可以在特定的应用服务器上使用。

5. Test

Test 模块支持使用 JUnit 和 TestNG 对 Spring 组件进行测试。

1.2 环境搭建

由于在上一版本中作者对环境搭建描述得比较粗糙，导致非常多的读者询问，作者在此表示歉意,并在这一版本中补充了非常详细的 Spring 环境搭建流程。如果是第一次接触 Spring 源码的环境搭建,确实还是比较麻烦的。

作者使用的编译器为目前流行的 IntelliJ IDEA,版本为 2018 旗舰版。Eclipse 用户还需要自己揣摩环境搭建方法,这里不再赘述。

1.2.1 源码链接获取

1. 输入 GitHub 官网网址并搜索 spring,如图 1-2 所示。

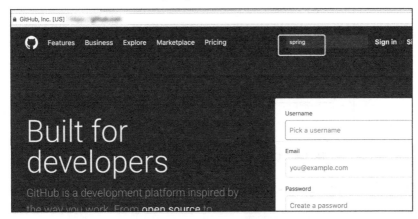

图 1-2　GitHub 上的 spring 搜索

2. 找到对应的 spring-framework 的工程,点击链接进入,如图 1-3 所示。

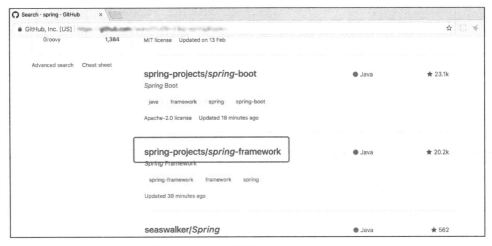

图 1-3　GitHub 上的 spring-framework

3. 切换为最新的 Spring 5.0.x 版本源码,如图 1-4 所示。

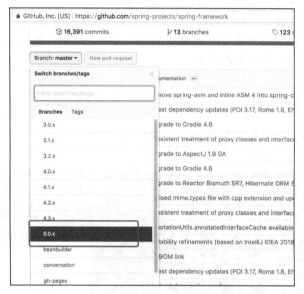

图 1-4　切换为最新的 Spring 5.0.x 版本源码

4. 获取 Git 分支链接，如图 1-5 所示。

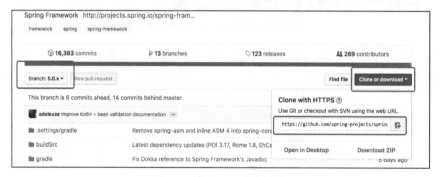

图 1-5　获取 Git 分支链接

1.2.2　源码下载及 IDEA 导入

1. IDEA 下 Spring Git 拉取分支，如图 1-6 所示。
2. 本地安装目录设置，如图 1-7 所示。
3. 拉取等待，如图 1-8 所示。
4. IDEA 导入，如图 1-9 所示。
5. Gradle 项目导入，如图 1-10 所示。

图 1-6　IDEA 下 Spring Git 拉取分支

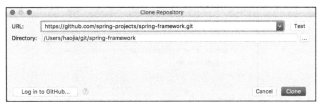
图 1-7　本地安装目录设置

图 1-8　拉取等待

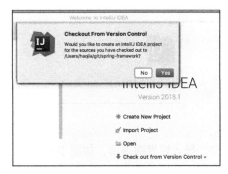
图 1-9　IDEA 导入

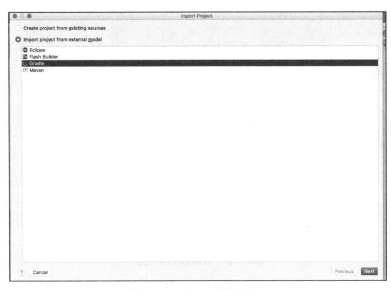
图 1-10　Gradle 项目导入

6. 工程属性设置，如图 1-11 所示。

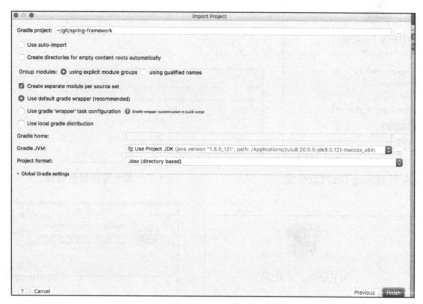

图 1-11　工程属性设置

7. 导入后界面展示，如图 1-12 所示。

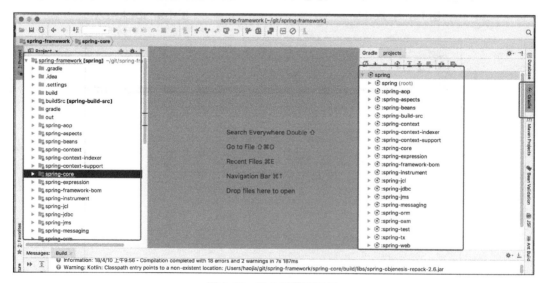

图 1-12　导入后界面展示

1.3 cglib 和 objenesis 的编译错误解决

1.3.1 问题发现及原因

错误信息获取，如图 1-13 所示。

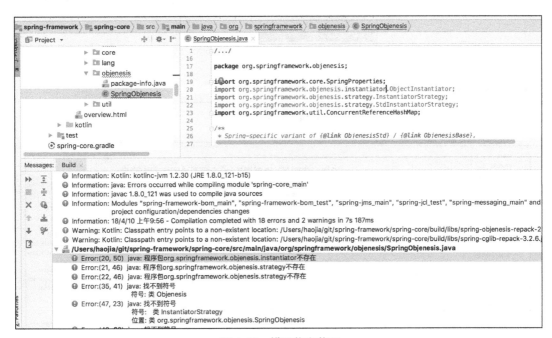

图 1-13 错误信息获取

为了避免第三方 class 的冲突，Spring 把最新的 cglib 和 objenesis 给重新打包（repack）了，它并没有在源码里提供这部分的代码，而是直接将其放在 jar 包当中，这也就导致了我们拉取代码后出现编译错误。那么为了通过编译，我们要把缺失的 jar 补回来。

1.3.2 问题解决

1. 缺失 jar 引入，如图 1-14 所示。
2. 新增 jar 在 Gradle 中生效，如图 1-15 所示。
   因为整个 Spring 都在 Gradle 环境中，所以要使得 jar 生效就必须更改 Gradle 配置文件：
   ```
   compile fileTree(dir: 'libs' ,include : '*.jar')
   ```

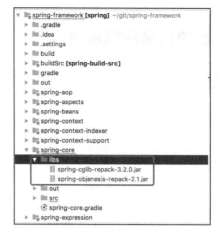

图 1-14 缺失 jar 引入

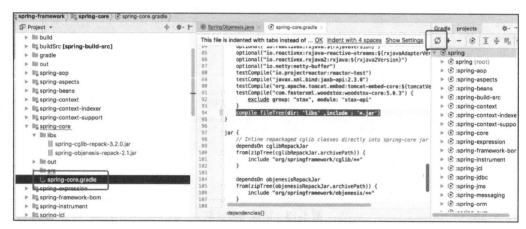

图 1-15 新增 jar 在 Gradle 中生效

1.4 AspectJ 编译问题解决

1.4.1 问题发现

当完成以上的 jar 包导入工作并进行重新编译后,发现还是有编译错误提醒,真是山路十八弯。查看编译错误原因,如图 1-16 所示,发现居然是类找不到。

aspect 关键字 Java 语法违背

可是我们明明能看到对应的类就在工程里面,为什么会找不到呢?于是打开对应的类查看,如图 1-17 所示。

1.4 AspectJ 编译问题解决

图 1-16 问题发现

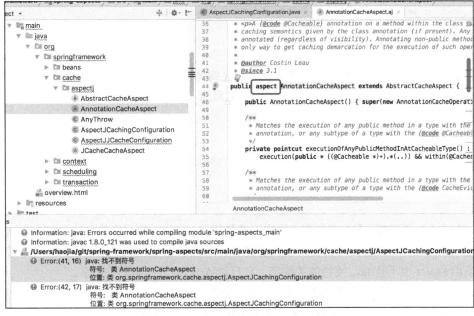

图 1-17 aspect 关键字 Java 语法违背

发现类的声明居然使用了 aspect 而不是 class，我瞬间吸了口凉气，太久没充电了，JDK 新出的语法变化这么大都没关注，经过几番周折后终于在 AspectJ 的使用上找到了答案。

1.4.2 问题原因

AOP（Aspect Orient Programming，面向切面编程）作为面向对象编程的一种补充，当前已经成为一种比较成熟的编程思想。其实 AOP 问世的时间并不长，甚至在国内的翻译还不太统一（另有人翻译为"面向方面编程"）。

而 AOP 在 Spring 中也占据着举足轻重的作用，可以说没有 AOP 就没有 Spring 现在的流行，当然 AOP 的实现有些时候也依赖于 AspectJ。

AspectJ 实现 AOP

脱离了 Spring，我们可以单独看看 AspectJ 的使用方法。

AspectJ 的用法很简单，就像我们使用 JDK 编译、运行 Java 程序一样。下面通过一个简单的程序来示范 AspectJ 的用法：

```
public class HelloWorld {
    public void sayHello(){
        System.out.println("Hello AspectJ!");
    }
    public static void main(String args[]) {
        HelloWorld h = new HelloWorld();
        h.sayHello();
    }
}
```

毫无疑问，结果将输出"Hello AspectJ!"字符串。假设现在客户需要在执行 sayHello 方法前启动事务，当该方法结束时关闭事务，则在传统编程模式下，我们必须手动修改 sayHello 方法——如果改为使用 AspectJ，则可以无须修改上面的 sayHello 方法。下面我们定义一个特殊的"类"：

```
public aspect TxAspect {
    void around():call(void sayHello()) {
        System.out.println("Transaction Begin");
        proceed();
        System.out.println("Transaction End");
    }
}
```

可能有人已经发现，在上面的类文件中不是使用 class、interface 或者 enum 来定义 Java 类，而是使用 aspect——难道 Java 语言又增加关键字了？No！上面的 TxAspect 根本不是一个 Java 类，所以 aspect 也不是 Java 支持的关键字，它只是 AspectJ 才认识的关键字。

上面 void around 中的内容也不是方法，它只是指定当程序执行 HelloWorld 对象的 sayHello 方法时，执行这个代码块，其中 proceed 表示调用原来的 sayHello 方法。正如前面提到的，因为 Java 无法识别 TxAspect.java 文件中的内容，所以我们需要使用 ajc.exe 来执行编译：

```
ajc HelloWorld.java TxAspect.java
```
我们可以把 ajc 命令理解为 javac 命令，两者都用于编译 Java 程序，区别是 ajc 命令可以识别 AspectJ 的语法。从这个角度看，我们可以将 ajc 命令当成增强版的 javac 命令。

运行该 HelloWorld 类依然无须任何改变，其结果如下：
```
Transaction Begin
Hello AspectJ!
Transaction End
```
从上面的运行结果来看，我们可以完全不修改 HelloWorld.java 文件，也不用修改执行 HelloWorld 的命令，就可以实现上文中的实现事务管理的需求。上面的 Transaction Begin 和 Transaction End 仅仅是模拟事务的事件，实际开发中，用代码替换掉这段输出即可实现事务管理。

1.4.3 问题解决

1. 下载 AspectJ 的最新稳定版本

安装 AspectJ 之前，请确保系统已经安装了 JDK。

下载下来后是一个 jar 包，如图 1-18 所示。

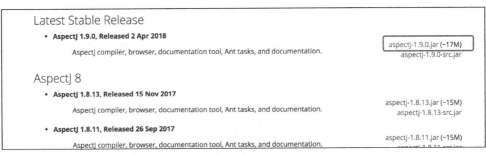

图 1-18　下载 AspectJ 的最新版本

2. AspectJ 安装

打开命令行，cd 到该 jar 包所在的文件夹，运行 java -jar aspectj-1.9.0.jar 命令，打开 AspectJ 的安装界面。第一个界面是欢迎界面，直接点击 Next，如图 1-19 所示。

在图 1-20 所示的第二个界面中选择 jre 的安装路径，继续点击 Next。

在图 1-21 所示的第三个界面中选择 AspectJ 的安装路径，点击 Install。因为安装过程的实质是解压一个压缩包，并不需要太多地依赖于系统，因此路径可以任意选择，这里选择和 Java 安装在一起。

第 1 章　Spring 整体架构和环境搭建

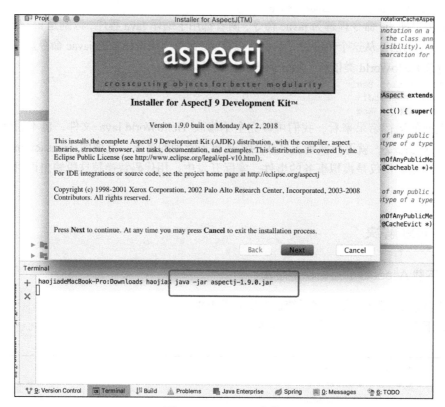

图 1-19　AspectJ 安装

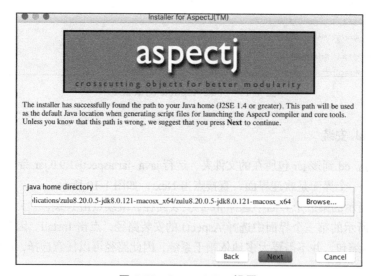

图 1-20　AspecJ JDK 设置

1.4 AspectJ 编译问题解决

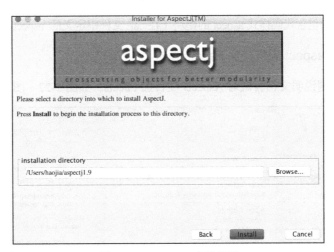

图 1-21 AspecJ 安装目录

至此，AspectJ 安装完成。

3. IDEA 对 Ajc 支持官方文档（使用 AspectJ 编译器）

此功能仅在 Ultimate 版本中得到支持。

默认情况下，IntelliJ IDEA 使用 Javac 编译器。要使用 AspectJ 编译器 Ajc（而不是与 javac 组合），应对相应的 IDE 设置进行更改。

项目级别指定的 Ajc 设置可以在各个模块进行微调。与模块相关的 AspectJ 用于此目的。

请注意，Ajc 不与 IntelliJ IDEA 捆绑在一起，它是 AspectJ 发行版的一部分，您可以从 AspectJ 网站下载。

将 Ajc 与 Javac 结合使用可以优化编译性能，IntelliJ IDEA 可把二者组合起来，而无须在 IDE 设置中切换编译器。

首先，您应该选择 Ajc 作为项目编译器（在 Java 编译器页面上的 Use 编译器字段）。

如果您想要同时使用 Javac，请打开 "Delegate to Javac" 选项。如果启用此选项，那么没有 aspect 的模块将被编译为 Javac（通常更快），并且包含 aspect 的模块将用 Ajc 编译（如果此选项为 off，则 Ajc 用于项目中的所有模块）。

您可以在各个模块级别对编译器（Ajc 和 Javac）之间的任务分配进行微调。对于只包含 @Aspect-annotated 的 Java 类（在 .java 文件中）的形式的模块，您可以指定 Ajc 仅应用于后编译的编织（weaving）。如果这样做，则 Javac 将用于编译所有源文件，然后 Ajc 将其应用于编译的类文件进行编织。因此，整个过程（编译+编织）(compilation + weaving) 将花费更少的时间。

如果打开了 "Javac 代理选项（Delegate to Javac）"，则通过在与模块关联的 AspectJ Facets 中打开相应的选项来启用 Ajc 的编译后编织模式。

请注意，不应为包含代码样式 aspect 的模块（在 .aj 文件中定义的 aspect）启用此选项。

4．为 spring-aspect 工程添加 Facets 属性

按照 IDEA 官网说明文档尝试对 AspectJ 项目加 Facets，如图 1-22～图 1-26 所示。

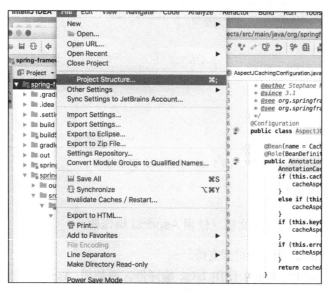

图 1-22 设置 Facets（1）

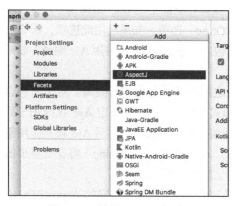

图 1-23 设置 Facets（2）

图 1-24 设置 Facets（3）

1.4　AspectJ 编译问题解决

图 1-25　设置 Facets（4）

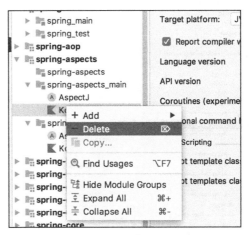

图 1-26　删除 Facets

5．更改编译器

编译器要改为 Ajc，同时要设置 Ajc 的安装目录，如图 1-27 所示。记住，要选择到 aspectjtools.jar 这个层面，同时，务必要选择 Delegate to Javac 选项，它的作用是只编译 AspectJ 的 Facets 项目，而其他则使用 JDK 代理，如图 1-28 所示。如果不勾选，则全部使用 Ajc 编译，那么会导致编译错误。如图 1-29 所示，编译器改为 Ajc。

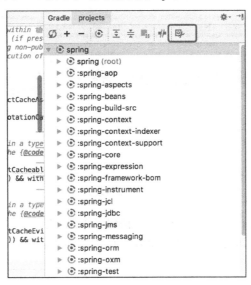

图 1-27　Gradle 入口更改编译器

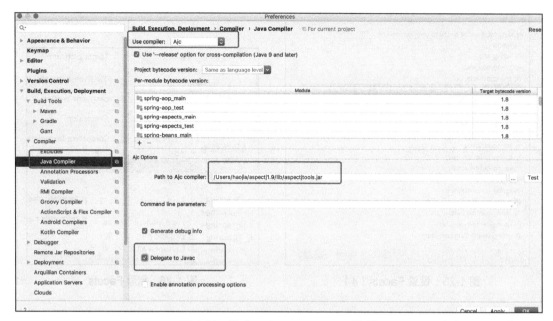

图 1-28　选中 Delegate to Javac 选项

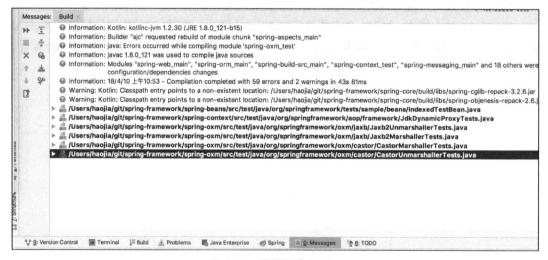

图 1-29　编译器改为 Ajc

至此，我们已经完成了整个 Spring 的环境搭建工作，还有一些单测类的错误已经不影响源码阅读，没有必要浪费时间去解决，删掉就好，有兴趣的读者可以自己解决。

第 2 章　容器的基本实现

源码分析是一件非常煎熬且极具挑战性的任务，你准备好开始战斗了吗？

在正式开始分析 Spring 源码之前，我们有必要先来回顾一下 Spring 中最简单的用法，尽管我相信您已经对这个例子非常熟悉了。

2.1　容器基本用法

bean 是 Spring 中最核心的东西，因为 Spring 就像是个大水桶，而 bean 就像是容器中的水，水桶脱离了水便也没什么用处了，那么我们先看看 bean 的定义。

```
public class MyTestBean {
    private String testStr = "testStr";

    public String getTestStr() {
        return testStr;
    }

    public void setTestStr(String testStr) {
        this.testStr = testStr;
    }
}
```

这么看来 bean 并没有任何特别之处，的确，Spring 的目的就是让我们的 bean 能成为一个纯粹的 POJO，这也是 Spring 所追求的。接下来看看配置文件：

```
<?xml version="1.0" encoding="UTF-8"?>
<beans xmlns="http://www.Springframework.org/schema/beans"
    xmlns:xsi="http://www.w3.org/2001/XMLSchema-instance"
    xsi:schemaLocation="http://www.Springframework.org/schema/beans http://www.Springframework.org/schema/beans/Spring-beans.xsd">

    <bean id="myTestBean" class="bean.MyTestBean"/>

</beans>
```

在上面的配置中我们看到了 bean 的声明方式，尽管 Spring 中 bean 的元素定义着 N 种属性来支撑我们业务的各种应用，但是我们只要声明成这样，基本上就已经可以满足我们的大多数应用了。好了，你可能觉得还有什么，但是，真没了，Spring 的入门示例到这里已经结束，我们可以写测试代码测试了。

```
@SuppressWarnings("deprecation")
public class BeanFactoryTest {

    @Test
    public void testSimpleLoad(){
        BeanFactory    bf = new XmlBeanFactory(new ClassPathResource("beanFactoryTest.xml"));
        MyTestBean bean=(MyTestBean) bf.getBean("myTestBean");
        assertEquals("testStr",bean.getTestStr());
    }
}
```

相信聪明的读者会很快看到我们期望的结果：在 Eclipse 中显示了 Green Bar。

直接使用 BeanFactory 作为容器对于 Spring 的使用来说并不多见，甚至是甚少使用，因为在企业级的应用中大多数都会使用的是 ApplicationContext（后续章节我们会介绍它们之间的区别），这里只是用于测试，让读者更快更好地分析 Spring 的内部原理。

OK，我们又复习了一遍 Spring，你是不是会很不屑呢？这样的小例子没有任何挑战性。嗯，确实，这样的使用是过于简单了，但是本书的目的并不是介绍如何使用 Spring，而是帮助您更好地了解 Spring 的内部原理。读者可以自己先想想，上面的一句简单代码都执行了什么样的逻辑呢？这样一句简单代码其实在 Spring 中执行了太多太多的逻辑，即使作者用半本书的文字也只能介绍它的大致原理。那么就让我们快速地进入分析状态吧。

2.2 功能分析

现在我们可以来好好分析一下上面测试代码的功能，来探索上面的测试代码中 Spring 究竟帮助我们完成了什么工作？不管之前你是否使用过 Spring，当然，你应该使用过的，毕竟本书面向的是对 Spring 有一定使用经验的读者，你都应该能猜出来，这段测试代码完成的功能无非就是以下几点。

- 读取配置文件 beanFactoryTest.xml。
- 根据 beanFactoryTest.xml 中的配置找到对应的类的配置，并实例化。
- 调用实例化后的实例。

为了更清楚地描述，作者临时画了设计类图，如图 2-1 所示，如果想完成我们预想的功能，至少需要 3 个类。

- ConfigReader：用于读取及验证配置文件。我们要用配置文件里面的东西，当然首先要做的就是读取，然后放置在内存中。
- ReflectionUtil：用于根据配置文件中的配置进行反射实例化。比如在上例中 beanFactoryTest.xml 出现的<bean id="myTestBean" class="bean.MyTestBean"/>，我们就

可以根据 bean.MyTestBean 进行实例化。
- App：用于完成整个逻辑的串联。

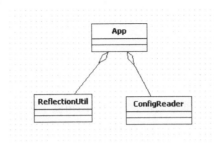

图 2-1　最简单的 Spring 功能架构

按照原始的思维方式，整个过程无非如此，但是作为一个风靡世界的优秀源码真的就这么简单吗？

2.3　工程搭建

不如我们首先大致看看 Spring 的源码。在 Spring 源码中，用于实现上面功能的是 org.Springframework.beans.jar，我们看源码的时候要打开这个工程，如果我们只使用上面的功能，那就没有必要引入 Spring 的其他更多的包，当然 Core 是必需的，还有些依赖的包如图 2-2 所示。

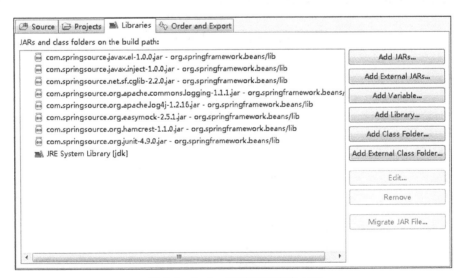

图 2-2　Spring 测试类依赖的 JAR

引入依赖的 JAR 消除掉所有编译错误后，终于可以看源码了。或许你已经知道了答案，Spring 居然用了 N 多代码实现了这个看似很简单的功能，那么这些代码都是做什么用的呢？Spring 在架构或者编码的时候又是如何考虑的呢？带着疑问，让我们踏上研读 Spring 源码的征程。

2.4　Spring 的结构组成

我们首先尝试梳理 Spring 的框架结构，从全局的角度了解 Spring 的结构组成。

2.4.1　beans 包的层级结构

作者认为阅读源码的最好方法是通过示例跟着操作一遍，虽然有时候或者说大多数时候会被复杂的代码绕来绕去，绕到最后已经不知道自己身在何处了，但是，如果配以 UML 还是可以搞定的。作者就是按照自己的思路进行分析，并配合必要的 UML，希望读者同样可以跟得上思路。

我们先看看整个 beans 工程的源码结构，如图 2-3 所示。

beans 包中的各个源码包的功能如下。

- src/main/java 用于展现 Spring 的主要逻辑。
- src/main/resources 用于存放系统的配置文件。
- src/test/java 用于对主要逻辑进行单元测试。
- src/test/resources 用于存放测试用的配置文件。

图 2-3　beans 工程的源码结构

2.4.2 核心类介绍

通过 beans 工程的结构介绍，我们现在对 beans 的工程结构有了初步的认识，但是在正式开始源码分析之前，有必要了解 Spring 中核心的两个类。

1．DefaultListableBeanFactory

XmlBeanFactory 继承自 DefaultListableBeanFactory，而 DefaultListableBeanFactory 是整个 bean 加载的核心部分，是 Spring 注册及加载 bean 的默认实现，而对于 XmlBeanFactory 与 DefaultListableBeanFactory 不同的地方其实是在 XmlBeanFactory 中使用了自定义的 XML 读取器 XmlBeanDefinitionReader，实现了个性化的 BeanDefinitionReader 读取，DefaultListableBeanFactory 继承了 AbstractAutowireCapableBeanFactory 并实现了 ConfigurableListableBeanFactory 以及 BeanDefinitionRegistry 接口。图 2-4 是 ConfigurableListableBeanFactory 的层次结构图，图 2-5 是相关类图。

图 2-4　ConfigurableListableBeanFactory 的层次结构图

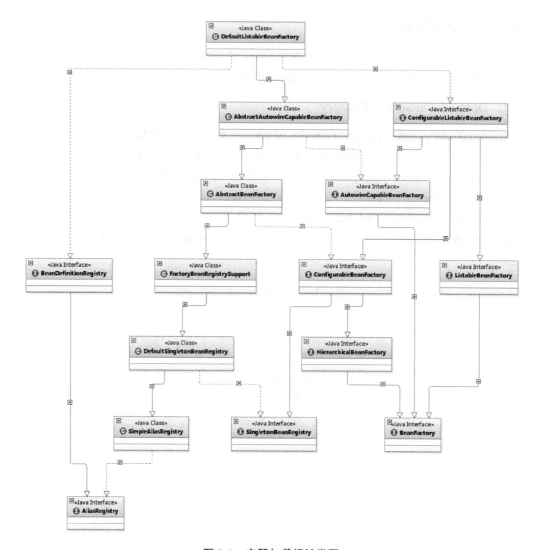

图 2-5 容器加载相关类图

从上面的类图以及层次结构图中,我们可以很清晰地从全局角度了解 DefaultListableBeanFactory 的脉络。如果读者没有了解过 Spring 源码可能对上面的类图不是很理解,不过没关系,通过后续的学习,你会逐渐了解每个类的作用。那么,让我们先简单地了解图 2-5 中各个类的作用。

- AliasRegistry:定义对 alias 的简单增删改等操作。
- SimpleAliasRegistry:主要使用 map 作为 alias 的缓存,并对接口 AliasRegistry 进行实现。

- SingletonBeanRegistry：定义对单例的注册及获取。
- BeanFactory：定义获取 bean 及 bean 的各种属性。
- DefaultSingletonBeanRegistry：对接口 SingletonBeanRegistry 各函数的实现。
- HierarchicalBeanFactory：继承 BeanFactory，也就是在 BeanFactory 定义的功能的基础上增加了对 parentFactory 的支持。
- BeanDefinitionRegistry：定义对 BeanDefinition 的各种增删改操作。
- FactoryBeanRegistrySupport：在 DefaultSingletonBeanRegistry 基础上增加了对 FactoryBean 的特殊处理功能。
- ConfigurableBeanFactory：提供配置 Factory 的各种方法。
- ListableBeanFactory：根据各种条件获取 bean 的配置清单。
- AbstractBeanFactory：综合 FactoryBeanRegistrySupport 和 ConfigurableBeanFactory 的功能。
- AutowireCapableBeanFactory：提供创建 bean、自动注入、初始化以及应用 bean 的后处理器。
- AbstractAutowireCapableBeanFactory：综合 AbstractBeanFactory 并对接口 Autowire Capable BeanFactory 进行实现。
- ConfigurableListableBeanFactory：BeanFactory 配置清单，指定忽略类型及接口等。
- DefaultListableBeanFactory：综合上面所有功能，主要是对 bean 注册后的处理。

XmlBeanFactory 对 DefaultListableBeanFactory 类进行了扩展，主要用于从 XML 文档中读取 BeanDefinition，对于注册及获取 bean 都是使用从父类 DefaultListableBeanFactory 继承的方法去实现，而唯独与父类不同的个性化实现就是增加了 XmlBeanDefinitionReader 类型的 reader 属性。在 XmlBeanFactory 中主要使用 reader 属性对资源文件进行读取和注册。

2. XmlBeanDefinitionReader

XML 配置文件的读取是 Spring 中重要的功能，因为 Spring 的大部分功能都是以配置作为切入点的，那么我们可以从 XmlBeanDefinitionReader 中梳理一下资源文件读取、解析及注册的大致脉络，首先我们看看各个类的功能。

- ResourceLoader：定义资源加载器，主要应用于根据给定的资源文件地址返回对应的 Resource。
- BeanDefinitionReader：主要定义资源文件读取并转换为 BeanDefinition 的各个功能。
- EnvironmentCapable：定义获取 Environment 方法。
- DocumentLoader：定义从资源文件加载到转换为 Document 的功能。
- AbstractBeanDefinitionReader：对 EnvironmentCapable、BeanDefinitionReader 类定义的功能进行实现。
- BeanDefinitionDocumentReader：定义读取 Docuemnt 并注册 BeanDefinition 功能。

- BeanDefinitionParserDelegate：定义解析 Element 的各种方法。

经过以上分析，我们可以梳理出整个 XML 配置文件读取的大致流程，如图 2-6 所示，在 XmlBeanDifinitionReader 中主要包含以下几步的处理。

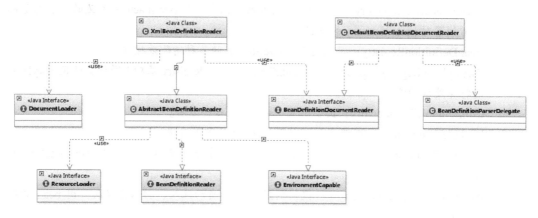

图 2-6　配置文件读取相关类图

1. 通过继承自 AbstractBeanDefinitionReader 中的方法，来使用 ResourLoader 将资源文件路径转换为对应的 Resource 文件。

2. 通过 DocumentLoader 对 Resource 文件进行转换，将 Resource 文件转换为 Document 文件。

3. 通过实现接口 BeanDefinitionDocumentReader 的 DefaultBeanDefinitionDocumentReader 类对 Document 进行解析，并使用 BeanDefinitionParserDelegate 对 Element 进行解析。

2.5　容器的基础 XmlBeanFactory

好了，到这里我们已经对 Spring 的容器功能有了大致的了解，尽管你可能还很迷糊，但是不要紧，接下来我们会详细探索每个步骤的实现。再次重申一下代码，我们接下来要深入分析以下功能的代码实现：

```
BeanFactory    bf = new XmlBeanFactory(new ClassPathResource("beanFactoryTest.xml"));
```

通过 XmlBeanFactory 初始化时序图（如图 2-7 所示）我们来看一看上面代码的执行逻辑。

时序图从 BeanFactoryTest 测试类开始，通过时序图我们可以一目了然地看到整个逻辑处理顺序。在测试的 BeanFactoryTest 中首先调用 ClassPathResource 的构造函数来构造 Resource 资源文件的实例对象，这样后续的资源处理就可以用 Resource 提供的各种服务来操作了，当我们有了 Resource 后就可以进行 XmlBeanFactory 的初始化了。那么 Resource 资源是如何封装的呢？

2.5 容器的基础 XmlBeanFactory

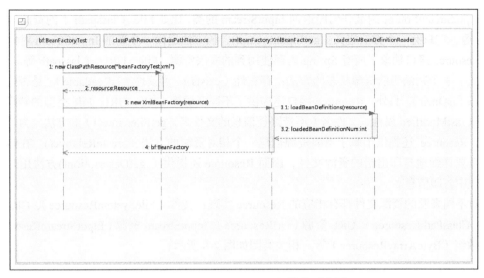

图 2-7 XmlBeanFactory 初始化时序图

2.5.1 配置文件封装

Spring 的配置文件读取是通过 ClassPathResource 进行封装的，如 new ClassPathResource("beanFactoryTest.xml")，那么 ClassPathResource 完成了什么功能呢？

在 Java 中，将不同来源的资源抽象成 URL，通过注册不同的 handler（URLStreamHandler）来处理不同来源的资源的读取逻辑，一般 handler 的类型使用不同前缀（协议，Protocol）来识别，如 "file:" "http:" "jar:" 等，然而 URL 没有默认定义相对 Classpath 或 ServletContext 等资源的 handler，虽然可以注册自己的 URLStreamHandler 来解析特定的 URL 前缀（协议），比如 "classpath:"，然而这需要了解 URL 的实现机制，而且 URL 也没有提供基本的方法，如检查当前资源是否存在、检查当前资源是否可读等方法。因而 Spring 对其内部使用到的资源实现了自己的抽象结构：Resource 接口封装底层资源。

```
public interface InputStreamSource {
    InputStream getInputStream() throws IOException;
}
public interface Resource extends InputStreamSource {
    boolean exists();
    boolean isReadable();
    boolean isOpen();
    URL getURL() throws IOException;
    URI getURI() throws IOException;
    File getFile() throws IOException;
    long lastModified() throws IOException;
    Resource createRelative(String relativePath) throws IOException;
    String getFilename();
    String getDescription();
}
```

InputStreamSource 封装任何能返回 InputStream 的类，比如 File、Classpath 下的资源和 Byte Array 等。它只有一个方法定义：getInputStream()，该方法返回一个新的 InputStream 对象。

Resource 接口抽象了所有 Spring 内部使用到的底层资源：File、URL、Classpath 等。首先，它定义了 3 个判断当前资源状态的方法：存在性（exists）、可读性（isReadable）、是否处于打开状态（isOpen）。另外，Resource 接口还提供了不同资源到 URL、URI、File 类型的转换，以及获取 lastModified 属性、文件名（不带路径信息的文件名，getFilename()）的方法。为了便于操作，Resource 还提供了基于当前资源创建一个相对资源的方法：createRelative()。在错误处理中需要详细地打印出错的资源文件，因而 Resource 还提供了 getDescription() 方法用来在错误处理中打印信息。

对不同来源的资源文件都有相应的 Resource 实现：文件（FileSystemResource）、Classpath 资源（ClassPathResource）、URL 资源（UrlResource）、InputStream 资源（InputStreamResource）、Byte 数组（ByteArrayResource）等。相关类图如图 2-8 所示。

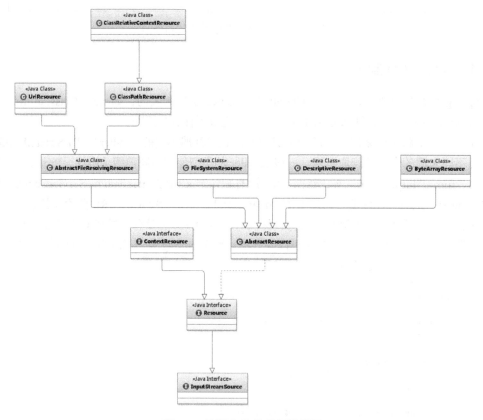

图 2-8　资源文件处理相关类图

在日常的开发工作中，资源文件的加载也是经常用到的，可以直接使用 Spring 提供的类，

2.5 容器的基础 XmlBeanFactory

比如在希望加载文件时可以使用以下代码：
```
Resource resource=new ClassPathResource("beanFactoryTest.xml");
InputStream inputStream=resource.getInputStream();
```
得到 inputStream 后，我们就可以按照以前的开发方式进行实现了，并且我们可以利用 Resource 及其子类为我们提供的诸多特性。

有了 Resource 接口便可以对所有资源文件进行统一处理。至于实现，其实是非常简单的，以 getInputStream 为例，ClassPathResource 中的实现方式便是通过 class 或者 classLoader 提供的底层方法进行调用，而对于 FileSystemResource 的实现其实更简单，直接使用 FileInputStream 对文件进行实例化。

ClassPathResource.java
```
if (this.clazz != null) {
        is = this.clazz.getResourceAsStream(this.path);
    }else {
        is = this.classLoader.getResourceAsStream(this.path);
    }
```

FileSystemResource.java
```
public InputStream getInputStream() throws IOException {
        return new FileInputStream(this.file);
    }
```

当通过 Resource 相关类完成了对配置文件进行封装后配置文件的读取工作就全权交给 XmlBeanDefinitionReader 来处理了。

了解了 Spring 中将配置文件封装为 Resource 类型的实例方法后，我们就可以继续探寻 XmlBeanFactory 的初始化过程了，XmlBeanFactory 的初始化有若干办法，Spring 中提供了很多的构造函数，在这里分析的是使用 Resource 实例作为构造函数参数的办法，代码如下：

XmlBeanFactory.java
```
public XmlBeanFactory(Resource resource) throws BeansException {
    //调用 XmlBeanFactory(Resource,BeanFactory)构造方法
        this(resource, null);
}
```
构造函数内部再次调用内部构造函数：
```
//parentBeanFactory为父类BeanFactory用于factory合并,可以为空
    public XmlBeanFactory(Resource resource, BeanFactory parentBeanFactory) throws BeansException {
        super(parentBeanFactory);
        this.reader.loadBeanDefinitions(resource);
}
```

上面函数中的代码 this.reader.loadBeanDefinitions(resource) 才是资源加载的真正实现，也是我们分析的重点之一。我们可以看到时序图中提到的 XmlBeanDefinitionReader 加载数据就是在这里完成的，但是在 XmlBeanDefinitionReader 加载数据前还有一个调用父类构造函数初始化的过程：

super(parentBeanFactory),跟踪代码到父类 AbstractAutowireCapableBeanFactory 的构造函数中:

```
AbstractAutowireCapableBeanFactory.java
public AbstractAutowireCapableBeanFactory() {
    super();
    ignoreDependencyInterface(BeanNameAware.class);
    ignoreDependencyInterface(BeanFactoryAware.class);
    ignoreDependencyInterface(BeanClassLoaderAware.class);
}
```

这里有必要提及 ignoreDependencyInterface 方法。ignoreDependencyInterface 的主要功能是忽略给定接口的自动装配功能,那么,这样做的目的是什么呢?会产生什么样的效果呢?

举例来说,当 A 中有属性 B,那么当 Spring 在获取 A 的 Bean 的时候如果其属性 B 还没有初始化,那么 Spring 会自动初始化 B,这也是 Spring 中提供的一个重要特性。但是,某些情况下,B 不会被初始化,其中的一种情况就是 B 实现了 BeanNameAware 接口。Spring 中是这样介绍的:自动装配时忽略给定的依赖接口,典型应用是通过其他方式解析 Application 上下文注册依赖,类似于 BeanFactory 通过 BeanFactoryAware 进行注入或者 ApplicationContext 通过 ApplicationContextAware 进行注入。

2.5.2 加载 Bean

之前提到的在 XmlBeanFactory 构造函数中调用了 XmlBeanDefinitionReader 类型的 reader 属性提供的方法 this.reader.loadBeanDefinitions(resource),而这句代码则是整个资源加载的切入点,我们先来看看这个方法的时序图,如图 2-9 所示。

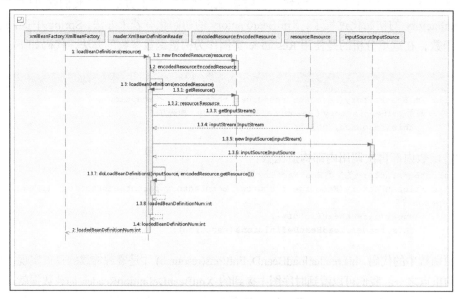

图 2-9　loadBeanDefinitions 函数执行时序图

2.5 容器的基础 XmlBeanFactory

看到图 2-9 我们才知道什么叫山路十八弯,绕了这么半天还没有真正地切入正题,比如加载 XML 文档和解析注册 Bean,一直还在做准备工作。我们根据上面的时序图来分析一下这里究竟在准备什么?从上面的时序图中我们尝试梳理整个的处理过程如下。

1. 封装资源文件。当进入 XmlBeanDefinitionReader 后首先对参数 Resource 使用 EncodedResource 类进行封装。
2. 获取输入流。从 Resource 中获取对应的 InputStream 并构造 InputSource。
3. 通过构造的 InputSource 实例和 Resource 实例继续调用函数 doLoadBeanDefinitions。

我们来看一下 loadBeanDefinitions 函数具体的实现过程。

```
public int loadBeanDefinitions(Resource resource) throws BeanDefinitionStoreException {
    return loadBeanDefinitions(new EncodedResource(resource));
}
```

那么 EncodedResource 的作用是什么呢?通过名称,我们可以大致推断这个类主要是用于对资源文件的编码进行处理的。其中的主要逻辑体现在 getReader()方法中,当设置了编码属性的时候 Spring 会使用相应的编码作为输入流的编码。

```
public Reader getReader() throws IOException {
    if (this.encoding != null) {
        return new InputStreamReader(this.resource.getInputStream(), this.encoding);
    }
    else {
        return new InputStreamReader(this.resource.getInputStream());
    }
}
```

上面代码构造了一个有编码(encoding)的 InputStreamReader。当构造好 encodedResource 对象后,再次转入了可复用方法 loadBeanDefinitions(new EncodedResource(resource))。

这个方法内部才是真正的数据准备阶段,也就是时序图所描述的逻辑:

```
public int loadBeanDefinitions(EncodedResource encodedResource) throws BeanDefinitionStoreException {
    Assert.notNull(encodedResource, "EncodedResource must not be null");
    if (logger.isInfoEnabled()) {
        logger.info("Loading XML bean definitions from " + encodedResource.getResource());
    }
    //通过属性来记录已经加载的资源
    Set<EncodedResource> currentResources = this.resourcesCurrentlyBeingLoaded.get();
    if (currentResources == null) {
        currentResources = new HashSet<EncodedResource>(4);
        this.resourcesCurrentlyBeingLoaded.set(currentResources);
    }
    if (!currentResources.add(encodedResource)) {
        throw new BeanDefinitionStoreException(
                "Detected cyclic loading of " + encodedResource + " - check your import definitions!");
    }
    try {
        //从encodedResource中获取已经封装的Resource对象并再次从Resource中获取其中的inputStream
        InputStream inputStream = encodedResource.getResource().getInputStream();
        try {
```

```
            //InputSource 这个类并不来自于 Spring，它的全路径是 org.xml.sax.InputSource
            InputSource inputSource = new InputSource(inputStream);
            if (encodedResource.getEncoding() != null) {
                inputSource.setEncoding(encodedResource.getEncoding());
            }
            //真正进入了逻辑核心部分
                return doLoadBeanDefinitions(inputSource, encodedResource.getResource());
            }
            finally {
                //关闭输入流
                inputStream.close();
            }
        }
        catch (IOException ex) {
            throw new BeanDefinitionStoreException(
                    "IOException parsing XML document from " + encodedResource.getResource(),ex);
        }
        finally {
            currentResources.remove(encodedResource);
            if (currentResources.isEmpty()) {
                this.resourcesCurrentlyBeingLoaded.remove();
            }
        }
    }
```

我们再次整理数据准备阶段的逻辑，首先对传入的 resource 参数做封装，目的是考虑到 Resource 可能存在编码要求的情况，其次，通过 SAX 读取 XML 文件的方式来准备 InputSource 对象，最后将准备的数据通过参数传入真正的核心处理部分 doLoadBeanDefinitions(inputSource, encodedResource.getResource())。

```
    protected int doLoadBeanDefinitions(InputSource inputSource, Resource resource)
                throws BeanDefinitionStoreException {
        try {
            int validationMode = getValidationModeForResource(resource);
            Document doc = this.documentLoader.loadDocument(
                    inputSource, getEntityResolver(), this.errorHandler, validationMode,
isNamespaceAware());
            return registerBeanDefinitions(doc, resource);
        }
        catch (BeanDefinitionStoreException ex) {
            throw ex;
        }
        catch (SAXParseException ex) {
            throw new XmlBeanDefinitionStoreException(resource.getDescription(),
                    "Line " + ex.getLineNumber() + " in XML document from " + resource
+ " is invalid", ex);
        }
        catch (SAXException ex) {
            throw new XmlBeanDefinitionStoreException(resource.getDescription(),
                    "XML document from " + resource + " is invalid", ex);
        }
        catch (ParserConfigurationException ex) {
            throw new BeanDefinitionStoreException(resource.getDescription(),
```

```
                        "Parser configuration exception parsing XML from " + resource,ex);
            }
            catch (IOException ex) {
                throw new BeanDefinitionStoreException(resource.getDescription(),
                        "IOException parsing XML document from " + resource, ex);
            }
            catch (Throwable ex) {
                throw new BeanDefinitionStoreException(resource.getDescription(),
                        "Unexpected exception parsing XML document from " + resource,ex);
            }
        }
```

在上面冗长的代码中假如不考虑异常类的代码，其实只做了三件事，这三件事的每一件都必不可少。

- 获取对 XML 文件的验证模式。
- 加载 XML 文件，并得到对应的 Document。
- 根据返回的 Document 注册 Bean 信息。

这 3 个步骤支撑着整个 Spring 容器部分的实现，尤其是第 3 步对配置文件的解析，逻辑非常的复杂，我们先从获取 XML 文件的验证模式讲起。

2.6 获取 XML 的验证模式

了解 XML 文件的读者都应该知道 XML 文件的验证模式保证了 XML 文件的正确性，而比较常用的验证模式有两种：DTD 和 XSD。它们之间有什么区别呢？

2.6.1 DTD 与 XSD 区别

DTD（Document Type Definition）即文档类型定义，是一种 XML 约束模式语言，是 XML 文件的验证机制，属于 XML 文件组成的一部分。DTD 是一种保证 XML 文档格式正确的有效方法，可以通过比较 XML 文档和 DTD 文件来看文档是否符合规范，元素和标签使用是否正确。一个 DTD 文档包含：元素的定义规则，元素间关系的定义规则，元素可使用的属性，可使用的实体或符号规则。

要使用 DTD 验证模式的时候需要在 XML 文件的头部声明，以下是在 Spring 中使用 DTD 声明方式的代码：

```
<?xml version="1.0" encoding="UTF-8"?>
<!DOCTYPE beans PUBLIC "-//Spring//DTD BEAN 2.0//EN" "http://www.Springframework.org/dtd/ Spring-beans-2.0.dtd">
<beans>
... ...
</beans>
```

而以 Spring 为例，具体的 Spring-beans-2.0.dtd 部分如下：

```
<!ELEMENT beans (
    description?,
```

```
        (import | alias | bean)*
)>
<!ATTLIST beans default-lazy-init (true | false) "false">
<!ATTLIST beans default-merge (true | false) "false">
<!ATTLIST beans default-autowire (no | byName | byType | constructor|autodetect)"no">
<!ATTLIST beans default-dependency-check (none | objects | simple | all) "none">
<!ATTLIST beans default-init-method CDATA #IMPLIED>
<!ATTLIST beans default-destroy-method CDATA #IMPLIED>
... ...
```

XML Schema 语言就是 XSD（XML Schemas Definition）。XML Schema 描述了 XML 文档的结构。可以用一个指定的 XML Schema 来验证某个 XML 文档，以检查该 XML 文档是否符合其要求。文档设计者可以通过 XML Schema 指定 XML 文档所允许的结构和内容，并可据此检查 XML 文档是否是有效的。XML Schema 本身是 XML 文档，它符合 XML 语法结构。可以用通用的 XML 解析器解析它。

在使用 XML Schema 文档对 XML 实例文档进行检验，除了要声明名称空间外（xmlns=http://www.Springframework.org/schema/beans），还必须指定该名称空间所对应的 XML Schema 文档的存储位置。通过 schemaLocation 属性来指定名称空间所对应的 XML Schema 文档的存储位置，它包含两个部分，一部分是名称空间的 URI，另一部分就是该名称空间所标识的 XML Schema 文件位置或 URL 地址（xsi:schemaLocation="http://www.springframework.org/schema/beans http://www.Springframework.org/schema/beans/Spring-beans.xsd）。

```
<?xml version="1.0" encoding="UTF-8"?>
<beans xmlns="http://www.Springframework.org/schema/beans"
    xmlns:xsi="http://www.w3.org/2001/XMLSchema-instance"
    xsi:schemaLocation="http://www.Springframework.org/schema/beans
     http://www.Springframework.org/schema/beans/Spring-beans.xsd">
    ... ...
</beans>
```

Spring-beans-3.0.xsd 部分代码如下：

```
<?xml version="1.0" encoding="UTF-8" standalone="no"?>

<xsd:schema xmlns="http://www.Springframework.org/schema/beans"
       xmlns:xsd="http://www.w3.org/2001/XMLSchema"
       targetNamespace="http://www.Springframework.org/schema/beans">

    <xsd:import namespace="http://www.w3.org/XML/1998/namespace"/>

    <xsd:annotation>
        <xsd:documentation><![CDATA[
    ... ...
        ]]></xsd:documentation>
    </xsd:annotation>

    <!-- base types -->
    <xsd:complexType name="identifiedType" abstract="true">
        <xsd:annotation>
            <xsd:documentation><![CDATA[
    The unique identifier for a bean. The scope of the identifier
```

```
                    is the enclosing bean factory.
                ]]></xsd:documentation>
            </xsd:annotation>
            <xsd:attribute name="id" type="xsd:ID">
                <xsd:annotation>
                    <xsd:documentation><![CDATA[
The unique identifier for a bean.
                    ]]></xsd:documentation>
                </xsd:annotation>
            </xsd:attribute>
    </xsd:complexType>
    ... ...

</xsd:schema>
```

我们只是简单地介绍一下 XML 文件的验证模式的相关知识，目的在于让读者对后续知识的理解能有连续性，如果对 XML 有兴趣的读者可以进一步查阅相关资料。

2.6.2 验证模式的读取

了解了 DTD 与 XSD 的区别后我们再去分析 Spring 中对于验证模式的提取就更容易理解了。通过之前的分析我们锁定了 Spring 通过 getValidationModeForResource 方法来获取对应资源的的验证模式。

```
protected int getValidationModeForResource(Resource resource) {
        int validationModeToUse = getValidationMode();
        //如果手动指定了验证模式则使用指定的验证模式
        if (validationModeToUse != VALIDATION_AUTO) {
            return validationModeToUse;
        }
        //如果未指定则使用自动检测
        int detectedMode = detectValidationMode(resource);
        if (detectedMode != VALIDATION_AUTO) {
            return detectedMode;
        }
        return VALIDATION_XSD;
}
```

方法的实现其实还是很简单的，无非是如果设定了验证模式则使用设定的验证模式（可以通过对调用 XmlBeanDefinitionReader 中的 setValidationMode 方法进行设定），否则使用自动检测的方式。而自动检测验证模式的功能是在函数 detectValidationMode 方法中实现的，在 detectValidationMode 函数中又将自动检测验证模式的工作委托给了专门处理类 XmlValidationMode-Detector，调用了 XmlValidationModeDetector 的 validationModeDetector 方法，具体代码如下：

```
protected int detectValidationMode(Resource resource) {
        if (resource.isOpen()) {
            throw new BeanDefinitionStoreException(
                    "Passed-in Resource [" + resource + "] contains an open stream: " +
                    "cannot determine validation mode automatically. Either pass in a Resource " +
                    "that is able to create fresh streams, or explicitly specify the validationMode " +
```

```
                    "on your XmlBeanDefinitionReader instance.");
        }

                InputStream inputStream;
                try {
                    inputStream = resource.getInputStream();
                }
                catch (IOException ex) {
                    throw new BeanDefinitionStoreException(
                        "Unable to determine validation mode for [" + resource + "]: cannot open InputStream. " +
                        "Did you attempt to load directly from a SAX InputSource without specifying the " +
                        "validationMode on your XmlBeanDefinitionReader instance?", ex);
                }

                try {
                    return this.validationModeDetector.detectValidationMode(inputStream);
                }
                catch (IOException ex) {
                    throw new BeanDefinitionStoreException("Unable to determine validation mode for [" +
                        resource + "]: an error occurred whilst reading from the InputStream.", ex);
                }
    }
```

XmlValidationModeDetector.java

```
            public int detectValidationMode(InputStream inputStream) throws IOException {
                BufferedReader reader=new BufferedReader(new InputStreamReader(inputStream));
                try {
                    boolean isDtdValidated = false;
                    String content;
                    while ((content = reader.readLine()) != null) {
                        content = consumeCommentTokens(content);
                        //如果读取的行是空或者是注释则略过
                        if (this.inComment || !StringUtils.hasText(content)) {
                            continue;
                        }
                        if (hasDoctype(content)) {
                            isDtdValidated = true;
                            break;
                        }
                        //读取到<开始符号,验证模式一定会在开始符号之前
                        if (hasOpeningTag(content)) {
                            break;
                        }
                    }
                    return (isDtdValidated ? VALIDATION_DTD : VALIDATION_XSD);
                }
                catch (CharConversionException ex) {
                    // Choked on some character encoding...
                    // Leave the decision up to the caller.
                    return VALIDATION_AUTO;
```

```java
            }
            finally {
                reader.close();
            }
        }

        private boolean hasDoctype(String content) {
            return (content.indexOf(DOCTYPE) > -1);
        }
```

只要我们理解了 XSD 与 DTD 的使用方法，理解上面的代码应该不会太难，Spring 用来检测验证模式的办法就是判断是否包含 DOCTYPE，如果包含就是 DTD，否则就是 XSD。

2.7 获取 Document

经过了验证模式准备的步骤就可以进行 Document 加载了，同样 XmlBeanFactoryReader 类对于文档读取并没有亲力亲为，而是委托给了 DocumentLoader 去执行，这里的 DocumentLoader 是个接口，而真正调用的是 DefaultDocumentLoader，解析代码如下：

DefaultDocumentLoader.java
```java
public Document loadDocument(InputSource inputSource, EntityResolver entityResolver,
        ErrorHandler errorHandler, int validationMode, boolean namespaceAware) throws
Exception {

        DocumentBuilderFactory factory = createDocumentBuilderFactory(validationMode,
namespaceAware);
        if (logger.isDebugEnabled()) {
            logger.debug("Using JAXP provider [" + factory.getClass().getName()+"]");
        }
        DocumentBuilder builder = createDocumentBuilder(factory, entityResolver,
errorHandler);
        return builder.parse(inputSource);
    }
```

对于这部分代码其实并没有太多可以描述的，因为通过 SAX 解析 XML 文档的套路大致都差不多，Spring 在这里并没有什么特殊的地方，同样首先创建 DocumentBuilderFactory，再通过 DocumentBuilderFactory 创建 DocumentBuilder，进而解析 inputSource 来返回 Document 对象。对此感兴趣的读者可以在网上获取更多的资料。这里有必要提及一下 EntityResolver，对于参数 entityResolver，传入的是通过 getEntityResolver()函数获取的返回值，如下代码：

```java
protected EntityResolver getEntityResolver() {
    if (this.entityResolver == null) {
        // Determine default EntityResolver to use.
        ResourceLoader resourceLoader = getResourceLoader();
        if (resourceLoader != null) {
            this.entityResolver = new ResourceEntityResolver(resourceLoader);
        }
        else {
            this.entityResolver=new DelegatingEntityResolver (getBeanClassLoader());
```

```
            }
        }
        return this.entityResolver;
}
```

那么，EntityResolver 到底是做什么用的呢？

2.7.1 EntityResolver 用法

在 loadDocument 方法中涉及一个参数 EntityResolver，何为 EntityResolver？官网这样解释：如果 SAX 应用程序需要实现自定义处理外部实体，则必须实现此接口并使用 setEntityResolver 方法向 SAX 驱动器注册一个实例。也就是说，对于解析一个 XML，SAX 首先读取该 XML 文档上的声明，根据声明去寻找相应的 DTD 定义，以便对文档进行一个验证。默认的寻找规则，即通过网络（实现上就是声明的 DTD 的 URI 地址）来下载相应的 DTD 声明，并进行认证。下载的过程是一个漫长的过程，而且当网络中断或不可用时，这里会报错，就是因为相应的 DTD 声明没有被找到的原因。

EntityResolver 的作用是项目本身就可以提供一个如何寻找 DTD 声明的方法，即由程序来实现寻找 DTD 声明的过程，比如我们将 DTD 文件放到项目中某处，在实现时直接将此文档读取并返回给 SAX 即可。这样就避免了通过网络来寻找相应的声明。

首先看 entityResolver 的接口方法声明：

```
InputSource resolveEntity(String publicId, String systemId)
```

这里，它接收两个参数 publicId 和 systemId，并返回一个 inputSource 对象。这里我们以特定配置文件来进行讲解。

1. 如果我们在解析验证模式为 XSD 的配置文件，代码如下：

```
<?xml version="1.0" encoding="UTF-8"?>
<beans xmlns="http://www.Springframework.org/schema/beans"
    xmlns:xsi="http://www.w3.org/2001/XMLSchema-instance"
    xsi:schemaLocation="http://www.Springframework.org/schema/beans
     http://www.springframework.org/schema/beans/Spring-beans.xsd">
    ... ...
</beans>
```

读取到以下两个参数。

- **publicId**：null
- **systemId**：http://www.springframework.org/schema/beans/Spring-beans.xsd

2. 如果我们在解析验证模式为 DTD 的配置文件，代码如下：

```
<?xml version="1.0" encoding="UTF-8"?>
<!DOCTYPE beans PUBLIC "-//Spring//DTD BEAN 2.0//EN" "http://www.Springframework.org/dtd/Spring-beans-2.0.dtd">
    <beans>
    ... ...
    </beans>
```

读取到以下两个参数。

- **publicId**:-//Spring//DTD BEAN 2.0//EN
- **systemId**:http://www.springframework.org/dtd/Spring-beans-2.0.dtd

之前已经提到过,验证文件默认的加载方式是通过 URL 进行网络下载获取,这样会造成延迟,用户体验也不好,一般的做法都是将验证文件放置在自己的工程里,那么怎么做才能将这个 URL 转换为自己工程里对应的地址文件呢?我们以加载 DTD 文件为例来看看 Spring 中是如何实现的。根据之前 Spring 中通过 getEntityResolver()方法对 EntityResolver 的获取,我们知道,Spring 中使用 DelegatingEntityResolver 类为 EntityResolver 的实现类,resolveEntity 实现方法如下:

DelegatingEntityResolver.java
```
    public InputSource resolveEntity(String publicId, String systemId) throws
SAXException, IOException {
        if (systemId != null) {
            if (systemId.endsWith(DTD_SUFFIX)) {
            //如果是dtd从这里解析
                return this.dtdResolver.resolveEntity(publicId, systemId);
            }
            else if (systemId.endsWith(XSD_SUFFIX)) {
                //通过调用META-INF/Spring.schemas解析
                return this.schemaResolver.resolveEntity(publicId, systemId);
            }
        }
        return null;
    }
```

我们可以看到,对不同的验证模式,Spring 使用了不同的解析器解析。这里简单描述一下原理,比如加载 DTD 类型的 BeansDtdResolver 的 resolveEntity 是直接截取 systemId 最后的 xx.dtd 然后去当前路径下寻找,而加载 XSD 类型的 PluggableSchemaResolver 类的 resolveEntity 是默认到 META-INF/Spring.schemas 文件中找到 systemid 所对应的 XSD 文件并加载。

BeansDtdResolver.java
```
    public InputSource resolveEntity(String publicId, String systemId)throws IOException {
        if (logger.isTraceEnabled()) {
            logger.trace("Trying to resolve XML entity with public ID [" + publicId +
                "] and system ID [" + systemId + "]");
        }
        // DTD_EXTENSION = ".dtd";
        if (systemId != null && systemId.endsWith(DTD_EXTENSION)) {
            int lastPathSeparator = systemId.lastIndexOf("/");
            for (String DTD_NAME : DTD_NAMES) {
                // DTD_NAMES = {"Spring-beans-2.0", "Spring-beans"};
                int dtdNameStart = systemId.indexOf(DTD_NAME);
                if (dtdNameStart > lastPathSeparator) {
                    String dtdFile = systemId.substring(dtdNameStart);
                    if (logger.isTraceEnabled()) {
                        logger.trace("Trying to locate [" + dtdFile +"]in Spring jar");
                    }
```

```
                try {
                    Resource resource = new ClassPathResource(dtdFile, getClass());
                    InputSource source = new InputSource(resource.getInputStream());
                    source.setPublicId(publicId);
                    source.setSystemId(systemId);
                    if (logger.isDebugEnabled()) {
                        logger.debug("Found beans DTD [" + systemId + "] in classpath: " + dtdFile);
                    }
                    return source;
                }
                catch (IOException ex) {
                    if (logger.isDebugEnabled()) {
                        logger.debug("Could not resolve beans DTD [" + systemId + "]: not found in class path", ex);
                    }
                }
            }
        }
    }
    return null;
}
```

2.8 解析及注册 BeanDefinitions

当把文件转换为 Document 后，接下来的提取及注册 bean 就是我们的重头戏。继续上面的分析，当程序已经拥有 XML 文档文件的 Document 实例对象时，就会被引入下面这个方法。

XmlBeanDefinitionReader.java
```
public int registerBeanDefinitions(Document doc, Resource resource) throws BeanDefinitionStoreException {
        //使用 DefaultBeanDefinitionDocumentReader 实例化 BeanDefinitionDocumentReader
        BeanDefinitionDocumentReader documentReader=createBeanDefinitionDocumentReader();
        //将环境变量设置其中
        documentReader.setEnvironment(this.getEnvironment());
        //在实例化 BeanDefinitionReader 时候会将 BeanDefinitionRegistry 传入，默认使用继承自 DefaultListableBeanFactory 的子类
        //记录统计前 BeanDefinition 的加载个数
        int countBefore = getRegistry().getBeanDefinitionCount();
    //加载及注册 bean
        documentReader.registerBeanDefinitions(doc, createReaderContext(resource));
        //记录本次加载的 BeanDefinition 个数
        return getRegistry().getBeanDefinitionCount() - countBefore;
}
```

其中的参数 doc 是通过上一节 loadDocument 加载转换出来的。在这个方法中很好地应用了面向对象中单一职责的原则，将逻辑处理委托给单一的类进行处理，而这个逻辑处理类就是 BeanDefinitionDocumentReader。BeanDefinitionDocumentReader 是一个接口，而实例化的工作是在 createBeanDefinitionDocumentReader()中完成的，而通过此方法，BeanDefinitionDocumentReader 真正

的类型其实已经是 DefaultBeanDefinitionDocumentReader 了，进入 DefaultBeanDefinitionDocument-Reader 后，发现这个方法的重要目的之一就是提取 root，以便于再次将 root 作为参数继续 BeanDefinition 的注册。

```
public void registerBeanDefinitions(Document doc, XmlReaderContext readerContext){
    this.readerContext = readerContext;

    logger.debug("Loading bean definitions");
    Element root = doc.getDocumentElement();

    doRegisterBeanDefinitions(root);
}
```

经过艰难险阻，磕磕绊绊，我们终于到了核心逻辑的底部 doRegisterBeanDefinitions(root)，至少我们在这个方法中看到了希望。

如果说以前一直是 XML 加载解析的准备阶段，那么 doRegisterBeanDefinitions 算是真正地开始进行解析了，我们期待的核心部分真正开始了。

```
protected void doRegisterBeanDefinitions(Element root) {
    //处理 profile 属性
    String profileSpec = root.getAttribute(PROFILE_ATTRIBUTE);
    if (StringUtils.hasText(profileSpec)) {
        Assert.state(this.environment !=null, "environment property must not be null");
        String[] specifiedProfiles = StringUtils.tokenizeToStringArray(profileSpec,
BeanDefinitionParserDelegate.MULTI_VALUE_ATTRIBUTE_DELIMITERS);
        if (!this.environment.acceptsProfiles(specifiedProfiles)) {
            return;
        }
    }
    //专门处理解析
    BeanDefinitionParserDelegate parent = this.delegate;
    this.delegate = createHelper(readerContext, root, parent);

    //解析前处理，留给子类实现
    preProcessXml(root);
    parseBeanDefinitions(root, this.delegate);
    //解析后处理，留给子类实现
    postProcessXml(root);

    this.delegate = parent;
}
```

通过上面的代码我们看到了处理流程，首先是对 profile 的处理，然后开始进行解析，可是当我们跟进 preProcessXml(root)或者 postProcessXml(root)发现代码是空的，既然是空的写着还有什么用呢？就像面向对象设计方法学中常说的一句话，一个类要么是面向继承的设计的，要么就用 final 修饰。在 DefaultBeanDefinitionDocumentReader 中并没有用 final 修饰，所以它是面向继承而设计的。这两个方法正是为子类而设计的，如果读者有了解过设计模式，可以很快速地反映出这是模版方法模式，如果继承自 DefaultBeanDefinitionDocumentReader 的子类需要在 Bean 解析前后做一些处理的话，那么只需要重写这两个方法就可以了。

2.8.1 profile 属性的使用

我们注意到在注册 Bean 的最开始是对 PROFILE_ATTRIBUTE 属性的解析，可能对于我们来说，profile 属性并不是很常用。让我们先了解一下这个属性。

分析 profile 前我们先了解下 profile 的用法，官方示例代码片段如下：

```
<beans xmlns="http://www.Springframework.org/schema/beans"
    xmlns:xsi="http://www.w3.org/2001/XMLSchema-instance" xmlns:jdbc="http://www.
Springframework.org/schema/jdbc"
    xmlns:jee="http://www.springframework.org/schema/jee"
    xsi:schemaLocation="...">
        ... ...
<beans profile="dev">
        ... ...
 </beans>
        <beans profile="production">
            ... ...
        </beans>
</beans>
```

集成到 Web 环境中时，在 web.xml 中加入以下代码：

```
  <context-param>
    <param-name>Spring.profiles.active</param-name>
    <param-value>dev</param-value>
</context-param>
```

有了这个特性我们就可以同时在配置文件中部署两套配置来适用于生产环境和开发环境，这样可以方便的进行切换开发、部署环境，最常用的就是更换不同的数据库。

了解了 profile 的使用再来分析代码会清晰得多，首先程序会获取 beans 节点是否定义了 profile 属性，如果定义了则会需要到环境变量中去寻找，所以这里首先断言 environment 不可能为空，因为 profile 是可以同时指定多个的，需要程序对其拆分，并解析每个 profile 是都符合环境变量中所定义的，不定义则不会浪费性能去解析。

2.8.2 解析并注册 BeanDefinition

处理了 profile 后就可以进行 XML 的读取了，跟踪代码进入 parseBeanDefinitions(root, this.delegate)。

```
protected void parseBeanDefinitions(Element root, BeanDefinitionParserDelegate delegate) {
        //对 beans 的处理
        if (delegate.isDefaultNamespace(root)) {
            NodeList nl = root.getChildNodes();
            for (int i = 0; i < nl.getLength(); i++) {
                Node node = nl.item(i);
                if (node instanceof Element) {
                    Element ele = (Element) node;
                    if (delegate.isDefaultNamespace(ele)) {
                        //对 bean 的处理
```

2.8 解析及注册 BeanDefinitions

```
                        parseDefaultElement(ele, delegate);
                    }
                    else {
                        //对 bean 的处理
                        delegate.parseCustomElement(ele);
                    }
                }
            }
        }
        else {
            delegate.parseCustomElement(root);
        }
    }
```

上面的代码看起来逻辑还是蛮清晰的，因为在 Spring 的 XML 配置里面有两大类 Bean 声明，一个是默认的，如：

`<bean id="test" class="test.TestBean"/>`

另一类就是自定义的，如：

`<tx:annotation-driven/>`

而两种方式的读取及解析差别是非常大的，如果采用 Spring 默认的配置，Spring 当然知道该怎么做，但是如果是自定义的，那么就需要用户实现一些接口及配置了。对于根节点或者子节点如果是默认命名空间的话则采用 parseDefaultElement 方法进行解析，否则使用 delegate.parseCustomElement 方法对自定义命名空间进行解析。而判断是否默认命名空间还是自定义命名空间的办法其实是使用 node.getNamespaceURI() 获取命名空间，并与 Spring 中固定的命名空间 http://www.springframework.org/schema/beans 进行比对。如果一致则认为是默认，否则就认为是自定义。而对于默认标签解析与自定义标签解析我们将会在下一章中进行讨论。

第 3 章 默认标签的解析

之前提到过 Spring 中的标签包括默认标签和自定义标签两种,而两种标签的用法以及解析方式存在着很大的不同,本章节重点带领读者详细分析默认标签的解析过程。

默认标签的解析是在 parseDefaultElement 函数中进行的,函数中的功能逻辑一目了然,分别对 4 种不同标签(import、alias、bean 和 beans)做了不同的处理。

```
private void parseDefaultElement(Element ele, BeanDefinitionParserDelegate delegate) {
    //对 import 标签的处理
        if (delegate.nodeNameEquals(ele, IMPORT_ELEMENT)) {
            importBeanDefinitionResource(ele);
        }
    //对 alias 标签的处理
        else if (delegate.nodeNameEquals(ele, ALIAS_ELEMENT)) {
            processAliasRegistration(ele);
        }
    //对 bean 标签的处理
        else if (delegate.nodeNameEquals(ele, BEAN_ELEMENT)) {
            processBeanDefinition(ele, delegate);
        }
    //对 beans 标签的处理
        else if (delegate.nodeNameEquals(ele, NESTED_BEANS_ELEMENT)) {
            doRegisterBeanDefinitions(ele);
        }
}
```

3.1 bean 标签的解析及注册

在 4 种标签的解析中,对 bean 标签的解析最为复杂也最为重要,所以我们从此标签开始深入分析,如果能理解此标签的解析过程,其他标签的解析自然会迎刃而解。首先我们进入函数 processBeanDefinition(ele, delegate)。

```
protected void processBeanDefinition(Element ele, BeanDefinitionParserDelegate delegate) {
    BeanDefinitionHolder bdHolder = delegate.parseBeanDefinitionElement(ele);
    if (bdHolder != null) {
```

```
            bdHolder = delegate.decorateBeanDefinitionIfRequired(ele, bdHolder);
            try {
                BeanDefinitionReaderUtils.registerBeanDefinition(bdHolder, getReaderContext().
getRegistry());
            }
            catch (BeanDefinitionStoreException ex) {
                getReaderContext().error("Failed to register bean definition with name '" +
                        bdHolder.getBeanName() + "'", ele, ex);
            }
            getReaderContext().fireComponentRegistered(new BeanComponentDefinition(bdHolder));
        }
    }
```

乍一看，似乎一头雾水，没有以前的函数那样清晰的逻辑。大致的逻辑总结如下。

1. 首先委托 BeanDefinitionDelegate 类的 parseBeanDefinitionElement 方法进行元素解析，返回 BeanDefinitionHolder 类型的实例 bdHolder，经过这个方法后，bdHolder 实例已经包含我们配置文件中配置的各种属性了，例如 class、name、id、alias 之类的属性。

2. 当返回的 bdHolder 不为空的情况下若存在默认标签的子节点下再有自定义属性，还需要再次对自定义标签进行解析。

3. 解析完成后，需要对解析后的 bdHolder 进行注册，同样，注册操作委托给了 BeanDefinitionReaderUtils 的 registerBeanDefinition 方法。

4. 最后发出响应事件，通知相关的监听器，这个 bean 已经加载完成了。

配合时序图（见图 3-1），可能会更容易理解。

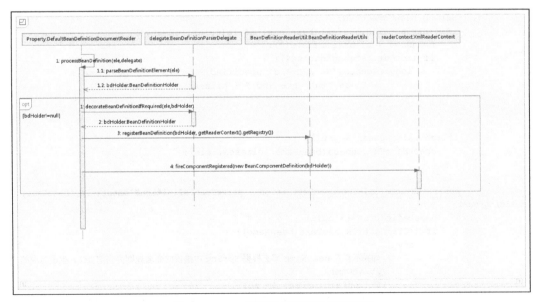

图 3-1　bean 标签的解析及注册时序图

3.1.1 解析 BeanDefinition

下面我们就针对各个操作做具体分析。首先我们从元素解析及信息提取开始,也就是 BeanDefinitionHolder bdHolder = delegate.parseBeanDefinitionElement(ele),进入 BeanDefinition-Delegate 类的 parseBeanDefinitionElement 方法。

BeanDefinitionDelegate.java

```java
public BeanDefinitionHolder parseBeanDefinitionElement(Element ele) {
    return parseBeanDefinitionElement(ele, null);
}

public BeanDefinitionHolder parseBeanDefinitionElement(Element ele, BeanDefinition containingBean) {
    //解析 id 属性
    String id = ele.getAttribute(ID_ATTRIBUTE);
    //解析 name 属性
    String nameAttr = ele.getAttribute(NAME_ATTRIBUTE);

    //分割 name 属性
    List<String> aliases = new ArrayList<String>();
    if (StringUtils.hasLength(nameAttr)) {
        String[] nameArr = StringUtils.tokenizeToStringArray(nameAttr, MULTI_VALUE_ATTRIBUTE_DELIMITERS);
        aliases.addAll(Arrays.asList(nameArr));
    }

    String beanName = id;
    if (!StringUtils.hasText(beanName) && !aliases.isEmpty()) {
        beanName = aliases.remove(0);
        if (logger.isDebugEnabled()) {
            logger.debug("No XML 'id' specified - using '" + beanName +
                "' as bean name and " + aliases + " as aliases");
        }
    }

    if (containingBean == null) {
        checkNameUniqueness(beanName, aliases, ele);
    }

    AbstractBeanDefinition beanDefinition = parseBeanDefinitionElement(ele, beanName, containingBean);
    if (beanDefinition != null) {
        if (!StringUtils.hasText(beanName)) {
            try {
                //如果不存在 beanName 那么根据 Spring 中提供的命名规则为当前 bean 生成对应的
                //beanName
                if (containingBean != null) {
                    beanName = BeanDefinitionReaderUtils.generateBeanName(
                        beanDefinition, this.readerContext.getRegistry(), true);
                }
```

```
            else {
                beanName = this.readerContext.generateBeanName(beanDefinition);

                String beanClassName = beanDefinition.getBeanClassName();
                if (beanClassName != null &&
                        beanName.startsWith(beanClassName) && beanName.length() > beanClassName.length() &&
                        !this.readerContext.getRegistry().IsBeanNameInUse(beanClassName)) {
                    aliases.add(beanClassName);
                }
            }
            if (logger.isDebugEnabled()) {
                logger.debug("Neither XML 'id' nor 'name' specified - " +
                        "using generated bean name [" + beanName + "]");
            }
        }
        catch (Exception ex) {
            error(ex.getMessage(), ele);
            return null;
        }
        String[] aliasesArray = StringUtils.toStringArray(aliases);
        return new BeanDefinitionHolder(beanDefinition, beanName, aliasesArray);
    }

    return null;
}
```

以上便是对默认标签解析的全过程了。当然，对 Spring 的解析犹如洋葱剥皮一样，一层一层地进行，尽管现在只能看到对属性 id 以及 name 的解析，但是很庆幸，思路我们已经了解了。在开始对属性展开全面解析前，Spring 在外层又做了一个当前层的功能架构，在当前层完成的主要工作包括如下内容。

1. 提取元素中的 id 以及 name 属性。
2. 进一步解析其他所有属性并统一封装至 GenericBeanDefinition 类型的实例中。
3. 如果检测到 bean 没有指定 beanName，那么使用默认规则为此 Bean 生成 beanName。
4. 将获取到的信息封装到 BeanDefinitionHolder 的实例中。

我们进一步地查看步骤 2 中对标签其他属性的解析过程。

```
public AbstractBeanDefinition parseBeanDefinitionElement(
        Element ele, String beanName, BeanDefinition containingBean) {

    this.parseState.push(new BeanEntry(beanName));

    String className = null;
    //解析 class 属性
    if (ele.hasAttribute(CLASS_ATTRIBUTE)) {
        className = ele.getAttribute(CLASS_ATTRIBUTE).trim();
    }

    try {
```

```java
            String parent = null;
            //解析parent属性
            if (ele.hasAttribute(PARENT_ATTRIBUTE)) {
                parent = ele.getAttribute(PARENT_ATTRIBUTE);
            }
            //创建用于承载属性的AbstractBeanDefinition类型的GenericBeanDefinition
            AbstractBeanDefinition bd = createBeanDefinition(className, parent);

            //硬编码解析默认bean的各种属性
            parseBeanDefinitionAttributes(ele, beanName, containingBean, bd);
            //提取description
            bd.setDescription(DomUtils.getChildElementValueByTagName(ele, DESCRIPTION_ELEMENT));

            //解析元数据
            parseMetaElements(ele, bd);
            //解析lookup-method属性
            parseLookupOverrideSubElements(ele, bd.getMethodOverrides());
            //解析replaced-method属性
            parseReplacedMethodSubElements(ele, bd.getMethodOverrides());

            //解析构造函数参数
            parseConstructorArgElements(ele, bd);
            //解析property子元素
            parsePropertyElements(ele, bd);
            //解析qualifier子元素
            parseQualifierElements(ele, bd);

            bd.setResource(this.readerContext.getResource());
            bd.setSource(extractSource(ele));

            return bd;
        }
        catch (ClassNotFoundException ex) {
            error("Bean class [" + className + "] not found", ele, ex);
        }
        catch (NoClassDefFoundError err) {
            error("Class that bean class [" +className +"] depends on not found", ele, err);
        }
        catch (Throwable ex) {
            error("Unexpected failure during bean definition parsing", ele, ex);
        }
        finally {
            this.parseState.pop();
        }

        return null;
    }
```

终于，bean标签的所有属性，不论常用的还是不常用的我们都看到了，尽管有些复杂的属性还需要进一步的解析，不过丝毫不会影响我们兴奋的心情。接下来，我们继续一些复杂标签属性的解析。

1. 创建用于属性承载的 BeanDefinition

BeanDefinition 是一个接口，在 Spring 中存在三种实现：RootBeanDefinition、ChildBeanDefinition 以及 GenericBeanDefinition。三种实现均继承了 AbstractBeanDefiniton，其中 BeanDefinition 是配置文件<bean>元素标签在容器中的内部表示形式。<bean>元素标签拥有 class、scope、lazy-init 等配置属性，BeanDefinition 则提供了相应的 beanClass、scope、lazyInit 属性，BeanDefinition 和<bean>中的属性是一一对应的。其中 RootBeanDefinition 是最常用的实现类，它对应一般性的<bean>元素标签，GenericBeanDefinition 是自 2.5 版本以后新加入的 bean 文件配置属性定义类，是一站式服务类。

在配置文件中可以定义父<bean>和子<bean>，父<bean>用 RootBeanDefinition 表示，而子<bean>用 ChildBeanDefiniton 表示，而没有父<bean>的<bean>就使用 RootBeanDefinition 表示。AbstractBeanDefinition 对两者共同的类信息进行抽象。

Spring 通过 BeanDefinition 将配置文件中的<bean>配置信息转换为容器的内部表示，并将这些 BeanDefiniton 注册到 BeanDefinitonRegistry 中。Spring 容器的 BeanDefinitionRegistry 就像是 Spring 配置信息的内存数据库，主要是以 map 的形式保存，后续操作直接从 BeanDefinitionRegistry 中读取配置信息。它们之间的关系如图 3-2 所示。

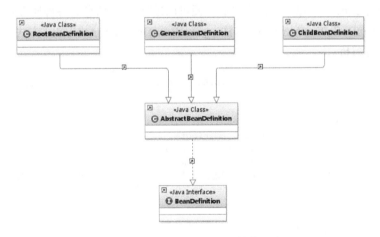

图 3-2　BeanDefinition 及其实现类

由此可知，要解析属性首先要创建用于承载属性的实例，也就是创建 GenericBeanDefinition 类型的实例。而代码 createBeanDefinition(className, parent) 的作用就是实现此功能。

```
protected AbstractBeanDefinition createBeanDefinition(String className, String parentName)
        throws ClassNotFoundException {

    return BeanDefinitionReaderUtils.createBeanDefinition(
            parentName, className, this.readerContext.getBeanClassLoader());
}
```

```
BeanDefinitionReaderUtils.java
    public static AbstractBeanDefinition createBeanDefinition(
            String parentName, String className, ClassLoader classLoader) throws
ClassNotFoundException {

            GenericBeanDefinition bd = new GenericBeanDefinition();
        //parentName 可能为空
        bd.setParentName(parentName);
            if (className != null) {
                if (classLoader != null) {
                //如果 classLoader 不为空,则使用以传入的 classLoader 同一虚拟机加载类对象,否则只是
                //记录 className
                    bd.setBeanClass(ClassUtils.forName(className, classLoader));
                }
                else {
                    bd.setBeanClassName(className);
                }
            }
            return bd;
    }
```

2. 解析各种属性

当我们创建了 bean 信息的承载实例后,便可以进行 bean 信息的各种属性解析了,首先我们进入 parseBeanDefinitionAttributes 方法。parseBeanDefinitionAttributes 方法是对 element 所有元素属性进行解析:

```
    public AbstractBeanDefinition parseBeanDefinitionAttributes(Element ele, String beanName,
            BeanDefinition containingBean, AbstractBeanDefinition bd) {

                //解析 scope 属性
                if (ele.hasAttribute(SCOPE_ATTRIBUTE)) {
                    // Spring 2.x "scope" attribute
                    bd.setScope(ele.getAttribute(SCOPE_ATTRIBUTE));
                    if (ele.hasAttribute(SINGLETON_ATTRIBUTE)) {
                //scope 与 singleton 两个属性只能指定其中之一,不可以同时出现,否则 Spring 将会报出异常
                        error("Specify either 'scope' or 'singleton', not both", ele);
                    }
                }
                //解析 singleton 属性
                else if (ele.hasAttribute(SINGLETON_ATTRIBUTE)) {
                    // Spring 1.x "singleton" attribute
                    bd.setScope(TRUE_VALUE.equals(ele.getAttribute(SINGLETON_ATTRIBUTE)) ?
                            BeanDefinition.SCOPE_SINGLETON : BeanDefinition.SCOPE_PROTOTYPE);
                }
                else if (containingBean != null) {
                // Take default from containing bean in case of an inner bean definition.
                //在嵌入 beanDifinition 情况下且没有单独指定 scope 属性则使用父类默认的属性
                    bd.setScope(containingBean.getScope());
                }
                //解析 abstract 属性
```

```java
        if (ele.hasAttribute(ABSTRACT_ATTRIBUTE)) {
            bd.setAbstract(TRUE_VALUE.equals(ele.getAttribute(ABSTRACT_ATTRIBUTE)));
        }
        //解析 lazy-init 属性
        String lazyInit = ele.getAttribute(LAZY_INIT_ATTRIBUTE);
        if (DEFAULT_VALUE.equals(lazyInit)) {
            lazyInit = this.defaults.getLazyInit();
        }
        //若没有设置或设置成其他字符都会被设置为 false
        bd.setLazyInit(TRUE_VALUE.equals(lazyInit));

        //解析 autowire 属性
        String autowire = ele.getAttribute(AUTOWIRE_ATTRIBUTE);
        bd.setAutowireMode(getAutowireMode(autowire));

        //解析 dependency-check 属性
        String dependencyCheck = ele.getAttribute(DEPENDENCY_CHECK_ATTRIBUTE);
        bd.setDependencyCheck(getDependencyCheck(dependencyCheck));

        //解析 depends-on 属性
        if (ele.hasAttribute(DEPENDS_ON_ATTRIBUTE)) {
            String dependsOn = ele.getAttribute(DEPENDS_ON_ATTRIBUTE);
            bd.setDependsOn(StringUtils.tokenizeToStringArray(dependsOn, MULTI_VALUE_ATTRIBUTE_DELIMITERS));
        }

        //解析 autowire-candidate 属性
        String autowireCandidate = ele.getAttribute(AUTOWIRE_CANDIDATE_ATTRIBUTE);
        if ("".equals(autowireCandidate) || DEFAULT_VALUE.equals(autowireCandidate)) {
            String candidatePattern = this.defaults.getAutowireCandidates();
            if (candidatePattern != null) {
                String[] patterns = StringUtils.commaDelimitedListToStringArray(candidatePattern);
                bd.setAutowireCandidate(PatternMatchUtils.simpleMatch(patterns, beanName));
            }
        }
        else {
            bd.setAutowireCandidate(TRUE_VALUE.equals(autowireCandidate));
        }

        //解析 primary 属性
        if (ele.hasAttribute(PRIMARY_ATTRIBUTE)) {
            bd.setPrimary(TRUE_VALUE.equals(ele.getAttribute(PRIMARY_ATTRIBUTE)));
        }
        //解析 init-method 属性
        if (ele.hasAttribute(INIT_METHOD_ATTRIBUTE)) {
            String initMethodName = ele.getAttribute(INIT_METHOD_ATTRIBUTE);
            if (!"".equals(initMethodName)) {
                bd.setInitMethodName(initMethodName);
            }
        }
        else {
            if (this.defaults.getInitMethod() != null) {
                bd.setInitMethodName(this.defaults.getInitMethod());
```

```
                    bd.setEnforceInitMethod(false);
                }
            }
            //解析 destroy-method 属性
            if (ele.hasAttribute(DESTROY_METHOD_ATTRIBUTE)) {
                String destroyMethodName = ele.getAttribute(DESTROY_METHOD_ATTRIBUTE);
                if (!"".equals(destroyMethodName)) {
                    bd.setDestroyMethodName(destroyMethodName);
                }
            }
            else {
                if (this.defaults.getDestroyMethod() != null) {
                    bd.setDestroyMethodName(this.defaults.getDestroyMethod());
                    bd.setEnforceDestroyMethod(false);
                }
            }
            //解析 factory-method 属性
            if (ele.hasAttribute(FACTORY_METHOD_ATTRIBUTE)) {
                bd.setFactoryMethodName(ele.getAttribute(FACTORY_METHOD_ATTRIBUTE));
            }
            //解析 factory-bean 属性
            if (ele.hasAttribute(FACTORY_BEAN_ATTRIBUTE)) {
                bd.setFactoryBeanName(ele.getAttribute(FACTORY_BEAN_ATTRIBUTE));
            }

            return bd;
    }
```

我们可以清楚地看到 Spring 完成了对所有 bean 属性的解析，这些属性中有很多是我们经常使用的，同时我相信也一定会有或多或少的属性是读者不熟悉或者是没有使用过的，有兴趣的读者可以查阅相关资料进一步了解每个属性。

3. 解析子元素 meta

在开始解析元数据的分析前，我们先回顾一下元数据 meta 属性的使用。

```
<bean id="myTestBean" class="bean.MyTestBean">
    <meta key="testStr" value="aaaaaaaa"/>
</bean>
```

这段代码并不会体现在 MyTestBean 的属性当中，而是一个额外的声明，当需要使用里面的信息的时候可以通过 BeanDefinition 的 getAttribute(key)方法进行获取。

对 meta 属性的解析代码如下：
```
    public void parseMetaElements(Element ele, BeanMetadataAttributeAccessor attributeAccessor) {
        //获取当前节点的所有子元素
        NodeList nl = ele.getChildNodes();
        for (int i = 0; i < nl.getLength(); i++) {
            Node node = nl.item(i);
            //提取 meta
            if (isCandidateElement(node) && nodeNameEquals(node, META_ELEMENT)) {
                Element metaElement = (Element) node;
                String key = metaElement.getAttribute(KEY_ATTRIBUTE);
```

```
                    String value = metaElement.getAttribute(VALUE_ATTRIBUTE);
                    //使用 key、value 构造 BeanMetadataAttribute
                    BeanMetadataAttribute attribute = new BeanMetadataAttribute(key, value);
                    attribute.setSource(extractSource(metaElement));
                    //记录信息
                    attributeAccessor.addMetadataAttribute(attribute);
                }
            }
        }
    }
```

4. 解析子元素 lookup-method

同样，子元素 lookup-method 似乎并不是很常用，但是在某些时候它的确是非常有用的属性，通常我们称它为获取器注入。引用 *Spring in Action* 中的一句话：获取器注入是一种特殊的方法注入，它是把一个方法声明为返回某种类型的 bean，但实际要返回的 bean 是在配置文件里面配置的，此方法可用在设计有些可插拔的功能上，解除程序依赖。我们看看具体的应用。

1. 首先我们创建一个父类。

```
package test.lookup.bean;

public class User {

    public void showMe(){
        System.out.println("i am user");
    }
}
```

2. 创建其子类并覆盖 showMe 方法。

```
package test.lookup.bean;

public class Teacher extends User{
    public void showMe(){
            System.out.println("i am Teacher");
    }
}
```

3. 创建调用方法。

```
public abstract class GetBeanTest {

    public void showMe(){
        this.getBean().showMe();
    }
    public abstract User getBean();
}
```

4. 创建测试方法。

```
package test.lookup;

import org.Springframework.context.ApplicationContext;
import org.Springframework.context.support.ClassPathXmlApplicationContext;
import test.lookup.app.GetBeanTest;
```

```java
public class Main {
    public static void main(String[] args) {
        ApplicationContext bf =
            new ClassPathXmlApplicationContext("test/lookup/lookupTest.xml");
        GetBeanTest test=(GetBeanTest) bf.getBean("getBeanTest");
        test.showMe();
    }
}
```

到现在为止,除了配置文件外,整个测试方法就完成了,如果之前没有接触过获取器注入的读者们可能会有疑问:抽象方法还没有被实现,怎么可以直接调用呢?答案就在 Spring 为我们提供的获取器中,我们看看配置文件是怎么配置的。

```xml
<?xml version="1.0" encoding="UTF-8"?>
<beans xmlns="http://www.Springframework.org/schema/beans"
    xmlns:xsi="http://www.w3.org/2001/XMLSchema-instance"
    xsi:schemaLocation="http://www.springframework.org/schema/beans http://www.springframework.org/schema/beans/Spring-beans.xsd">

            <bean id="getBeanTest" class="test.lookup.app.GetBeanTest">
                <lookup-method name="getBean" bean="teacher"/>
            </bean>

<bean id="teacher" class="test.lookup.bean.Teacher"/>
</beans>
```

在配置文件中,我们看到了源码解析中提到的 lookup-method 子元素,这个配置完成的功能是动态地将 teacher 所代表的 bean 作为 getBean 的返回值,运行测试方法我们会看到控制台上的输出:

```
i am Teacher
```

当我们的业务变更或者在其他情况下,teacher 里面的业务逻辑已经不再符合我们的业务要求,需要进行替换怎么办呢?这是我们需要增加新的逻辑类:

```java
package test.lookup.bean;

public class Student extends User {

    public void showMe(){
        System.out.println("i am student");
    }
}
```

同时修改配置文件:

```xml
<?xml version="1.0" encoding="UTF-8"?>
<beans xmlns="http://www.springframework.org/schema/beans"
    xmlns:xsi="http://www.w3.org/2001/XMLSchema-instance"
    xsi:schemaLocation="http://www.Springframework.org/schema/beans http://www.springframework.org/schema/beans/Spring-beans.xsd">

            <bean id="getBeanTest" class="test.lookup.app.GetBeanTest">
                <lookup-method name="getBean" bean="student"/>
            </bean>

<bean id="teacher" class="test.lookup.bean.Teacher"/>
```

```
    <bean id="student" class="test.lookup.bean.Student"/>
</beans>
```
再次运行测试类,你会发现不一样的结果:

i am Student

至此,我们已经初步了解了 lookup-method 子元素所提供的大致功能,相信这时再次去看它的属性提取源码会觉得更有针对性。

```
public void parseLookupOverrideSubElements(Element beanEle, MethodOverrides overrides) {
    NodeList nl = beanEle.getChildNodes();
    for (int i = 0; i < nl.getLength(); i++) {
        Node node = nl.item(i);
        //仅当在 Spring 默认 bean 的子元素下且为    <lookup-method 时有效
        if (isCandidateElement(node) && nodeNameEquals(node, LOOKUP_METHOD_ELEMENT)) {
            Element ele = (Element) node;
            //获取要修饰的方法
            String methodName = ele.getAttribute(NAME_ATTRIBUTE);
            //获取配置返回的 bean
            String beanRef = ele.getAttribute(BEAN_ELEMENT);
            LookupOverride override = new LookupOverride(methodName, beanRef);
            override.setSource(extractSource(ele));
            overrides.addOverride(override);
        }
    }
}
```

上面的代码很眼熟,似乎与 parseMetaElements 的代码大同小异,最大的区别就是在 if 判断中的节点名称在这里被修改为 LOOKUP_METHOD_ELEMENT。还有,在数据存储上面通过使用 LookupOverride 类型的实体类来进行数据承载并记录在 AbstractBeanDefinition 中的 methodOverrides 属性中。

5. 解析子元素 replaced-method

这个方法主要是对 bean 中 replaced-method 子元素的提取,在开始提取分析之前我们还是预先介绍下这个元素的用法。

方法替换:可以在运行时用新的方法替换现有的方法。与之前的 look-up 不同的是,replaced-method 不但可以动态地替换返回实体 bean,而且还能动态地更改原有方法的逻辑。我们来看看使用示例。

1. 在 changeMe 中完成某个业务逻辑。

```
public class TestChangeMethod {

    public void changeMe(){
        System.out.println("changeMe");
    }
}
```

2. 在运营一段时间后需要改变原有的业务逻辑。

```
public class TestMethodReplacer implements MethodReplacer{
@Override
```

```java
public Object reimplement(Object obj, Method method, Object[] args)throws Throwable {
        System.out.println("我替换了原有的方法");
        return null;
    }
}
```

3. 使替换后的类生效。

```xml
<?xml version="1.0" encoding="UTF-8"?>
<beans xmlns="http://www.Springframework.org/schema/beans"
       xmlns:xsi="http://www.w3.org/2001/XMLSchema-instance"
    xsi:schemaLocation="http://www.Springframework.org/schema/beans http://www.Springframework.org/schema/beans/Spring-beans.xsd">

    <bean id="testChangeMethod" class="test.replacemethod.TestChangeMethod">
        <replaced-method name="changeMe" replacer="replacer"/>
    </bean>

    <bean id="replacer" class="test.replacemethod.TestMethodReplacer"/>
</beans>
```

4. 测试。

```java
public static void main(String[] args) {
    ApplicationContext bf =
        new ClassPathXmlApplicationContext("test/replacemethod/replaceMethodTest.xml");
    TestChangeMethod test=(TestChangeMethod) bf.getBean("testChangeMethod");
    test.changeMe();
}
```

好了，运行测试类就可以看到预期的结果了，控制台成功打印出"我替换了原有的方法"，也就是说我们做到了动态替换原有方法，知道了这个元素的用法，我们再次来看元素的提取过程：

```java
public void parseReplacedMethodSubElements(Element beanEle, MethodOverrides overrides) {
    NodeList nl = beanEle.getChildNodes();
    for (int i = 0; i < nl.getLength(); i++) {
        Node node = nl.item(i);
        //仅当在Spring默认bean的子元素下且为<replaced-method时有效
        if (isCandidateElement(node) && nodeNameEquals(node, REPLACED_METHOD_ELEMENT)) {
            Element replacedMethodEle = (Element) node;
            //提取要替换的旧的方法
            String name = replacedMethodEle.getAttribute(NAME_ATTRIBUTE);

            //提取对应的新的替换方法
            String callback = replacedMethodEle.getAttribute(REPLACER_ATTRIBUTE);
            ReplaceOverride replaceOverride = new ReplaceOverride(name, callback);

            List<Element> argTypeEles = DomUtils.getChildElementsByTagName(replacedMethodEle, ARG_TYPE_ELEMENT);
            for (Element argTypeEle : argTypeEles) {
                //记录参数
                String match = argTypeEle.getAttribute (ARG_TYPE_MATCH_ATTRIBUTE);
                match = (StringUtils.hasText(match) ? match : DomUtils.getTextValue (argTypeEle));
```

```
                if (StringUtils.hasText(match)) {
                    replaceOverride.addTypeIdentifier(match);
                }
            }
            replaceOverride.setSource(extractSource(replacedMethodEle));
            overrides.addOverride(replaceOverride);
        }
    }
}
```

我们可以看到无论是 look-up 还是 replaced-method 都是构造了一个 MethodOverride，并最终记录在了 AbstractBeanDefinition 中的 methodOverrides 属性中。而这个属性如何使用以完成它所提供的功能我们会在后续的章节进行详细的介绍。

6. 解析子元素 constructor-arg

对构造函数的解析是非常常用的，同时也是非常复杂的，也相信大家对构造函数的配置都不陌生，举个简单的小例子：

```
... ...
<beans>
<!-- 默认的情况下是按照参数的顺序注入，当指定 index 索引后就可以改变注入参数的顺序 -->
<bean id="helloBean" class="com.HelloBean">
    <constructor-arg index="0">
        <value>郝佳</value>
    </constructor-arg>
    <constructor-arg index="1">
        <value>你好</value>
    </constructor-arg>
</bean>
... ...
</beans>
```

上面的配置是 Spring 构造函数配置中最基础的配置，实现的功能就是对 HelloBean 自动寻找对应的构造函数，并在初始化的时候将设置的参数传入进去。那么让我们来看看具体的 XML 解析过程。

对于 constructor-arg 子元素的解析，Spring 是通过 parseConstructorArgElements 函数来实现的，具体的代码如下：

```
public void parseConstructorArgElements(Element beanEle, BeanDefinition bd) {
    NodeList nl = beanEle.getChildNodes();
    for (int i = 0; i < nl.getLength(); i++) {
        Node node = nl.item(i);
        if (isCandidateElement(node) && nodeNameEquals(node, CONSTRUCTOR_ARG_ELEMENT)) {
            //解析 constructor-arg
            parseConstructorArgElement((Element) node, bd);
        }
    }
}
```

这个结构似乎我们可以想象得到，遍历所有子元素，也就是提取所有 constructor-arg，然后进行解析，但是具体的解析却被放置在了另个函数 parseConstructorArgElement 中，具体代码如下：

```java
public void parseConstructorArgElement(Element ele, BeanDefinition bd) {
    //提取 index 属性
    String indexAttr = ele.getAttribute(INDEX_ATTRIBUTE);
    //提取 type 属性
    String typeAttr = ele.getAttribute(TYPE_ATTRIBUTE);
    //提取 name 属性
    String nameAttr = ele.getAttribute(NAME_ATTRIBUTE);
    if (StringUtils.hasLength(indexAttr)) {
        try {
            int index = Integer.parseInt(indexAttr);
            if (index < 0) {
                error("'index' cannot be lower than 0", ele);
            }else {
                try {
                    this.parseState.push(new ConstructorArgumentEntry(index));
                    //解析 ele 对应的属性元素
                    Object value = parsePropertyValue(ele, bd, null);
                    ConstructorArgumentValues.ValueHolder valueHolder = new ConstructorArgumentValues.ValueHolder(value);
                    if (StringUtils.hasLength(typeAttr)) {
                        valueHolder.setType(typeAttr);
                    }
                    if (StringUtils.hasLength(nameAttr)) {
                        valueHolder.setName(nameAttr);
                    }
                    valueHolder.setSource(extractSource(ele));
                    //不允许重复指定相同参数
                    if (bd.getConstructorArgumentValues().hasIndexedArgumentValue(index)) {
                        error("Ambiguous constructor-arg entries for index " + index, ele);
                    }else {
                        bd.getConstructorArgumentValues().AddIndexedArgumentValue(index, valueHolder);
                    }
                }finally {
                    this.parseState.pop();
                }
            }
        }catch (NumberFormatException ex) {
            error("Attribute 'index' of tag 'constructor-arg' must be an integer", ele);
        }
    }else {
        //没有 index 属性则忽略去属性，自动寻找
        try {
            this.parseState.push(new ConstructorArgumentEntry());
            Object value = parsePropertyValue(ele, bd, null);
            ConstructorArgumentValues.ValueHolder valueHolder = new ConstructorArgumentValues.ValueHolder(value);
            if (StringUtils.hasLength(typeAttr)) {
                valueHolder.setType(typeAttr);
            }
```

```java
            if (StringUtils.hasLength(nameAttr)) {
                valueHolder.setName(nameAttr);
            }
            valueHolder.setSource(extractSource(ele));
            bd.getConstructorArgumentValues().**addGenericArgumentValue**(valueHolder);
        }
        finally {
            this.parseState.pop();
        }
    }
}
```

上面一段看似复杂的代码让很多人失去了耐心，但是，涉及的逻辑其实并不复杂，首先是提取 constructor-arg 上必要的属性（index、type、name）。

● 如果配置中指定了 index 属性，那么操作步骤如下。

1. 解析 Constructor-arg 的子元素。
2. 使用 ConstructorArgumentValues.ValueHolder 类型来封装解析出来的元素。
3. 将 type、name 和 index 属性一并封装在 ConstructorArgumentValues.ValueHolder 类型中并添加至当前 BeanDefinition 的 constructorArgumentValues 的 indexedArgumentValues 属性中。

● 如果没有指定 index 属性，那么操作步骤如下。

1. 解析 constructor-arg 的子元素。
2. 使用 ConstructorArgumentValues.ValueHolder 类型来封装解析出来的元素。
3. 将 type、name 和 index 属性一并封装在 ConstructorArgumentValues.ValueHolder 类型中并添加至当前 BeanDefinition 的 constructorArgumentValues 的 genericArgumentValues 属性中。

可以看到，对于是否制定 index 属性来讲，Spring 的处理流程是不同的，关键在于属性信息被保存的位置。

那么了解了整个流程后，我们尝试着进一步了解解析构造函数配置中子元素的过程，进入 parsePropertyValue：

```java
public Object parsePropertyValue(Element ele, BeanDefinition bd, String propertyName) {
    String elementName = (propertyName != null) ?
                    "<property> element for property '" + propertyName + "'" :
                    "<constructor-arg> element";

    //一个属性只能对应一种类型: ref、value、list 等
    NodeList nl = ele.getChildNodes();
    Element subElement = null;
    for (int i = 0; i < nl.getLength(); i++) {
        Node node = nl.item(i);
        //对应 description 或者 meta 不处理
        if (node instanceof Element && !nodeNameEquals(node, DESCRIPTION_ELEMENT) &&
                !nodeNameEquals(node, META_ELEMENT)) {
            if (subElement != null) {
                error(elementName + " must not contain more than one sub-element", ele);
            }
            else {
                subElement = (Element) node;
```

```java
                        }
                    }
                }
                //解析 constructor-arg 上的 ref 属性
                boolean hasRefAttribute = ele.hasAttribute(REF_ATTRIBUTE);
                //解析 constructor-arg 上的 value 属性
                boolean hasValueAttribute = ele.hasAttribute(VALUE_ATTRIBUTE);
                if ((hasRefAttribute && hasValueAttribute) ||
                        ((hasRefAttribute || hasValueAttribute) && subElement != null)) {
                    /*
                     * 在 constructor-arg 上不存在:
                     *     1. 同时既有 ref 属性又有 value 属性
                     *     2. 存在 ref 属性或者 value 属性且又有子元素
                     */
                    error(elementName +
                            " is only allowed to contain either 'ref' attribute OR 'value' attribute OR sub-element", ele);
                }

                if (hasRefAttribute) {
                    //ref 属性的处理,使用 RuntimeBeanReference 封装对应的 ref 名称
                    String refName = ele.getAttribute(REF_ATTRIBUTE);
                    if (!StringUtils.hasText(refName)) {
                        error(elementName + " contains empty 'ref' attribute", ele);
                    }
                    RuntimeBeanReference ref = new RuntimeBeanReference(refName);
                    ref.setSource(extractSource(ele));
                    return ref;
                }else if (hasValueAttribute) {
                    //value 属性的处理,使用 TypedStringValue 封装
                    TypedStringValue valueHolder = new TypedStringValue (ele.getAttribute(VALUE_ATTRIBUTE));
                    valueHolder.setSource(extractSource(ele));
                    return valueHolder;
                }else if (subElement != null) {
                    //解析子元素
                    return parsePropertySubElement(subElement, bd);
                }else {
                    //既没有 ref 也没有 value 也没有子元素,Spring 蒙圈了
                    error(elementName + " must specify a ref or value", ele);
                    return null;
                }
            }
```

从代码上来看,对构造函数中属性元素的解析,经历了以下几个过程。

1. 略过 description 或者 meta。

2. 提取 constructor-arg 上的 ref 和 value 属性,以便于根据规则验证正确性,其规则为在 constructor-arg 上不存在以下情况。

- 同时既有 ref 属性又有 value 属性。
- 存在 ref 属性或者 value 属性且又有子元素。

3. ref 属性的处理。使用 RuntimeBeanReference 封装对应的 ref 名称,如:

```
<constructor-arg ref="a" >
```
4. value 属性的处理。使用 TypedStringValue 封装,如:
```
<constructor-arg value="a" >
```
5. 子元素的处理,如:
```
            <constructor-arg>
                <map>
                    <entry key="key" value="value" />
                </map>
            </constructor-arg>
```
而对于子元素的处理,例如这里提到的在构造函数中又嵌入了子元素 map 是怎么实现的呢? parsePropertySubElement 中实现了对各种子元素的分类处理。
```
public Object parsePropertySubElement(Element ele, BeanDefinition bd) {
        return parsePropertySubElement(ele, bd, null);
}

public Object parsePropertySubElement(Element ele, BeanDefinition bd, String defaultValueType) {
        if (!isDefaultNamespace(ele)) {
            return parseNestedCustomElement(ele, bd);
        }
        else if (nodeNameEquals(ele, BEAN_ELEMENT)) {
            BeanDefinitionHolder nestedBd = parseBeanDefinitionElement(ele, bd);
            if (nestedBd != null) {
                nestedBd = decorateBeanDefinitionIfRequired(ele, nestedBd, bd);
            }
            return nestedBd;
        }
        else if (nodeNameEquals(ele, REF_ELEMENT)) {
            // A generic reference to any name of any bean.
            String refName = ele.getAttribute(BEAN_REF_ATTRIBUTE);
            boolean toParent = false;
            if (!StringUtils.hasLength(refName)) {
                //解析 local
                refName = ele.getAttribute(LOCAL_REF_ATTRIBUTE);
                if (!StringUtils.hasLength(refName)) {
                    //解析 parent
                    refName = ele.getAttribute(PARENT_REF_ATTRIBUTE);
                    toParent = true;
                    if (!StringUtils.hasLength(refName)) {
                        error("'bean', 'local' or 'parent' is required for <ref> element", ele);
                        return null;
                    }
                }
            }
            if (!StringUtils.hasText(refName)) {
                error("<ref> element contains empty target attribute", ele);
                return null;
            }
            RuntimeBeanReference ref = new RuntimeBeanReference(refName, toParent);
            ref.setSource(extractSource(ele));
            return ref;
        }
        //对 idref 元素的解析
```

```
        else if (nodeNameEquals(ele, IDREF_ELEMENT)) {
            return parseIdRefElement(ele);
        }
        //对 value 子元素的解析
        else if (nodeNameEquals(ele, VALUE_ELEMENT)) {
            return parseValueElement(ele, defaultValueType);
        }
        //对 null 子元素的解析
        else if (nodeNameEquals(ele, NULL_ELEMENT)) {
            // It's a distinguished null value. Let's wrap it in a TypedStringValue
            // object in order to preserve the source location.
            TypedStringValue nullHolder = new TypedStringValue(null);
            nullHolder.setSource(extractSource(ele));
            return nullHolder;
        }
        else if (nodeNameEquals(ele, ARRAY_ELEMENT)) {
            //解析 array 子元素
            return parseArrayElement(ele, bd);
        }
        else if (nodeNameEquals(ele, LIST_ELEMENT)) {
            //解析 list 子元素
            return parseListElement(ele, bd);
        }
        else if (nodeNameEquals(ele, SET_ELEMENT)) {
            //解析 set 子元素
            return parseSetElement(ele, bd);
        }
        else if (nodeNameEquals(ele, MAP_ELEMENT)) {
            //解析 map 子元素
            return parseMapElement(ele, bd);
        }
        else if (nodeNameEquals(ele, PROPS_ELEMENT)) {
            //解析 props 子元素
            return parsePropsElement(ele);
        }
        else {
            error("Unknown property sub-element: [" + ele.getNodeName() + "]", ele);
            return null;
        }
}
```

可以看到，在上面的函数中实现了所有可支持的子类的分类处理，到这里，我们已经大致理清构造函数的解析流程，至于更深入的解析读者有兴趣可以自己去探索。

7. 解析子元素 property

parsePropertyElement 函数完成了对 property 属性的提取，property 使用方式如下：

```
<bean id="test" class="test.TestClass">
    <property name="testStr" value="aaa"/>
</bean>
```

或者

```
<bean id="a">
    <property name="p">
```

```
            <list>
                <value>aa</value>
                <value>bb</value>
            </list>
        </property>
</bean>
```
而具体的解析过程如下:
```
public void parsePropertyElements(Element beanEle, BeanDefinition bd) {
        NodeList nl = beanEle.getChildNodes();
        for (int i = 0; i < nl.getLength(); i++) {
            Node node = nl.item(i);
            if (isCandidateElement(node) && nodeNameEquals(node, PROPERTY_ELEMENT)) {
                parsePropertyElement((Element) node, bd);
            }
        }
}
```
有了之前分析构造函数的经验,这个函数我们并不难理解,无非是提取所有 property 的子元素,然后调用 parsePropertyElement 处理, parsePropertyElement 代码如下:
```
public void parsePropertyElement(Element ele, BeanDefinition bd) {
        //获取配置元素中 name 的值
        String propertyName = ele.getAttribute(NAME_ATTRIBUTE);
        if (!StringUtils.hasLength(propertyName)) {
            error("Tag 'property' must have a 'name' attribute", ele);
            return;
        }
        this.parseState.push(new PropertyEntry(propertyName));
        try {
            //不允许多次对同一属性配置
            if (bd.getPropertyValues().contains(propertyName)) {
                error("Multiple 'property' definitions for property '" + propertyName + "'", ele);
                return;
            }
            Object val = parsePropertyValue(ele, bd, propertyName);
            PropertyValue pv = new PropertyValue(propertyName, val);
            parseMetaElements(ele, pv);
            pv.setSource(extractSource(ele));
            bd.getPropertyValues().addPropertyValue(pv);
        }
        finally {
            this.parseState.pop();
        }
}
```
可以看到上面函数与构造函数注入方式不同的是将返回值使用 PropertyValue 进行封装,并记录在了 BeanDefinition 中的 propertyValues 属性中。

8. 解析子元素 qualifier

对于 qualifier 元素的获取,我们接触更多的是注解的形式,在使用 Spring 框架中进行自动注入时, Spring 容器中匹配的候选 Bean 数目必须有且仅有一个。当找不到一个匹配的 Bean 时, Spring 容器将抛出 BeanCreationException 异常,并指出必须至少拥有一个匹配的 Bean。

Spring 允许我们通过 Qualifier 指定注入 Bean 的名称，这样歧义就消除了，而对于配置方式使用如：

```
<bean id="myTestBean" class="bean.MyTestBean">
    <qualifier type="org.Springframework.beans.factory.annotation.Qualifier" value="qf"/>
</bean>
```

其解析过程与之前大同小异，这里不再重复叙述。

3.1.2 AbstractBeanDefinition 属性

至此我们便完成了对 XML 文档到 GenericBeanDefinition 的转换，也就是说到这里，XML 中所有的配置都可以在 GenericBeanDefinition 的实例类中找到对应的配置。

GenericBeanDefinition 只是子类实现，而大部分的通用属性都保存在了 AbstractBeanDefinition 中，那么我们再次通过 AbstractBeanDefinition 的属性来回顾一下我们都解析了哪些对应的配置。

```java
public abstract class AbstractBeanDefinition extends BeanMetadataAttributeAccessor
        implements BeanDefinition, Cloneable {

//此处省略静态变量以及final常量

    /**
     * bean 的作用范围,对应 bean 属性 scope
     */
    private String scope = SCOPE_DEFAULT;

    /**
     * 是否是单例,来自 bean 属性 scope
     */
    private boolean singleton = true;

    /**
     * 是否是原型,来自 bean 属性 scope
     */
    private boolean prototype = false;

    /**
     * 是否是抽象,对应 bean 属性 abstract
     */
    private boolean abstractFlag = false;

    /**
     * 是否延迟加载,对应 bean 属性 lazy-init
     */
    private boolean lazyInit = false;

    /**
     * 自动注入模式,对应 bean 属性 autowire
     */
    private int autowireMode = AUTOWIRE_NO;
```

```java
/**
 * 依赖检查, Spring 3.0 后弃用这个属性
 */
private int dependencyCheck = DEPENDENCY_CHECK_NONE;
/**
 * 用来表示一个 bean 的实例化依靠另一个 bean 先实例化, 对应 bean 属性 depend-on
 */
private String[] dependsOn;

/**
 * autowire-candidate 属性设置为 false, 这样容器在查找自动装配对象时,
 * 将不考虑该 bean, 即它不会被考虑作为其他 bean 自动装配的候选者, 但是该 bean 本身还是可以使用自
 * 动装配来注入其他 bean 的。
 *    对应 bean 属性 autowire-candidate
 */
private boolean autowireCandidate = true;

/**
 * 自动装配时当出现多个 bean 候选者时, 将作为首选者, 对应 bean 属性 primary
 */
private boolean primary = false;

/**
 * 用于记录 Qualifier, 对应子元素 qualifier
 */
private final Map<String, AutowireCandidateQualifier> qualifiers =
        new LinkedHashMap<String, AutowireCandidateQualifier>(0);

/**
 * 允许访问非公开的构造器和方法, 程序设置
 */
private boolean nonPublicAccessAllowed = true;

/**
 * 是否以一种宽松的模式解析构造函数, 默认为 true,
 * 如果为 false, 则在如下情况
 * interface ITest{}
 * class  ITestImpl implements ITest{};
 * class Main{
 *     Main(ITest i){}
 *     Main(ITestImpl i){}
 * }
 * 抛出异常, 因为 Spring 无法准确定位哪个构造函数
 * 程序设置
 */
private boolean lenientConstructorResolution = true;

/**
 * 记录构造函数注入属性, 对应 bean 属性 constructor-arg
 */
private ConstructorArgumentValues constructorArgumentValues;

/**
```

```java
     * 普通属性集合
     */
    private MutablePropertyValues propertyValues;

    /**
     * 方法重写的持有者，记录lookup-method、replaced-method 元素
     */
    private MethodOverrides methodOverrides = new MethodOverrides();

    /**
     * 对应bean属性factory-bean,用法：
     * <bean id="instanceFactoryBean" class="example.chapter3.InstanceFactoryBean"/>
     * <bean id="currentTime" factory-bean="instanceFactoryBean" factory-method="createTime"/>
     */
    private String factoryBeanName;

    /**
     * 对应bean属性factory-method
     */
    private String factoryMethodName;

    /**
     * 初始化方法，对应bean属性init-method
     */
    private String initMethodName;

    /**
     * 销毁方法，对应bean属性destory-method
     */
    private String destroyMethodName;

    /**
     * 是否执行init-method,程序设置
     */
    private boolean enforceInitMethod = true;

    /**
     * 是否执行destory-method,程序设置
     */
    private boolean enforceDestroyMethod = true;

    /**
     * 是否是用户定义的而不是应用程序本身定义的,创建AOP时候为true,程序设置
     */
    private boolean synthetic = false;

    /**
     * 定义这个bean的应用 ,APPLICATION:用户,INFRASTRUCTURE:完全内部使用,与用户无关,SUPPORT:某些复杂配置的一部分
     * 程序设置
     */
    private int role = BeanDefinition.ROLE_APPLICATION;
```

```
    /**
     * bean 的描述信息
     */
    private String description;
    /**
     * 这个 bean 定义的资源
     */
    private Resource resource;

    //此处省略 set/get 方法
}
```

3.1.3 解析默认标签中的自定义标签元素

到这里我们已经完成了分析默认标签的解析与提取过程,或许涉及的内容太多,我们已经忘了我们从哪个函数开始的了,我们再次回顾下默认标签解析函数的起始函数:

```
protected void processBeanDefinition(Element ele, BeanDefinitionParserDelegate delegate) {
        BeanDefinitionHolder bdHolder = delegate.parseBeanDefinitionElement(ele);
        if (bdHolder != null) {
            bdHolder = delegate.decorateBeanDefinitionIfRequired(ele, bdHolder);
            try {
                // Register the final decorated instance.
                BeanDefinitionReaderUtils.registerBeanDefinition(bdHolder, getReaderContext().getRegistry());
            }
            catch (BeanDefinitionStoreException ex) {
                getReaderContext().error("Failed to register bean definition with name '" +
                        bdHolder.getBeanName() + "'", ele, ex);
            }
            // Send registration event.
            getReaderContext().fireComponentRegistered(new BeanComponentDefinition(bdHolder));
        }
}
```

我们已经用了大量的篇幅分析了 BeanDefinitionHolder bdHolder = delegate.parseBeanDefinitionElement(ele)这句代码,接下来,我们要进行 bdHolder = delegate.decorateBeanDefinitionIfRequired(ele, bdHolder)代码的分析,首先大致了解下这句代码的作用,其实我们可以从语义上分析:如果需要的话就对 beanDefinition 进行装饰,那这句代码到底是什么功能呢?其实这句代码适用于这样的场景,如:

```
<bean id="test" class="test.MyClass">
        <mybean:user username="aaa"/>
</bean>
```

当 Spring 中的 bean 使用的是默认的标签配置,但是其中的子元素却使用了自定义的配置时,这句代码便会起作用了。可能有人会有疑问,之前讲过,对 bean 的解析分为两种类型,一种是默认类型的解析,另一种是自定义类型的解析,这不正是自定义类型的解析吗?为什么会在默认类型解析中单独添加一个方法处理呢?确实,这个问题很让人迷惑,但是,不知道聪明的读者是否有发现，这个自定义类型并不是以 Bean 的形式出现的呢？我们之前讲过的两种类型的不同处理只是针对 Bean 的，这里我们看到，这个自定义类型其实是属性。好了，我们继续分析下这段代码的逻辑。

```
public BeanDefinitionHolder decorateBeanDefinitionIfRequired(Element ele, BeanDefinitionHolder definitionHolder) {
        return decorateBeanDefinitionIfRequired(ele, definitionHolder, null);
}
```

这里将函数中第三个参数设置为空，那么第三个参数是做什么用的呢？什么情况下不为空呢？其实这第三个参数是父类 bean，当对某个嵌套配置进行分析时，这里需要传递父类 beanDefinition。分析源码得知这里传递的参数其实是为了使用父类的 scope 属性，以备子类若没有设置 scope 时默认使用父类的属性，这里分析的是顶层配置，所以传递 null。将第三个参数设置为空后进一步跟踪函数：

```
public BeanDefinitionHolder decorateBeanDefinitionIfRequired(
        Element ele, BeanDefinitionHolder definitionHolder, BeanDefinition containingBd){

            BeanDefinitionHolder finalDefinition = definitionHolder;

            NamedNodeMap attributes = ele.getAttributes();
    //遍历所有的属性，看看是否有适用于修饰的属性
            for (int i = 0; i < attributes.getLength(); i++) {
                Node node = attributes.item(i);
                finalDefinition = decorateIfRequired(node, finalDefinition, containingBd);
            }

            NodeList children = ele.getChildNodes();
    //遍历所有的子节点，看看是否有适用于修饰的子元素
            for (int i = 0; i < children.getLength(); i++) {
                Node node = children.item(i);
                if (node.getNodeType() == Node.ELEMENT_NODE) {
                    finalDefinition = decorateIfRequired(node, finalDefinition, containingBd);
                }
            }
            return finalDefinition;
}
```

上面的代码，我们看到函数分别对元素的所有属性以及子节点进行了 decorateIfRequired 函数的调用，我们继续跟踪代码：

```
private BeanDefinitionHolder decorateIfRequired(
        Node node, BeanDefinitionHolder originalDef, BeanDefinition containingBd) {

    //获取自定义标签的命名空间
        String namespaceUri = getNamespaceURI(node);
    //对于非默认标签进行修饰
        if (!isDefaultNamespace(namespaceUri)) {
    //根据命名空间找到对应的处理器
            NamespaceHandler handler=this.readerContext. getNamespaceHandler Resolver(). resolve(namespaceUri);
            if (handler != null) {
```

```
                //进行修饰
                    return handler.decorate(node, originalDef, new ParserContext(this.readerContext,
this, containingBd));
            }
            else if (namespaceUri != null && namespaceUri.startsWith("http://www.
springframework.org/")) {
                    error("Unable to locate Spring NamespaceHandler for XML schema
namespace [" + namespaceUri + "]", node);
            }
            else {
                // A custom namespace, not to be handled by Spring - maybe "xml:...".
                if (logger.isDebugEnabled()) {
                    logger.debug("No Spring NamespaceHandler found for XML schema
namespace [" + namespaceUri + "]");
                }
            }
        }
        return originalDef;
    }

    public String getNamespaceURI(Node node) {
        return node.getNamespaceURI();
    }

    public boolean isDefaultNamespace(String namespaceUri) {
        //BEANS_NAMESPACE_URI = "http://www.springframework.org/schema/beans";
        return (!StringUtils.hasLength(namespaceUri) || BEANS_NAMESPACE_URI.equals
(namespaceUri));
    }
```

程序走到这里，条理其实已经非常清楚了，首先获取属性或者元素的命名空间，以此来判断该元素或者属性是否适用于自定义标签的解析条件，找出自定义类型所对应的 NamespaceHandler 并进行进一步解析。在自定义标签解析的章节我们会重点讲解，这里暂时先略过。

我们总结下 decorateBeanDefinitionIfRequired 方法的作用，在 decorateBeanDefinitionIfRequired 中我们可以看到对于程序默认的标签的处理其实是直接略过的，因为默认的标签到这里已经被处理完了，这里只对自定义的标签或者说对 bean 的自定义属性感兴趣。在方法中实现了寻找自定义标签并根据自定义标签寻找命名空间处理器，并进行进一步的解析。

3.1.4 注册解析的 BeanDefinition

对于配置文件，解析也解析完了，装饰也装饰完了，对于得到的 beanDinition 已经可以满足后续的使用要求了，唯一还剩下的工作就是注册了，也就是 processBeanDefinition 函数中的 BeanDefinitionReaderUtils.registerBeanDefinition(bdHolder, getReaderContext().getRegistry())代码的解析了。

```
    public static void registerBeanDefinition(
            BeanDefinitionHolder definitionHolder, BeanDefinitionRegistry registry)
            throws BeanDefinitionStoreException {
```

```java
//使用 beanName 做唯一标识注册
String beanName = definitionHolder.getBeanName();
registry.registerBeanDefinition(beanName, definitionHolder.getBeanDefinition());

//注册所有的别名
    String[] aliases = definitionHolder.getAliases();
    if (aliases != null) {
        for (String aliase : aliases) {
            registry.registerAlias(beanName, aliase);
        }
    }
}
```

从上面的代码可以看出，解析的 beanDefinition 都会被注册到 BeanDefinitionRegistry 类型的实例 registry 中，而对于 beanDefinition 的注册分成了两部分：通过 beanName 的注册以及通过别名的注册。

1. 通过 beanName 注册 BeanDefinition

对于 beanDefinition 的注册，或许很多人认为的方式就是将 beanDefinition 直接放入 map 中就好了，使用 beanName 作为 key。确实，Spring 就是这么做的，只不过除此之外，它还做了点别的事情。

```java
public void registerBeanDefinition(String beanName, BeanDefinition beanDefinition)
        throws BeanDefinitionStoreException {

    Assert.hasText(beanName, "Bean name must not be empty");
    Assert.notNull(beanDefinition, "BeanDefinition must not be null");

    if (beanDefinition instanceof AbstractBeanDefinition) {
        try {
            /*
             * 注册前的最后一次校验，这里的校验不同于之前的 XML 文件校验，
             * 主要是对于 AbstractBeanDefinition 属性中的 methodOverrides 校验，
             * 校验 methodOverrides 是否与工厂方法并存或者 methodOverrides 对应的方法根本不存在
             */
            ((AbstractBeanDefinition) beanDefinition).validate();
        }
        catch (BeanDefinitionValidationException ex) {
            throw new BeanDefinitionStoreException (beanDefinition. getResourceDescription(), beanName,
                    "Validation of bean definition failed", ex);
        }
    }
    //因为 beanDefinitionMap 是全局变量，这里定会存在并发访问的情况
    synchronized (this.beanDefinitionMap) {
        Object oldBeanDefinition = this.beanDefinitionMap.get(beanName);
        //处理注册已经注册的 beanName 情况
        if (oldBeanDefinition != null) {
            //如果对应的 BeanName 已经注册且在配置中配置了 bean 不允许被覆盖，则抛出异常
            if (!this.allowBeanDefinitionOverriding) {
```

```
                    throw new BeanDefinitionStoreException(beanDefinition.getResource
Description(), beanName,
                        "Cannot register bean definition [" + beanDefinition + "] 
 for bean '" + beanName +
                        "': There is already [" + oldBeanDefinition + "] bound.");
                }else {
                    if (this.logger.isInfoEnabled()) {
                        this.logger.info("Overriding bean definition for bean '" + beanName +
                                "': replacing [" + oldBeanDefinition + "] with [" +
beanDefinition + "]");
                    }
                }
            }else {
                //记录 beanName
                this.beanDefinitionNames.add(beanName);
                this.frozenBeanDefinitionNames = null;
            }
            //注册 beanDefinition
            this.beanDefinitionMap.put(beanName, beanDefinition);
        }
        //重置所有 beanName 对应的缓存
        resetBeanDefinition(beanName);
    }
```

上面的代码中我们看到，在对于 bean 的注册处理方式上，主要进行了几个步骤。

1. 对 AbstractBeanDefinition 的校验。在解析 XML 文件的时候我们提过校验，但是此校验非彼校验，之前的校验时针对于 XML 格式的校验，而此时的校验时针是对于 AbstractBean-Definition 的 methodOverrides 属性的。

2. 对 beanName 已经注册的情况的处理。如果设置了不允许 bean 的覆盖，则需要抛出异常，否则直接覆盖。

3. 加入 map 缓存。

4. 清除解析之前留下的对应 beanName 的缓存。

2. 通过别名注册 BeanDefinition

在理解了注册 bean 的原理后，理解注册别名的原理就容易多了。

```
public void registerAlias(String name, String alias) {
    Assert.hasText(name, "'name' must not be empty");
    Assert.hasText(alias, "'alias' must not be empty");
    //如果 beanName 与 alias 相同的话不记录 alias,并删除对应的 alias
    if (alias.equals(name)) {
        this.aliasMap.remove(alias);
    }else {
        //如果 alias 不允许被覆盖则抛出异常
        if (!allowAliasOverriding()) {
            String registeredName = this.aliasMap.get(alias);
            if (registeredName != null && !registeredName.equals(name)) {
                throw new IllegalStateException("Cannot register alias '" + alias
 + "' for name '" +
                        name + "': It is already registered for name '" +
registeredName + "'.");
```

```
            }
        }
        //当A->B存在时，若再次出现A->C->B时候则会抛出异常
        checkForAliasCircle(name, alias);
        this.aliasMap.put(alias, name);
    }
}
```

由以上代码中可以得知，注册 alias 的步骤如下。

1. alias 与 beanName 相同情况处理。若 alias 与 beanName 并名称相同则不需要处理并删除掉原有 alias。

2. alias 覆盖处理。若 aliasName 已经使用并已经指向了另一 beanName 则需要用户的设置进行处理。

3. alias 循环检查。当 A->B 存在时，若再次出现 A->C->B 时候则会抛出异常。

4. 注册 alias。

3.1.5 通知监听器解析及注册完成

通过代码 getReaderContext().fireComponentRegistered(new BeanComponentDefinition(bdHolder)) 完成此工作，这里的实现只为扩展，当程序开发人员需要对注册 BeanDefinition 事件进行监听时可以通过注册监听器的方式并将处理逻辑写入监听器中，目前在 Spring 中并没有对此事件做任何逻辑处理。

3.2 alias 标签的解析

通过上面较长的篇幅我们终于分析完了默认标签中对 bean 标签的处理，那么我们之前提到过，对配置文件的解析包括对 import 标签、alias 标签、bean 标签、beans 标签的处理，现在我们已经完成了最重要也是最核心的功能，其他的解析步骤也都是围绕第 3 个解析而进行的。在分析了第 3 个解析步骤后，再回过头来看看对 alias 标签的解析。

在对 bean 进行定义时，除了使用 id 属性来指定名称之外，为了提供多个名称，可以使用 alias 标签来指定。而所有的这些名称都指向同一个 bean，在某些情况下提供别名非常有用，比如为了让应用的每一个组件能更容易地对公共组件进行引用。

然而，在定义 bean 时就指定所有的别名并不是总是恰当的。有时我们期望能在当前位置为那些在别处定义的 bean 引入别名。在 XML 配置文件中，可用单独的<alias/>元素来完成 bean 别名的定义。如配置文件中定义了一个 JavaBean：

```
<bean id="testBean" class="com.test"/>
```

要给这个 JavaBean 增加别名，以方便不同对象来调用。我们就可以直接使用 bean 标签中的 name 属性：

```
<bean id="testBean" name="testBean,testBean2" class="com.test"/>
```

同样，Spring 还有另外一种声明别名的方式：
```
<bean id="testBean" class="com.test"/>
<alias name="testBean" alias="testBean,testBean2"/>
```
考虑一个更为具体的例子，组件 A 在 XML 配置文件中定义了一个名为 componentA 的 DataSource 类型的 bean，但组件 B 却想在其 XML 文件中以 componentB 命名来引用此 bean。而且在主程序 MyApp 的 XML 配置文件中，希望以 myApp 的名字来引用此 bean。最后容器加载 3 个 XML 文件来生成最终的 ApplicationContext。在此情形下，可通过在配置文件中添加下列 alias 元素来实现：
```
<alias name="componentA" alias="componentB"/>
<alias name="componentA" alias="myApp" />
```
这样一来，每个组件及主程序就可通过唯一名字来引用同一个数据源而互不干扰。

在之前的章节已经讲过了对于 bean 中 name 元素的解析，那么我们现在再来深入分析下对于 alias 标签的解析过程。

```java
protected void processAliasRegistration(Element ele) {
    //获取 beanName
    String name = ele.getAttribute(NAME_ATTRIBUTE);
    //获取 alias
    String alias = ele.getAttribute(ALIAS_ATTRIBUTE);
    boolean valid = true;
    if (!StringUtils.hasText(name)) {
        getReaderContext().error("Name must not be empty", ele);
        valid = false;
    }
    if (!StringUtils.hasText(alias)) {
        getReaderContext().error("Alias must not be empty", ele);
        valid = false;
    }
    if (valid) {
        try {
            //注册 alias
            getReaderContext().getRegistry().registerAlias(name, alias);
        }
        catch (Exception ex) {
            getReaderContext().error("Failed to register alias '" + alias +
                "' for bean with name '" + name + "'", ele, ex);
        }
        //别名注册后通知监听器做相应处理
        getReaderContext().fireAliasRegistered(name, alias, extractSource(ele));
    }
}
```

可以发现，跟之前讲过的 bean 中的 alias 解析大同小异，都是将别名与 beanName 组成一对注册至 registry 中。这里不再赘述。

3.3　import 标签的解析

对于 Spring 配置文件的编写，我想，经历过庞大项目的人，都有那种恐惧的心理，太多的配置文件了。不过，分模块是大多数人能想到的方法，但是，怎么分模块，那就仁者见仁，智者见智了。使用 import 是个好办法，例如我们可以构造这样的 Spring 配置文件：

applicationContext.xml
```xml
        <?xml version="1.0" encoding="gb2312"?>
<!DOCTYPE beans PUBLIC "-//Spring//DTD BEAN//EN" "http://www.springframework.org/dtd/Spring-beans.dtd">
    <beans>

        <import resource="customerContext.xml" />
        <import resource="systemContext.xml" />
        ... ...

</beans>
```

applicationContext.xml 文件中使用 import 的方式导入有模块配置文件，以后若有新模块的加入，那就可以简单修改这个文件了。这样大大简化了配置后期维护的复杂度，并使配置模块化，易于管理。我们来看看 Spring 是如何解析 import 配置文件的呢？

```java
protected void importBeanDefinitionResource(Element ele) {
            //获取 resource 属性
            String location = ele.getAttribute(RESOURCE_ATTRIBUTE);
            //如果不存在 resource 属性则不做任何处理
            if (!StringUtils.hasText(location)) {
                getReaderContext().error("Resource location must not be empty", ele);
                return;
            }

            //解析系统属性，格式如： "${user.dir}"
            location = environment.resolveRequiredPlaceholders(location);

            Set<Resource> actualResources = new LinkedHashSet<Resource>(4);

            //判定 location 是决定 URI 还是相对 URI
            boolean absoluteLocation = false;
            try {
                absoluteLocation = ResourcePatternUtils.isUrl(location) || ResourceUtils.toURI(location).isAbsolute();
            }
            catch (URISyntaxException ex) {
                // cannot convert to an URI, considering the location relative
                // unless it is the well-known Spring prefix "classpath*:"
            }

            // Absolute or relative?
            //如果是绝对 URI 则直接根据地址加载对应的配置文件
            if (absoluteLocation) {
                try {
                    int importCount = getReaderContext().getReader().loadBeanDefinitions(location, actualResources);
                    if (logger.isDebugEnabled()) {
                        logger.debug("Imported " + importCount + " bean definitions from URL location [" + location + "]");
```

3.3 import 标签的解析

```
                    }
                }
                catch (BeanDefinitionStoreException ex) {
                    getReaderContext().error(
                        "Failed to import bean definitions from URL location [" +
location + "]", ele, ex);
                }
            }
            else {
                //如果是相对地址则根据相对地址计算出绝对地址
                try {
                    int importCount;
                    //Resource存在多个子实现类，如VfsResource、FileSystemResource等，
                    //而每个resource的createRelative方式实现都不一样，所以这里先使用子类的方法尝
                    //试解析
                    Resource relativeResource = getReaderContext().getResource().Create
Relative(location);
                    if (relativeResource.exists()) {
                        importCount = getReaderContext().getReader().loadBeanDefinitions
(relativeResource);
                        actualResources.add(relativeResource);
                    }else {
                        //如果解析不成功，则使用默认的解析器ResourcePatternResolver进行解析
                        String baseLocation = getReaderContext().getResource().getURL().
toString();
                        importCount = getReaderContext().getReader().loadBeanDefinitions(
                            StringUtils.applyRelativePath(baseLocation, location),
actualResources);
                    }
                    if (logger.isDebugEnabled()) {
                        logger.debug("Imported " + importCount + " bean definitions from
relative location [" + location + "]");
                    }
                }
                catch (IOException ex) {
                    getReaderContext().error("Failed to resolve current resource location", ele, ex);
                }
                catch (BeanDefinitionStoreException ex) {
                    getReaderContext().error("Failed to import bean definitions from
relative location [" + location + "]",
                        ele, ex);
                }
            }
            //解析后进行监听器激活处理
            Resource[] actResArray = actualResources.toArray(new Resource[actualResources.
size()]);
            getReaderContext().fireImportProcessed(location, actResArray, extractSource(ele));
    }
```

上面的代码不难，相信配合注释会很好理解，我们总结一下大致流程便于读者更好地梳理，在解析<import 标签时，Spring 进行解析的步骤大致如下。

1. 获取 resource 属性所表示的路径。
2. 解析路径中的系统属性，格式如 "${user.dir}"。

3. 判定 location 是绝对路径还是相对路径。
4. 如果是绝对路径则递归调用 bean 的解析过程，进行另一次的解析。
5. 如果是相对路径则计算出绝对路径并进行解析。
6. 通知监听器，解析完成。

3.4 嵌入式 beans 标签的解析

对于嵌入式的 beans 标签，相信大家使用过或者至少接触过，非常类似于 import 标签所提供的功能，使用如下：

```
<?xml version="1.0" encoding="UTF-8"?>
<beans xmlns="http://www.Springframework.org/schema/beans"
       xmlns:xsi="http://www.w3.org/2001/XMLSchema-instance"
       xsi:schemaLocation="http://www.Springframework.org/schema/beans http://www.Springframework. org/schema/beans/Spring-beans.xsd">

    <bean id="aa" class="test.aa"/>
    <beans>
    </beans>
</beans>
```

对于嵌入式 beans 标签来讲，并没有太多可讲，与单独的配置文件并没有太大的差别，无非是递归调用 beans 的解析过程，相信读者根据之前讲解过的内容已经有能力理解其中的奥秘了。

第 4 章　自定义标签的解析

在之前的章节中，我们提到了在 Spring 中存在默认标签与自定义标签两种，而在上一章中我们分析了 Spring 中对默认标签的解析过程，相信大家一定已经有所感悟。那么，现在将开始新的里程，分析 Spring 中自定义标签的加载过程。同样，我们还是先再次回顾一下，当完成从配置文件到 Document 的转换并提取对应的 root 后，将开始了所有元素的解析，而在这一过程中便开始了默认标签与自定义标签两种格式的区分，函数如下：

```
protected void parseBeanDefinitions(Element root, BeanDefinitionParserDelegate delegate) {
    if (delegate.isDefaultNamespace(root)) {
        NodeList nl = root.getChildNodes();
        for (int i = 0; i < nl.getLength(); i++) {
            Node node = nl.item(i);
            if (node instanceof Element) {
                Element ele = (Element) node;
                if (delegate.isDefaultNamespace(ele)) {
                    parseDefaultElement(ele, delegate);
                }
                else {
                    delegate.parseCustomElement(ele);
                }
            }
        }
    }
    else {
        delegate.parseCustomElement(root);
    }
}
```

在本章中，所有的功能都是围绕其中的一句代码 delegate.parseCustomElement(root)开展的。从上面的函数我们可以看出，当 Spring 拿到一个元素时首先要做的是根据命名空间进行解析，如果是默认的命名空间，则使用 parseDefaultElement 方法进行元素解析，否则使用 parseCustom-Element 方法进行解析。在分析自定义标签的解析过程前，我们先了解一下自定义标签的使用过程。

4.1 自定义标签使用

在很多情况下，我们需要为系统提供可配置化支持，简单的做法可以直接基于 Spring 的标准 bean 来配置，但配置较为复杂或者需要更多丰富控制的时候，会显得非常笨拙。一般的做法会用原生态的方式去解析定义好的 XML 文件，然后转化为配置对象。这种方式当然可以解决所有问题，但实现起来比较烦琐，特别是在配置非常复杂的时候，解析工作是一个不得不考虑的负担。Spring 提供了可扩展 Schema 的支持，这是一个不错的折中方案，扩展 Spring 自定义标签配置大致需要以下几个步骤（前提是要把 Spring 的 Core 包加入项目中）。

- 创建一个需要扩展的组件。
- 定义一个 XSD 文件描述组件内容。
- 创建一个文件，实现 BeanDefinitionParser 接口，用来解析 XSD 文件中的定义和组件定义。
- 创建一个 Handler 文件，扩展自 NamespaceHandlerSupport，目的是将组件注册到 Spring 容器。
- 编写 Spring.handlers 和 Spring.schemas 文件。

现在我们就按照上面的步骤带领读者一步步地体验自定义标签的过程。

1. 首先我们创建一个普通的 POJO，这个 POJO 没有任何特别之处，只是用来接收配置文件。

```
package test.customtag;

public class User {
    private String userName;
    private String email;
    //省略 set/get 方法
}
```

2. 定义一个 XSD 文件描述组件内容。

```xml
<?xml version="1.0" encoding="UTF-8"?>
<schema xmlns="http://www.w3.org/2001/XMLSchema"
targetNamespace="http://www.lexueba.com/schema/user"
xmlns:tns="http://www.lexueba.com/schema/user"
elementFormDefault="qualified">

<element name="user">
        <complexType>
            <attribute name="id" type="string"/>
            <attribute name="userName" type="string"/>
            <attribute name="email" type="string"/>
        </complexType>
    </element>
</schema>
```

在上面的 XSD 文件中描述了一个新的 targetNamespace，并在这个空间中定义了一个 name 为 user 的 element，user 有 3 个属性 id、userName 和 email，其中 email 的类型为 string。这 3 个类主要用于验证 Spring 配置文件中自定义格式。XSD 文件是 XML DTD 的替代者，使用 XML Schema

语言进行编写,这里对 XSD Schema 不做太多解释,有兴趣的读者可以参考相关的资料。

3. 创建一个文件,实现 BeanDefinitionParser 接口,用来解析 XSD 文件中的定义和组件定义。

```
package test.customtag;
Import org.Springframework.beans.factory.support.BeanDefinitionBuilder;
import org.Springframework.beans.factory.xml.AbstractSingleBeanDefinitionParser;
import org.Springframework.util.StringUtils;
import org.w3c.dom.Element;

public class UserBeanDefinitionParser extends AbstractSingleBeanDefinitionParser {
    //Element 对应的类
    protected Class getBeanClass(Element element) {
        return User.class;
    }
    //从 element 中解析并提取对应的元素
    protected void doParse(Element element, BeanDefinitionBuilder bean) {
        String userName = element.getAttribute("userName");
        String email = element.getAttribute("email");
        //将提取的数据放入到 BeanDefinitionBuilder 中,待到完成所有 bean 的解析后统一注册到 beanFactory 中
        if (StringUtils.hasText(userName)) {
            bean.addPropertyValue("userName", userName);
        }
        if (StringUtils.hasText(email)) {
            bean.addPropertyValue("email", email);
        }
    }
}
```

4. 创建一个 Handler 文件,扩展自 NamespaceHandlerSupport,目的是将组件注册到 Spring 容器。

```
package test.customtag;

import org.Springframework.beans.factory.xml.NamespaceHandlerSupport;

public class MyNamespaceHandler extends NamespaceHandlerSupport {
    public void init() {
        registerBeanDefinitionParser("user", new UserBeanDefinitionParser());
    }
}
```

以上代码很简单,无非是当遇到自定义标签<user:aaa 这样类似于以 user 开头的元素,就会把这个元素扔给对应的 UserBeanDefinitionParser 去解析。

5. 编写 Spring.handlers 和 Spring.schemas 文件,默认位置是在工程的/META-INF/文件夹下,当然,你可以通过 Spring 的扩展或者修改源码的方式改变路径。

- Spring.handlers。

http\://www.lexueba.com/schema/user=test.customtag.MyNamespaceHandler

- Spring.schemas。

http\://www.lexueba.com/schema/user.xsd=META-INF/Spring-test.xsd

到这里,自定义的配置就结束了,而 Spring 加载自定义的大致流程是遇到自定义标签然后就去 Spring.handlers 和 Spring.schemas 中去找对应的 handler 和 XSD,默认位置是/META-INF/下,进而有找到对应的 handler 以及解析元素的 Parser,从而完成了整个自定义元素的解析,也

就是说自定义与 Spring 中默认的标准配置不同在于 Spring 将自定义标签解析的工作委托给了用户去实现。

6. 创建测试配置文件，在配置文件中引入对应的命名空间以及 XSD 后，便可以直接使用自定义标签了。

```
<beans xmlns="http://www.Springframework.org/schema/beans"
       xmlns:xsi="http://www.w3.org/2001/XMLSchema-instance"
       xmlns:myname="http://www.lexueba.com/schema/user"
       xsi:schemaLocation="http://www.Springframework.org/schema/beans http://www.Springframework.org/schema/beans/Spring-beans-2.0.xsd
       http://www.lexueba.com/schema/user http://www.lexueba.com/schema/user.xsd">

    <myname:user id="testbean" userName="aaa" email="bbb"/>

</beans>
```

7. 测试。

```
public static void main(String[] args) {
    ApplicationContext bf = new ClassPathXmlApplicationContext ("test/customtag/test.xml");
    User user=(User) bf.getBean("testbean");
    System.out.println(user.getUserName()+","+user.getEmail());
}
```

不出意外的话，你应该看到了我们期待的结果，控制台上打印出了：

aaa,bbb

在上面的例子中，我们实现了通过自定义标签实现了通过属性的方式将 user 类型的 Bean 赋值，在 Spring 中自定义标签非常常用，例如我们熟知的事务标签：tx(<tx:annotation-driven>)。

4.2 自定义标签解析

了解了自定义标签的使用后，我们带着强烈的好奇心来探究一下自定义标签的解析过程。

```
public BeanDefinition parseCustomElement(Element ele) {
    return parseCustomElement(ele, null);
}

//containingBd 为父类 bean, 对顶层元素的解析应设置为 null
public BeanDefinition parseCustomElement(Element ele, BeanDefinition containingBd) {
    //获取对应的命名空间
    String namespaceUri = getNamespaceURI(ele);
    //根据命名空间找到对应的 NamespaceHandler
    NamespaceHandler handler = this.readerContext.getNamespaceHandlerResolver().resolve(namespaceUri);
    if (handler == null) {
        error("Unable to locate Spring NamespaceHandler for XML schema namespace [" + namespaceUri + "]", ele);
        return null;
    }
    //调用自定义的 NamespaceHandler 进行解析
    return handler.parse(ele, new ParserContext(this.readerContext, this, containingBd));
}
```

相信了解了自定义标签的使用方法后，或多或少会对自定义标签的实现过程有一个自己的想法。其实思路非常的简单，无非是根据对应的 bean 获取对应的命名空间，根据命名空间解析对应的处理器，然后根据用户自定义的处理器进行解析。可是有些事情说起来简单做起来难，我们先看看如何获取命名空间吧。

4.2.1 获取标签的命名空间

标签的解析是从命名空间的提起开始的，无论是区分 Spring 中默认标签和自定义标签还是区分自定义标签中不同标签的处理器都是以标签所提供的命名空间为基础的，而至于如何提取对应元素的命名空间其实并不需要我们亲自去实现，在 org.w3c.dom.Node 中已经提供了方法供我们直接调用：

```java
public String getNamespaceURI(Node node) {
    return node.getNamespaceURI();
}
```

4.2.2 提取自定义标签处理器

有了命名空间，就可以进行 NamespaceHandler 的提取了，继续之前的 parseCustomElement 函数的跟踪，分析 NamespaceHandler handler = this.readerContext.getNamespaceHandlerResolver().resolve(namespaceUri)，在 readerContext 初始化的时候其属性 namespaceHandlerResolver 已经被初始化为了 DefaultNamespaceHandlerResolver 的实例，所以，这里调用的 resolve 方法其实调用的是 DefaultNamespaceHandlerResolver 类中的方法。我们进入 DefaultNamespaceHandlerResolver 的 resolve 方法进行查看。

DefaultNamespaceHandlerResolver.java
```java
public NamespaceHandler resolve(String namespaceUri) {
            //获取所有已经配置的handler映射
            Map<String, Object> handlerMappings = getHandlerMappings();
            //根据命名空间找到对应的信息
            Object handlerOrClassName = handlerMappings.get(namespaceUri);
            if (handlerOrClassName == null) {
                return null;
            }else if (handlerOrClassName instanceof NamespaceHandler) {
                //已经做过解析的情况，直接从缓存读取
                return (NamespaceHandler) handlerOrClassName;
            }else {
                //没有做过解析，则返回的是类路径
                String className = (String) handlerOrClassName;
                try {
                    //使用反射将类路径转化为类
                    Class<?> handlerClass = ClassUtils.forName(className, this.classLoader);
                    if (!NamespaceHandler.class.isAssignableFrom(handlerClass)) {
                        throw new FatalBeanException("Class [" + className + "] for namespace [" + namespaceUri +
                            "] does not implement the [" + NamespaceHandler.class.getName() + "] interface");
                    }
                    //初始化类
```

```
                    NamespaceHandler namespaceHandler = (NamespaceHandler) BeanUtils.
instantiateClass(handlerClass);
                    //调用自定义的 NamespaceHandler 的初始化方法
                    namespaceHandler.init();
                    //记录在缓存
                    handlerMappings.put(namespaceUri, namespaceHandler);
                    return namespaceHandler;
                }catch (ClassNotFoundException ex) {
                    throw new FatalBeanException("NamespaceHandler class [" + className +
"] for namespace [" +
                            namespaceUri + "] not found", ex);
                }catch (LinkageError err) {
                    throw new FatalBeanException("Invalid NamespaceHandler class [" +
className + "] for namespace [" +
                            namespaceUri + "]: problem with handler class file or dependent
class", err);
                }
            }
        }
```

上面的函数清晰地阐述了解析自定义 NamespaceHandler 的过程，通过之前的示例程序我们了解到如果要使用自定义标签，那么其中一项必不可少的操作就是在 Spring.handlers 文件中配置命名空间与命名空间处理器的映射关系。只有这样，Spring 才能根据映射关系找到匹配的处理器，而寻找匹配的处理器就是在上面函数中实现的，当获取到自定义的 NamespaceHandler 之后就可以进行处理器初始化并解析了。我们不妨再次回忆一下示例中对于命名空间处理器的内容：

```
public class MyNamespaceHandler extends NamespaceHandlerSupport {
    public void init() {
        registerBeanDefinitionParser("user", new UserBeanDefinitionParser());
    }
}
```

当得到自定义命名空间处理后会马上执行 namespaceHandler.init()来进行自定义 BeanDefinitionParser 的注册。在这里，你可以注册多个标签解析器，当前示例中只有支持<myname:user 的写法，你也可以在这里注册多个解析器，如<myname:A、<myname:B 等，使得 myname 的命名空间中可以支持多种标签解析。

注册后，命名空间处理器就可以根据标签的不同来调用不同的解析器进行解析。那么，根据上面的函数与之前介绍过的例子，我们基本上可以推断 getHandlerMappings 的主要功能就是读取 Spring.handlers 配置文件并将配置文件缓存在 map 中。

```
        private Map<String, Object> getHandlerMappings() {
            //如果没有被缓存则开始进行缓存
            if (this.handlerMappings == null) {
                synchronized (this) {
                    if (this.handlerMappings == null) {
                        try {
                            //this.handlerMappingsLocation 在构造函数中已经被初始化为:META- INF/
Spring.handlers
                            Properties mappings =
                                    PropertiesLoaderUtils.loadAllProperties (this.
handlerMappingsLocation, this.classLoader);
                            if (logger.isDebugEnabled()) {
                                logger.debug("Loaded NamespaceHandler mappings: " + mappings);
```

```
                            }
                            Map<String, Object> handlerMappings = new ConcurrentHashMap<
String, Object>();
                            //将 Properties 格式文件合并到 Map 格式的 handlerMappings 中
                            CollectionUtils.mergePropertiesIntoMap(mappings, handlerMappings);
                            this.handlerMappings = handlerMappings;
                        }
                        catch (IOException ex) {
                            throw new IllegalStateException(
                                    "Unable to load NamespaceHandler mappings from
location [" + this.handlerMappingsLocation + "]", ex);
                        }
                    }
                }
            }
            return this.handlerMappings;
        }
```

同我们想象的一样，借助了工具类 PropertiesLoaderUtils 对属性 handlerMappingsLocation 进行了配置文件的读取，handlerMappingsLocation 被默认初始化为 "META-INF/Spring.handlers"。

4.2.3 标签解析

得到了解析器以及要分析的元素后，Spring 就可以将解析工作委托给自定义解析器去解析了。在 Spring 中的代码为：

```
return handler.parse(ele, new ParserContext(this.readerContext, this, containingBd))
```

以之前提到的示例进行分析，此时的 handler 已经被实例化成我们自定义的 MyNamespaceHandler 了，而 MyNamespaceHandler 也已经完成了初始化的工作，但是在我们实现的自定义命名空间处理器中并没有实现 parse 方法，所以推断，这个方法是父类中的实现，查看父类 NamespaceHandlerSupport 中的 parse 方法。

NamespaceHandlerSupport.java
```
public BeanDefinition parse(Element element, ParserContext parserContext) {
    //寻找解析器并进行解析操作
            return findParserForElement(element, parserContext).parse(element, parserContext);
        }
```

解析过程中首先是寻找元素对应的解析器，进而调用解析器中的 parse 方法，那么结合示例来讲，其实就是首先获取在 MyNameSpaceHandler 类中的 init 方法中注册的对应的 UserBeanDefinitionParser 实例，并调用其 parse 方法进行进一步解析。

```
private BeanDefinitionParser findParserForElement(Element element, ParserContext parser
Context) {
            //获取元素名称，也就是<myname:user 中的 user,若在示例中，此时 localName 为 user
            String localName = parserContext.getDelegate().getLocalName(element);
        //根据 user 找到对应的解析器，也就是在
//registerBeanDefinitionParser("user", new UserBeanDefinitionParser());
//注册的解析器
            BeanDefinitionParser parser = this.parsers.get(localName);
            if (parser == null) {
```

```
            parserContext.getReaderContext().fatal(
                "Cannot locate BeanDefinitionParser for element [" + localName +
"]", element);
        }
        return parser;
    }
```

而对于 parse 方法的处理：
```
public final BeanDefinition parse(Element element, ParserContext parserContext) {
    AbstractBeanDefinition definition = parseInternal(element, parserContext);
    if (definition != null && !parserContext.isNested()) {
        try {
            String id = resolveId(element, definition, parserContext);
            if (!StringUtils.hasText(id)) {
                parserContext.getReaderContext().error(
                    "Id is required for element '" + parserContext. getDelegate(). getLocalName(element)
                    + "' when used as a top-level tag", element);
            }
            String[] aliases = new String[0];
            String name = element.getAttribute(NAME_ATTRIBUTE);
            if (StringUtils.hasLength(name)) {
                aliases = StringUtils.trimArrayElements(StringUtils.commaDelimitedListToStringArray(name));
            }
            //将 AbstractBeanDefinition 转换为 BeanDefinitionHolder 并注册
            BeanDefinitionHolder holder = new BeanDefinitionHolder(definition, id, aliases);
            registerBeanDefinition(holder, parserContext.getRegistry());
            if (shouldFireEvents()) {
                //需要通知监听器则进行处理
                BeanComponentDefinition componentDefinition = new BeanComponentDefinition(holder);
                postProcessComponentDefinition(componentDefinition);
                parserContext.registerComponent(componentDefinition);
            }
        }
        catch (BeanDefinitionStoreException ex) {
            parserContext.getReaderContext().error(ex.getMessage(), element);
            return null;
        }
    }
    return definition;
}
```

虽说是对自定义配置文件的解析，但是，我们可以看到，在这个函数中大部分的代码是用来处理将解析后的 AbstractBeanDefinition 转化为 BeanDefinitionHolder 并注册的功能，而真正去做解析的事情委托给了函数 parseInternal，正是这句代码调用了我们自定义的解析函数。

在 parseInternal 中并不是直接调用自定义的 doParse 函数，而是进行了一系列的数据准备，包括对 beanClass、scope、lazyInit 等属性的准备。
```
protected final AbstractBeanDefinition parseInternal(Element element, ParserContext parserContext) {
    BeanDefinitionBuilder builder = BeanDefinitionBuilder.genericBeanDefinition();
    String parentName = getParentName(element);
```

```
        if (parentName != null) {
            builder.getRawBeanDefinition().setParentName(parentName);
        }
//获取自定义标签中的class,此时会调用自定义解析器如UserBeanDefinitionParser中的getBeanClass方法
        Class<?> beanClass = getBeanClass(element);
        if (beanClass != null) {
            builder.getRawBeanDefinition().setBeanClass(beanClass);
        }
        else {
//若子类没有重写getBeanClass方法则尝试检查子类是否重写getBeanClassName方法
            String beanClassName = getBeanClassName(element);
            if (beanClassName != null) {
                builder.getRawBeanDefinition().setBeanClassName(beanClassName);
            }
        }
        builder.getRawBeanDefinition().setSource(parserContext.extractSource(element));
        if (parserContext.isNested()) {
//若存在父类则使用父类的scope属性
            builder.setScope(parserContext.getContainingBeanDefinition().getScope());
        }
        if (parserContext.isDefaultLazyInit()) {
            // Default-lazy-init applies to custom bean definitions as well.
//配置延迟加载
            builder.setLazyInit(true);
        }
//调用子类重写的doParse方法进行解析
        doParse(element, parserContext, builder);
        return builder.getBeanDefinition();
    }

    protected void doParse(Element element, ParserContext parserContext, BeanDefinition
Builder builder) {
        doParse(element, builder);
    }
```

回顾一下全部的自定义标签处理过程，虽然在实例中我们定义 UserBeanDefinitionParser，但是在其中我们只是做了与自己业务逻辑相关的部分。不过我们没做但是并不代表没有，在这个处理过程中同样也是按照 Spring 中默认标签的处理方式进行，包括创建 BeanDefinition 以及进行相应默认属性的设置，对于这些工作 Spring 都默默地帮我们实现了，只是暴露出一些接口来供用户实现个性化的业务。通过对本章的了解，相信读者对 Spring 中自定义标签的使用以及在解析自定义标签过程中 Spring 为我们做了哪些工作会有一个全面的了解。到此为止我们已经完成了 Spring 中全部的解析工作，也就是说到现在为止我们已经理解了 Spring 将 bean 从配置文件到加载到内存中的全过程，而接下来的任务便是如何使用这些 bean，下一章将介绍 bean 的加载。

第 5 章 bean 的加载

经过前面的分析,我们终于结束了对 XML 配置文件的解析,接下来将会面临更大的挑战,就是对 bean 加载的探索。bean 加载的功能实现远比 bean 的解析要复杂得多,同样,我们还是以本书开篇的示例为基础,对于加载 bean 的功能,在 Spring 中的调用方式为:

```
MyTestBean bean=(MyTestBean) bf.getBean("myTestBean")
```

这句代码实现了什么样的功能呢?我们可以先快速体验一下 Spring 中代码是如何实现的。

```java
public Object getBean(String name) throws BeansException {
    return doGetBean(name, null, null, false);
}

protected <T> T doGetBean(
        final String name, final Class<T> requiredType, final Object[] args,
boolean typeCheckOnly) throws BeansException {
        //提取对应的beanName
        final String beanName = transformedBeanName(name);
        Object bean;

        /*
         * 检查缓存中或者实例工厂中是否有对应的实例
         * 为什么首先会使用这段代码呢,
         * 因为在创建单例bean的时候会存在依赖注入的情况,而在创建依赖的时候为了避免循环依赖,
         * Spring创建bean的原则是不等bean创建完成就会将创建bean的ObjectFactory提早曝光
         * 也就是将ObjectFactory加入到缓存中,一旦下个bean创建时候需要依赖上个bean则直接使用ObjectFactory
         */
        //直接尝试从缓存获取或者singletonFactories中的ObjectFactory中获取
        Object sharedInstance = getSingleton(beanName);
        if (sharedInstance != null && args == null) {
            if (logger.isDebugEnabled()) {
                if (isSingletonCurrentlyInCreation(beanName)) {
                    logger.debug("Returning eagerly cached instance of singleton bean '" + beanName +
                            "' that is not fully initialized yet - a consequence of a circular reference");
                }
                else {
                    logger.debug("Returning cached instance of singleton bean '" + beanName + "'");
```

第 5 章 bean 的加载

```
        }
    }
//返回对应的实例，有时候存在诸如 BeanFactory 的情况并不是直接返回实例本身而是返回指定方法返回的实例
    bean = getObjectForBeanInstance(sharedInstance, name, beanName, null);
}else {
    //只有在单例情况才会尝试解决循环依赖，原型模式情况下，如果存在
    //A 中有 B 的属性，B 中有 A 的属性，那么当依赖注入的时候，就会产生当 A 还未创建完的时候因为
    //对于 B 的创建再次返回创建 A，造成循环依赖，也就是下面的情况
    //isPrototypeCurrentlyInCreation(beanName) 为 true
    if (isPrototypeCurrentlyInCreation(beanName)) {
        throw new BeanCurrentlyInCreationException(beanName);
    }

    BeanFactory parentBeanFactory = getParentBeanFactory();
    //如果 beanDefinitionMap 中也就是在所有已经加载的类中不包括 beanName 则尝试从
    //parentBeanFactory 中检测
    if (parentBeanFactory != null && !containsBeanDefinition(beanName)) {
        String nameToLookup = originalBeanName(name);
        //递归到 BeanFactory 中寻找
        if (args != null) {
            return (T) parentBeanFactory.getBean(nameToLookup, args);
        }
        else {
            return parentBeanFactory.getBean(nameToLookup, requiredType);
        }
    }
    //如果不是仅仅做类型检查则是创建 bean，这里要进行记录
    if (!typeCheckOnly) {
        markBeanAsCreated(beanName);
    }

    //将存储 XML 配置文件的 GernericBeanDefinition 转换为 RootBeanDefinition,如果指定
    //BeanName 是子 Bean 的话同时会合并父类的相关属性
    final RootBeanDefinition mbd = getMergedLocalBeanDefinition(beanName);
    checkMergedBeanDefinition(mbd, beanName, args);

    String[] dependsOn = mbd.getDependsOn();
    //若存在依赖则需要递归实例化依赖的 bean
    if (dependsOn != null) {
        for (String dependsOnBean : dependsOn) {
            getBean(dependsOnBean);
            //缓存依赖调用
            registerDependentBean(dependsOnBean, beanName);
        }
    }

    //实例化依赖的 bean 后便可以实例化 mbd 本身了
    //singleton 模式的创建
    if (mbd.isSingleton()) {
        sharedInstance = getSingleton(beanName, new ObjectFactory<Object>() {
            public Object getObject() throws BeansException {
                try {
                    return createBean(beanName, mbd, args);
                }
                catch (BeansException ex) {
```

```java
                            destroySingleton(beanName);
                            throw ex;
                        }
                    }
                });
                bean = getObjectForBeanInstance(sharedInstance, name, beanName, mbd);
            }else if (mbd.isPrototype()) {
                //prototype 模式的创建(new)
                Object prototypeInstance = null;
                try {
                    beforePrototypeCreation(beanName);
                    prototypeInstance = createBean(beanName, mbd, args);
                }
                finally {
                    afterPrototypeCreation(beanName);
                }
                bean = getObjectForBeanInstance(prototypeInstance, name, beanName, mbd);
            }else {
                //指定的 scope 上实例化 bean
                String scopeName = mbd.getScope();
                final Scope scope = this.scopes.get(scopeName);
                if (scope == null) {
                    throw new IllegalStateException("No Scope registered for scope '" + scopeName + "'");
                }
                try {
                    Object scopedInstance = scope.get(beanName, new ObjectFactory<Object>() {
                        public Object getObject() throws BeansException {
                            beforePrototypeCreation(beanName);
                            try {
                                return createBean(beanName, mbd, args);
                            }
                            finally {
                                afterPrototypeCreation(beanName);
                            }
                        }
                    });
                    bean = getObjectForBeanInstance(scopedInstance, name, beanName, mbd);
                }
                catch (IllegalStateException ex) {
                    throw new BeanCreationException(beanName,
                            "Scope '" + scopeName + "' is not active for the current thread; " +
                            "consider defining a scoped proxy for this bean if you intend to refer to it from a singleton",
                            ex);
                }
            }
        }

        //检查需要的类型是否符合 bean 的实际类型
        if (requiredType != null && bean != null && !requiredType.isAssignableFrom(bean.getClass())) {
            try {
                return getTypeConverter().convertIfNecessary(bean, requiredType);
```

```
                }
                catch (TypeMismatchException ex) {
                    if (logger.isDebugEnabled()) {
                        logger.debug("Failed to convert bean '" + name + "' to required type [" +
                                ClassUtils.getQualifiedName(requiredType) + "]", ex);
                    }
                    throw new BeanNotOfRequiredTypeException(name, requiredType, bean.getClass());
                }
            }
        }
        return (T) bean;
    }
```

仅从代码量上就能看出来 bean 的加载经历了一个相当复杂的过程，其中涉及各种各样的考虑。相信读者细心阅读上面的代码，并参照部分代码注释，是可以粗略地了解整个 Spring 加载 bean 的过程。对于加载过程中所涉及的步骤大致如下。

1. 转换对应 beanName

或许很多人不理解转换对应 beanName 是什么意思，传入的参数 name 不就是 beanName 吗？其实不是，这里传入的参数可能是别名，也可能是 FactoryBean，所以需要进行一系列的解析，这些解析内容包括如下内容。

- 去除 FactoryBean 的修饰符，也就是如果 name="&aa"，那么会首先去除&而使 name="aa"。
- 取指定 alias 所表示的最终 beanName，例如别名 A 指向名称为 B 的 bean 则返回 B；若别名 A 指向别名 B，别名 B 又指向名称为 C 的 bean 则返回 C。

2. 尝试从缓存中加载单例

单例在 Spring 的同一个容器内只会被创建一次，后续再获取 bean，就直接从单例缓存中获取了。当然这里也只是尝试加载，首先尝试从缓存中加载，如果加载不成功则再次尝试从 singletonFactories 中加载。因为在创建单例 bean 的时候会存在依赖注入的情况，而在创建依赖的时候为了避免循环依赖，在 Spring 中创建 bean 的原则是不等 bean 创建完成就会将创建 bean 的 ObjectFactory 提早曝光加入到缓存中，一旦下一个 bean 创建时候需要依赖上一个 bean 则直接使用 ObjectFactory（后面章节会对循环依赖重点讲解）。

3. bean 的实例化

如果从缓存中得到了 bean 的原始状态，则需要对 bean 进行实例化。这里有必要强调一下，缓存中记录的只是最原始的 bean 状态，并不一定是我们最终想要的 bean。举个例子，假如我们需要对工厂 bean 进行处理，那么这里得到的其实是工厂 bean 的初始状态，但是我们真正需要的是工厂 bean 中定义的 factory-method 方法中返回的 bean，而 getObjectForBeanInstance 就是完成这个工作的，后续会详细讲解。

4. 原型模式的依赖检查

只有在单例情况下才会尝试解决循环依赖，如果存在 A 中有 B 的属性，B 中有 A 的属性，那么当依赖注入的时候，就会产生当 A 还未创建完的时候因为对于 B 的创建再次返回创建 A，造成循环依赖，也就是情况：isPrototypeCurrentlyInCreation(beanName)判断 true。

5. 检测 parentBeanFactory

从代码上看，如果缓存没有数据的话直接转到父类工厂上去加载了，这是为什么呢？

可能读者忽略了一个很重要的判断条件：parentBeanFactory != null && !containsBean Definition (beanName)，parentBeanFactory != null。parentBeanFactory 如果为空，则其他一切都是浮云，这个没什么说的，但是!containsBeanDefinition(beanName)就比较重要了，它是在检测如果当前加载的 XML 配置文件中不包含 beanName 所对应的配置，就只能到 parentBeanFactory 去尝试下了，然后再去递归的调用 getBean 方法。

6. 将存储 XML 配置文件的 GernericBeanDefinition 转换为 RootBeanDefinition

因为从 XML 配置文件中读取到的 bean 信息是存储在 GernericBeanDefinition 中的，但是所有的 bean 后续处理都是针对于 RootBeanDefinition 的，所以这里需要进行一个转换，转换的同时如果父类 bean 不为空的话，则会一并合并父类的属性。

7. 寻找依赖

因为 bean 的初始化过程中很可能会用到某些属性，而某些属性很可能是动态配置的，并且配置成依赖于其他的 bean，那么这个时候就有必要先加载依赖的 bean，所以，在 Spring 的加载顺寻中，在初始化某一个 bean 的时候首先会初始化这个 bean 所对应的依赖。

8. 针对不同的 scope 进行 bean 的创建

我们都知道，在 Spring 中存在着不同的 scope，其中默认的是 singleton，但是还有些其他的配置诸如 prototype、request 之类的。在这个步骤中，Spring 会根据不同的配置进行不同的初始化策略。

9. 类型转换

程序到这里返回 bean 后已经基本结束了,通常对该方法的调用参数 requiredType 是为空的，但是可能会存在这样的情况，返回的 bean 其实是个 String，但是 requiredType 却传入 Integer 类型，那么这时候本步骤就会起作用了，它的功能是将返回的 bean 转换为 requiredType 所指定的类型。当然，String 转换为 Integer 是最简单的一种转换，在 Spring 中提供了各种各样的转换器，用户也可以自己扩展转换器来满足需求。

经过上面的步骤后 bean 的加载就结束了，这个时候就可以返回我们所需要的 bean 了，

图 5-1 直观地反映了整个过程。其中最重要的就是步骤 8，针对不同的 scope 进行 bean 的创建，你会看到各种常用的 Spring 特性在这里的实现。

在细化分析各个步骤提供的功能前，我们有必要先了解下 FactoryBean 的用法。

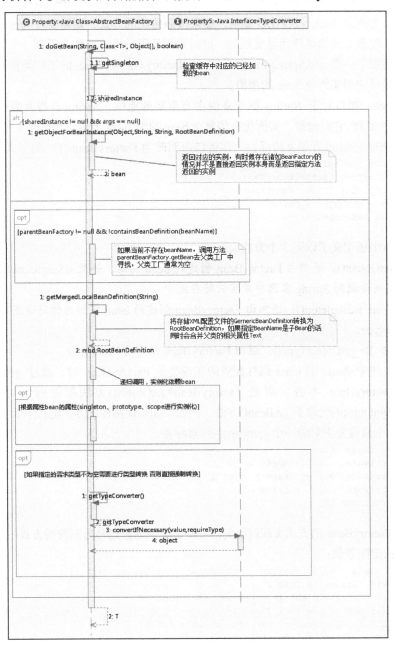

图 5-1 bean 的获取过程

5.1 FactoryBean 的使用

一般情况下，Spring 通过反射机制利用 bean 的 class 属性指定实现类来实例化 bean。在某些情况下，实例化 bean 过程比较复杂，如果按照传统的方式，则需要在<bean>中提供大量的配置信息，配置方式的灵活性是受限的，这时采用编码的方式可能会得到一个简单的方案。Spring 为此提供了一个 org.Springframework.bean.factory.FactoryBean 的工厂类接口，用户可以通过实现该接口定制实例化 bean 的逻辑。

FactoryBean 接口对于 Spring 框架来说占有重要的地位，Spring 自身就提供了 70 多个 FactoryBean 的实现。它们隐藏了实例化一些复杂 bean 的细节，给上层应用带来了便利。从 Spring 3.0 开始，FactoryBean 开始支持泛型，即接口声明改为 FactoryBean<T> 的形式：

```
package org.Springframework.beans.factory;
public interface FactoryBean<T> {
    T getObject() throws Exception;
    Class<?> getObjectType();
    boolean isSingleton();
}
```

在该接口中还定义了以下 3 个方法。

- T getObject()：返回由 FactoryBean 创建的 bean 实例，如果 isSingleton()返回 true，则该实例会放到 Spring 容器中单实例缓存池中。
- boolean isSingleton()：返回由 FactoryBean 创建的 bean 实例的作用域是 singleton 还是 prototype。
- Class<T> getObjectType()：返回 FactoryBean 创建的 bean 类型。

当配置文件中<bean>的 class 属性配置的实现类是 FactoryBean 时，通过 getBean()方法返回的不是 FactoryBean 本身，而是 FactoryBean#getObject()方法所返回的对象，相当于 FactoryBean#getObject()代理了 getBean()方法。例如：如果使用传统方式配置下面 Car 的<bean>时，Car 的每个属性分别对应一个<property>元素标签。

```
public class Car {
    private int maxSpeed ;
    private String brand ;
    private double price ;
    //get/set方法
}
```

如果用 FactoryBean 的方式实现就会灵活一些，下例通过逗号分割符的方式一次性地为 Car 的所有属性指定配置值：

```
public class CarFactoryBean implements FactoryBean<Car> {
    private String carInfo ;
    public Car getObject () throws Exception {
        Car car = new Car () ;
        String [] infos = carInfo .split ( "," ) ;
        car.setBrand ( infos [ 0 ]) ;
        car.setMaxSpeed ( Integer. valueOf ( infos [ 1 ])) ;
```

```java
            car.setPrice(Double.valueOf(infos[2]));
            return car;
        }
        public Class<Car> getObjectType() {
            return Car.class;
        }
        public boolean isSingleton() {
            return false;
        }
        public String getCarInfo() {
            return this.carInfo;
        }

        // 接受逗号分割符设置属性信息
        public void setCarInfo(String carInfo) {
            this.carInfo = carInfo;
        }
    }
```

有了这个 CarFactoryBean 后，就可以在配置文件中使用下面这种自定义的配置方式配置 Car Bean 了：

```xml
<bean id="car" class="com.test.factorybean.CarFactoryBean" carInfo="超级跑车,400,2000000"/>
```

当调用 getBean("car") 时，Spring 通过反射机制发现 CarFactoryBean 实现了 FactoryBean 的接口，这时 Spring 容器就调用接口方法 CarFactoryBean#getObject()方法返回。如果希望获取 CarFactoryBean 的实例，则需要在使用 getBean(beanName) 方法时在 beanName 前显示的加上 "&" 前缀，例如 getBean("&car")。

5.2　缓存中获取单例 bean

介绍过 FactoryBean 的用法后，我们就可以了解 bean 加载的过程了。前面已经提到过，单例在 Spring 的同一个容器内只会被创建一次，后续再获取 bean 直接从单例缓存中获取，当然这里也只是尝试加载，首先尝试从缓存中加载，然后再次尝试尝试从 singletonFactories 中加载。因为在创建单例 bean 的时候会存在依赖注入的情况，而在创建依赖的时候为了避免循环依赖，Spring 创建 bean 的原则是不等 bean 创建完成就会将创建 bean 的 ObjectFactory 提早曝光加入到缓存中，一旦下一个 bean 创建时需要依赖上个 bean，则直接使用 ObjectFactory。

```java
public Object getSingleton(String beanName) {
    //参数 true 设置标识允许早期依赖
    return getSingleton(beanName, true);
}

protected Object getSingleton(String beanName, boolean allowEarlyReference) {
        //检查缓存中是否存在实例
        Object singletonObject = this.singletonObjects.get(beanName);
        if (singletonObject == null) {
            //如果为空，则锁定全局变量并进行处理
            synchronized (this.singletonObjects) {
                //如果此 bean 正在加载则不处理
```

```
                    singletonObject = this.earlySingletonObjects.get(beanName);
                    if (singletonObject == null && allowEarlyReference) {
                        //当某些方法需要提前初始化的时候则会调用 addSingletonFactory 方法将对应的
                        //ObjectFactory 初始化策略存储在 singletonFactories
                        ObjectFactory singletonFactory = this.singletonFactories.get
(beanName);
                        if (singletonFactory != null) {
                            //调用预先设定的 getObject 方法
                            singletonObject = singletonFactory.getObject();
                            //记录在缓存中，earlySingletonObjects 和 singletonFactories 互斥
                            this.earlySingletonObjects.put(beanName, singletonObject);
                            this.singletonFactories.remove(beanName);
                        }
                    }
                }
            }
            return (singletonObject != NULL_OBJECT ? singletonObject : null);
}
```

这个方法因为涉及循环依赖的检测，以及涉及很多变量的记录存取，所以让很多读者摸不着头脑。这个方法首先尝试从 singletonObjects 里面获取实例，如果获取不到再从 earlySingletonObjects 里面获取，如果还获取不到，再尝试从 singletonFactories 里面获取 beanName 对应的 ObjectFactory，然后调用这个 ObjectFactory 的 getObject 来创建 bean，并放到 earlySingletonObjects 里面去，并且从 singletonFacotories 里面 remove 掉这个 ObjectFactory，而对于后续的所有内存操作都只为了循环依赖检测时候使用，也就是在 allowEarlyReference 为 true 的情况下才会使用。

这里涉及用于存储 bean 的不同的 map，可能让读者感到崩溃，简单解释如下。

- singletonObjects：用于保存 BeanName 和创建 bean 实例之间的关系，bean name --> bean instance。
- singletonFactories：用于保存 BeanName 和创建 bean 的工厂之间的关系，bean name --> ObjectFactory。
- earlySingletonObjects：也是保存 BeanName 和创建 bean 实例之间的关系，与 singletonObjects 的不同之处在于，当一个单例 bean 被放到这里面后，那么当 bean 还在创建过程中，就可以通过 getBean 方法获取到了，其目的是用来检测循环引用。
- registeredSingletons：用来保存当前所有已注册的 bean。

5.3 从 bean 的实例中获取对象

在 getBean 方法中，getObjectForBeanInstance 是个高频率使用的方法，无论是从缓存中获得 bean 还是根据不同的 scope 策略加载 bean。总之，我们得到 bean 的实例后要做的第一步就是调用这个方法来检测一下正确性，其实就是用于检测当前 bean 是否是 FactoryBean 类型的 bean，如果是，那么需要调用该 bean 对应的 FactoryBean 实例中的 getObject()作为

5.3 从 bean 的实例中获取对象

返回值。

无论是从缓存中获取到的bean还是通过不同的scope策略加载的bean都只是最原始的bean状态，并不一定是我们最终想要的 bean。举个例子，假如我们需要对工厂 bean 进行处理，那么这里得到的其实是工厂 bean 的初始状态，但是我们真正需要的是工厂 bean 中定义的 factory-method 方法中返回的 bean，而 getObjectForBeanInstance 方法就是完成这个工作的。

```
protected Object getObjectForBeanInstance(
        Object beanInstance, String name, String beanName, RootBeanDefinition mbd) {

    //如果指定的name是工厂相关(以&为前缀)且beanInstance又不是FactoryBean类型则验证不通过
    if (BeanFactoryUtils.isFactoryDereference(name) && !(beanInstance instanceof FactoryBean)) {
        throw new BeanIsNotAFactoryException(transformedBeanName(name), beanInstance.getClass());
    }

    //现在我们有了个bean的实例，这个实例可能会是正常的bean或者是FactoryBean
    //如果是FactoryBean我们使用它创建实例,但是如果用户想要直接获取工厂实例而不是工厂的
    //getObject方法对应的实例那么传入的name应该加入前缀&
    if (!(beanInstance instanceof FactoryBean) || BeanFactoryUtils. IsFactoryDereference(name)) {
        return beanInstance;
    }

    //加载FactoryBean
    Object object = null;
    if (mbd == null) {
        //尝试从缓存中加载bean
        object = getCachedObjectForFactoryBean(beanName);
    }
    if (object == null) {
        //到这里已经明确知道beanInstance一定是FactoryBean类型
        FactoryBean<?> factory = (FactoryBean<?>) beanInstance;
        //containsBeanDefinition检测beanDefinitionMap中也就是在所有已经加载的类中检测
        //是否定义beanName
        if (mbd == null && containsBeanDefinition(beanName)) {
            //将存储XML配置文件的GernericBeanDefinition转换为RootBeanDefinition,如
            //果指定BeanName是子Bean的话同时会合并父类的相关属性
            mbd = getMergedLocalBeanDefinition(beanName);
        }
        //是否是用户定义的而不是应用程序本身定义的
        boolean synthetic = (mbd != null && mbd.isSynthetic());
        object = **getObjectFromFactoryBean**(factory, beanName, !synthetic);
    }
    return object;
}
```

从上面的代码来看，其实这个方法并没有什么重要的信息，大多是些辅助代码以及一些功能性的判断，而真正的核心代码却委托给了 getObjectFromFactoryBean，我们来看看 getObjectForBeanInstance 中的所做的工作。

1. 对 FactoryBean 正确性的验证。

2. 对非 FactoryBean 不做任何处理。
3. 对 bean 进行转换。
4. 将从 Factory 中解析 bean 的工作委托给 getObjectFromFactoryBean。

```
protected Object getObjectFromFactoryBean(FactoryBean factory, String beanName, Boolean shouldPostProcess) {
    //如果是单例模式
    if (factory.isSingleton() && containsSingleton(beanName)) {
        synchronized (getSingletonMutex()) {
            Object object = this.factoryBeanObjectCache.get(beanName);
            if (object == null) {
                object = doGetObjectFromFactoryBean(factory, beanName, shouldPostProcess);
                this.factoryBeanObjectCache.put(beanName, (object != null ? object : NULL_OBJECT));
            }
            return (object != NULL_OBJECT ? object : null);
        }
    }
    else {
        return doGetObjectFromFactoryBean(factory, beanName, shouldPostProcess);
    }
}
```

很遗憾，在这个代码中我们还是没有看到想要看到的代码，在这个方法里只做了一件事情，就是返回的 bean 如果是单例的，那就必须要保证全局唯一，同时，也因为是单例的，所以不必重复创建，可以使用缓存来提高性能，也就是说已经加载过就要记录下来以便于下次复用，否则的话就直接获取了。

在 doGetObjectFromFactoryBean 方法中我们终于看到了我们想要看到的方法，也就是 object = factory.getObject()，是的，就是这句代码，我们的历程犹如剥洋葱一样，一层一层的直到最内部的代码实现，虽然很简单。

```
private Object doGetObjectFromFactoryBean(
        final FactoryBean factory, final String beanName, final boolean shouldPostProcess)
        throws BeanCreationException {

    Object object;
    try {
        //需要权限验证
        if (System.getSecurityManager() != null) {
            AccessControlContext acc = getAccessControlContext();
            try {
                object = AccessController.doPrivileged(new PrivilegedExceptionAction<Object>() {
                    public Object run() throws Exception {
                        return factory.getObject();
                    }
                }, acc);
            }
            catch (PrivilegedActionException pae) {
                throw pae.getException();
```

```
                    }
                }
                else {
                    //直接调用getObject方法
                    object = factory.getObject();
                }
            }
            catch (FactoryBeanNotInitializedException ex) {
                throw new BeanCurrentlyInCreationException(beanName, ex.toString());
            }
            catch (Throwable ex) {
                throw new BeanCreationException(beanName, "FactoryBean threw exception on
 object creation", ex);
            }
            if (object == null && isSingletonCurrentlyInCreation(beanName)) {
                throw new BeanCurrentlyInCreationException(
                        beanName, "FactoryBean which is currently in creation returned
 null from getObject");
            }

            if (object != null && shouldPostProcess) {
                try {
                    //调用ObjectFactory的后处理器
                    object = postProcessObjectFromFactoryBean(object, beanName);
                }
                catch (Throwable ex) {
                    throw new BeanCreationException(beanName, "Post-processing of the
 FactoryBean's object failed", ex);
                }
            }

            return object;
    }
```

上面我们已经讲述了FactoryBean的调用方法，如果bean声明为FactoryBean类型，则当提取bean时提取的并不是FactoryBean，而是FactoryBean中对应的getObject方法返回的bean，而doGetObjectFromFactoryBean正是实现这个功能的。但是，我们看到在上面的方法中除了调用object = factory.getObject()得到我们想要的结果后并没有直接返回，而是接下来又做了些后处理的操作，这个又是做什么用的呢？于是我们跟踪进入AbstractAutowireCapableBeanFactory类的postProcessObjectFromFactoryBean方法：

AbstractAutowireCapableBeanFactory.java
```
    protected Object postProcessObjectFromFactoryBean(Object object, String beanName) {
            return applyBeanPostProcessorsAfterInitialization(object, beanName);
    }
    public Object applyBeanPostProcessorsAfterInitialization(Object existingBean, String beanName)
            throws BeansException {

            Object result = existingBean;
            for (BeanPostProcessor beanProcessor : getBeanPostProcessors()) {
                result = beanProcessor.postProcessAfterInitialization(result, beanName);
```

```
            if (result == null) {
                return result;
            }
        }
        return result;
}
```

对于后处理器的使用我们还未过多接触，后续章节会使用大量篇幅介绍，这里，我们只需了解在 Spring 获取 bean 的规则中有这样一条：尽可能保证所有 bean 初始化后都会调用注册的 BeanPostProcessor 的 postProcessAfterInitialization 方法进行处理，在实际开发过程中大可以针对此特性设计自己的业务逻辑。

5.4 获取单例

之前我们讲解了从缓存中获取单例的过程，那么，如果缓存中不存在已经加载的单例 bean 就需要从头开始 bean 的加载过程了，而 Spring 中使用 getSingleton 的重载方法实现 bean 的加载过程。

```
public Object getSingleton(String beanName, ObjectFactory singletonFactory) {
            Assert.notNull(beanName, "'beanName' must not be null");
            //全局变量需要同步
            synchronized (this.singletonObjects) {
                //首先检查对应的 bean 是否已经加载过，因为 singleton 模式其实就是复用以创建的 bean,
                //所以这一步是必须的
                Object singletonObject = this.singletonObjects.get(beanName);
                //如果为空才可以进行 singleto 的 bean 的初始化
                if (singletonObject == null) {
                    if (this.singletonsCurrentlyInDestruction) {
                        throw new BeanCreationNotAllowedException(beanName,
                                "Singleton bean creation not allowed while the singletons of this factory are in destruction " +
                                "(Do not request a bean from a BeanFactory in a destroy method implementation!)");
                    }
                    if (logger.isDebugEnabled()) {
                        logger.debug("Creating shared instance of singleton bean '" + beanName + "'");
                    }
                    **beforeSingletonCreation**(beanName);
                    boolean recordSuppressedExceptions = (this.suppressedExceptions == null);
                    if (recordSuppressedExceptions) {
                        this.suppressedExceptions = new LinkedHashSet<Exception>();
                    }
                    try {
                        //初始化 bean
                        singletonObject = **singletonFactory.getObject()**;
                    }
                    catch (BeanCreationException ex) {
                        if (recordSuppressedExceptions) {
                            for (Exception suppressedException : this.suppressedExceptions) {
```

```
                    ex.addRelatedCause(suppressedException);
                }
            }
            throw ex;
        }
        finally {
            if (recordSuppressedExceptions) {
                this.suppressedExceptions = null;
            }
            afterSingletonCreation(beanName);
        }
        //加入缓存
        addSingleton(beanName, singletonObject);
    }
    return (singletonObject != NULL_OBJECT ? singletonObject : null);
}
```

上述代码中其实是使用了回调方法，使得程序可以在单例创建的前后做一些准备及处理操作，而真正的获取单例 bean 的方法其实并不是在此方法中实现的，其实现逻辑是在 ObjectFactory 类型的实例 singletonFactory 中实现的。而这些准备及处理操作包括如下内容。

1. 检查缓存是否已经加载过。
2. 若没有加载，则记录 beanName 的正在加载状态。
3. 加载单例前记录加载状态。

可能你会觉得 beforeSingletonCreation 方法是个空实现，里面没有任何逻辑，但其实不是，这个函数中做了一个很重要的操作：记录加载状态，也就是通过 this.singletonsCurrentlyInCreation.add(beanName)将当前正要创建的 bean 记录在缓存中，这样便可以对循环依赖进行检测。

```
protected void beforeSingletonCreation(String beanName) {
    if (!this.inCreationCheckExclusions.contains(beanName) && !this.singletons
CurrentlyInCreation.add(beanName)) {
        throw new BeanCurrentlyInCreationException(beanName);
    }
}
```

4. 通过调用参数传入的 ObjectFactory 的个体 Object 方法实例化 bean。
5. 加载单例后的处理方法调用。

同步骤 3 的记录加载状态相似，当 bean 加载结束后需要移除缓存中对该 bean 的正在加载状态的记录。

```
protected void afterSingletonCreation(String beanName) {
    if (!this.inCreationCheckExclusions.contains(beanName) && !this.singletons
CurrentlyInCreation.remove(beanName)) {
        throw new IllegalStateException("Singleton '" + beanName + "' isn't
currently in creation");
    }
}
```

6. 将结果记录至缓存并删除加载 bean 过程中所记录的各种辅助状态。

```java
protected void addSingleton(String beanName, Object singletonObject) {
    synchronized (this.singletonObjects) {
        this.singletonObjects.put(beanName, (singletonObject != null ? singletonObject : NULL_OBJECT));
        this.singletonFactories.remove(beanName);
        this.earlySingletonObjects.remove(beanName);
        this.registeredSingletons.add(beanName);
    }
}
```

7. 返回处理结果。

虽然我们已经从外部了解了加载 bean 的逻辑架构，但现在我们还并没有开始对 bean 加载功能的探索，之前提到过，bean 的加载逻辑其实是在传入的 ObjectFactory 类型的参数 singletonFactory 中定义的，我们反推参数的获取，得到如下代码：

```java
sharedInstance = getSingleton(beanName, new ObjectFactory<Object>() {
    public Object getObject() throws BeansException {
        try {
            return createBean(beanName, mbd, args);
        }
        catch (BeansException ex) {
            destroySingleton(beanName);
            throw ex;
        }
    }
});
```

ObjectFactory 的核心部分其实只是调用了 createBean 的方法，所以我们还需要到 createBean 方法中追寻真理。

5.5 准备创建 bean

我们不可能指望在一个函数中完成一个复杂的逻辑，而且我们跟踪了这么多 Spring 代码，经历了这么多函数，或多或少也发现了一些规律：一个真正干活的函数其实是以 do 开头的，比如 doGetObjectFromFactoryBean；而给我们错觉的函数，比如 getObjectFromFactoryBean，其实只是从全局角度去做些统筹的工作。这个规则对于 createBean 也不例外，那么让我们看看在 createBean 函数中做了哪些准备工作。

```java
protected Object createBean(final String beanName, final RootBeanDefinition mbd, final Object[] args)    throws BeanCreationException {

    if (logger.isDebugEnabled()) {
        logger.debug("Creating instance of bean '" + beanName + "'");
    }
    //锁定class,根据设置的class属性或者根据className来解析Class
    resolveBeanClass(mbd, beanName);

    //验证及准备覆盖的方法
    try {
        mbd.prepareMethodOverrides();
```

```
        }
        catch (BeanDefinitionValidationException ex) {
            throw new BeanDefinitionStoreException(mbd.getResourceDescription(),
                    beanName, "Validation of method overrides failed", ex);
        }

        try {
            //给 BeanPostProcessors 一个机会来返回代理来替代真正的实例
            Object bean = resolveBeforeInstantiation(beanName, mbd);
            if (bean != null) {
                return bean;
            }
        }
        catch (Throwable ex) {
            throw new BeanCreationException(mbd.getResourceDescription(), beanName,
                    "BeanPostProcessor before instantiation of bean failed", ex);
        }

        Object beanInstance = doCreateBean(beanName, mbd, args);
        if (logger.isDebugEnabled()) {
            logger.debug("Finished creating instance of bean '" + beanName + "'");
        }
        return beanInstance;
    }
```

从代码中我们可以总结出函数完成的具体步骤及功能。

1. 根据设置的 class 属性或者根据 className 来解析 Class。
2. 对 override 属性进行标记及验证。

很多读者可能会不知道这个方法的作用，因为在 Spring 的配置里面根本就没有诸如 override-method 之类的配置，那么这个方法到底是干什么用的呢？

其实在 Spring 中确实没有 override-method 这样的配置，但是如果读过前面的部分，可能会有所发现，在 Spring 配置中是存在 lookup-method 和 replace-method 的，而这两个配置的加载其实就是将配置统一存放在 BeanDefinition 中的 methodOverrides 属性里，而这个函数的操作其实也就是针对于这两个配置的。

3. 应用初始化前的后处理器，解析指定 bean 是否存在初始化前的短路操作。
4. 创建 bean。

我们首先查看下对 override 属性标记及验证的逻辑实现。

5.5.1 处理 override 属性

查看源码中 AbstractBeanDefinition 类的 prepareMethodOverrides 方法：

```
public void prepareMethodOverrides() throws BeanDefinitionValidationException {
    // Check that lookup methods exists.
    MethodOverrides methodOverrides = getMethodOverrides();
    if (!methodOverrides.isEmpty()) {
        for (MethodOverride mo : methodOverrides.getOverrides()) {
            prepareMethodOverride(mo);
```

```
            }
        }
    }
    protected void prepareMethodOverride(MethodOverride mo) throws BeanDefinitionValidationException {
        //获取对应类中对应方法名的个数
        int count = ClassUtils.getMethodCountForName(getBeanClass(), mo.getMethodName());
        if (count == 0) {
            throw new BeanDefinitionValidationException(
                    "Invalid method override: no method with name '" + mo.getMethodName() +
                    "' on class [" + getBeanClassName() + "]");
        }
        else if (count == 1) {
            //标记 MethodOverride 暂未被覆盖，避免参数类型检查的开销。
            mo.setOverloaded(false);
        }
    }
```

通过以上两个函数的代码你能体会到它所要实现的功能吗？之前反复提到过，在 Spring 配置中存在 lookup-method 和 replace-method 两个配置功能，而这两个配置的加载其实就是将配置统一存放在 BeanDefinition 中的 methodOverrides 属性里，这两个功能实现原理其实是在 bean 实例化的时候如果检测到存在 methodOverrides 属性，会动态地为当前 bean 生成代理并使用对应的拦截器为 bean 做增强处理，相关逻辑实现在 bean 的实例化部分详细介绍。

但是，这里要提到的是，对于方法的匹配来讲，如果一个类中存在若干个重载方法，那么，在函数调用及增强的时候还需要根据参数类型进行匹配，来最终确认当前调用的到底是哪个函数。但是，Spring 将一部分匹配工作在这里完成了，如果当前类中的方法只有一个，那么就设置重载该方法没有被重载，这样在后续调用的时候便可以直接使用找到的方法，而不需要进行方法的参数匹配验证了，而且还可以提前对方法存在性进行验证，正可谓一箭双雕。

5.5.2　实例化的前置处理

在真正调用 doCreate 方法创建 bean 的实例前使用了这样一个方法 resolveBeforeInstantiation(beanName, mbd)对 BeanDefinigiton 中的属性做些前置处理。当然，无论其中是否有相应的逻辑实现我们都可以理解，因为真正逻辑实现前后留有处理函数也是可扩展的一种体现，但是，这并不是最重要的，在函数中还提供了一个短路判断，这才是最为关键的部分。

```
    if (bean != null) {
        return bean;
    }
```

当经过前置处理后返回的结果如果不为空，那么会直接略过后续的 bean 的创建而直接返回结果。这一特性虽然很容易被忽略，但是却起着至关重要的作用，我们熟知的 AOP 功能就是基于这里的判断的。

```
    protected Object resolveBeforeInstantiation(String beanName, RootBeanDefinition mbd) {
        Object bean = null;
        //如果尚未被解析
        if (!Boolean.FALSE.equals(mbd.beforeInstantiationResolved)) {
            // Make sure bean class is actually resolved at this point.
```

```java
                if (mbd.hasBeanClass() && !mbd.isSynthetic() && hasInstantiationAware
BeanPostProcessors()) {
                    bean = applyBeanPostProcessorsBeforeInstantiation(mbd.getBeanClass(),
 beanName);
                    if (bean != null) {
                        bean = applyBeanPostProcessorsAfterInitialization(bean, beanName);
                    }
                }
                mbd.beforeInstantiationResolved = (bean != null);
            }
            return bean;
        }
```

此方法中最吸引我们的无疑是两个方法 applyBeanPostProcessorsBeforeInstantiation 以及 applyBeanPostProcessorsAfterInitialization。两个方法实现的非常简单，无非是对后处理器中的所有 InstantiationAwareBeanPostProcessor 类型的后处理器进行 postProcessBeforeInstantiation 方法和 BeanPostProcessor 的 postProcessAfterInitialization 方法的调用。

1. 实例化前的后处理器应用

bean 的实例化前调用，也就是将 AbsractBeanDefinition 转换为 BeanWrapper 前的处理。给子类一个修改 BeanDefinition 的机会，也就是说当程序经过这个方法后，bean 可能已经不是我们认为的 bean 了，而是或许成为了一个经过处理的代理 bean，可能是通过 cglib 生成的，也可能是通过其他技术生成的。这在第 7 章中会详细介绍，我们只需要知道，在 bean 的实例化前会调用后处理器的方法进行处理。

```java
protected Object applyBeanPostProcessorsBeforeInstantiation(Class beanClass, String beanName)
        throws BeansException {

    for (BeanPostProcessor bp : getBeanPostProcessors()) {
        if (bp instanceof InstantiationAwareBeanPostProcessor) {
            InstantiationAwareBeanPostProcessor ibp = (Instantiation AwareBean
PostProcessor) bp;
            Object result = ibp.postProcessBeforeInstantiation(beanClass, beanName);
            if (result != null) {
                return result;
            }
        }
    }
    return null;
}
```

2. 实例化后的后处理器应用

在讲解从缓存中获取单例 bean 的时候就提到过，Spring 中的规则是在 bean 的初始化后尽可能保证将注册的后处理器的 postProcessAfterInitialization 方法应用到该 bean 中，因为如果返回的 bean 不为空，那么便不会再次经历普通 bean 的创建过程，所以只能在这里应用后处理器的 postProcessAfterInitialization 方法。

```
public Object applyBeanPostProcessorsAfterInitialization(Object existingBean, String beanName)
        throws BeansException {
    Object result = existingBean;
    for (BeanPostProcessor beanProcessor : getBeanPostProcessors()) {
        result = beanProcessor.postProcessAfterInitialization(result, beanName);
        if (result == null) {
            return result;
        }
    }
    return result;
}
```

5.6 循环依赖

实例化 bean 是一个非常复杂的过程，而其中比较难以理解的就是对循环依赖的解决，不管之前读者有没有对循环依赖方面的研究，这里有必要先对此知识点稍作回顾。

5.6.1 什么是循环依赖

循环依赖就是循环引用，就是两个或多个 bean 相互之间的持有对方，比如 CircleA 引用 CircleB，CircleB 引用 CircleC，CircleC 引用 CircleA，则它们最终反映为一个环。此处不是循环调用，循环调用是方法之间的环调用，如图 5-2 所示。

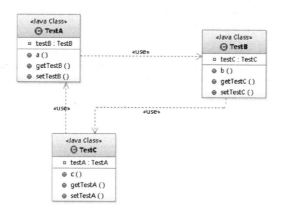

图 5-2　循环依赖

循环调用是无法解决的，除非有终结条件，否则就是死循环，最终导致内存溢出错误。

5.6.2　Spring 如何解决循环依赖

Spring 容器循环依赖包括构造器循环依赖和 setter 循环依赖，那 Spring 容器如何解决循环依赖呢？首先让我们来定义循环引用类：

```java
public class TestA {

    private TestB testB;

    public void a() {
        testB.b();
    }

    public TestB getTestB() {
        return testB;
    }

    public void setTestB(TestB testB) {
        this.testB = testB;
    }
}

public class TestB {
    private TestC testC;

    public void b() {
        testC.c();
    }

    public TestC getTestC() {
        return testC;
    }

    public void setTestC(TestC testC) {
        this.testC = testC;
    }
}


public class TestC {
    private TestA testA;

    public void c() {
        testA.a();
    }

    public TestA getTestA() {
        return testA;
    }

    public void setTestA(TestA testA) {
        this.testA = testA;
    }
}
```

在 Spring 中将循环依赖的处理分成了 3 种情况。

1. 构造器循环依赖

表示通过构造器注入构成的循环依赖，此依赖是无法解决的，只能抛出 BeanCurrentlyIn-CreationException 异常表示循环依赖。

如在创建 TestA 类时，构造器需要 TestB 类，那将去创建 TestB，在创建 TestB 类时又发现需要 TestC 类，则又去创建 TestC，最终在创建 TestC 时发现又需要 TestA，从而形成一个环，没办法创建。

Spring 容器将每一个正在创建的 bean 标识符放在一个"当前创建 bean 池"中，bean 标识符在创建过程中将一直保持在这个池中，因此如果在创建 bean 过程中发现自己已经在"当前创建 bean 池"里时，将抛出 BeanCurrentlyInCreationException 异常表示循环依赖；而对于创建完毕的 bean 将从"当前创建 bean 池"中清除掉。

我们通过一个直观的测试用例来进行分析。

1. 创建配置文件。

```xml
<bean id="testA" class="com.bean.TestA">
    <constructor-arg index="0" ref="testB"/>
</bean>
<bean id="testB" class="com.bean.TestB">
    <constructor-arg index="0" ref="testC"/>
</bean>
<bean id="testC" class="com.bean.TestC">
    <constructor-arg index="0" ref="testA"/>
</bean>
```

2. 创建测试用例。

```java
@Test(expected = BeanCurrentlyInCreationException.class)
public void testCircleByConstructor() throws Throwable {
    try {
            new ClassPathXmlApplicationContext("test.xml");
    } catch (Exception e) {
      //因为要在创建testC时抛出;
      Throwable e1 = e.getCause().getCause().getCause();
      throw e1;
    }
}
```

针对以上代码的分析如下。

- Spring 容器创建"testA" bean，首先去"当前创建 bean 池"查找是否当前 bean 正在创建，如果没发现，则继续准备其需要的构造器参数"testB"，并将"testA"标识符放到"当前创建 bean 池"。

- Spring 容器创建"testB" bean，首先去"当前创建 bean 池"查找是否当前 bean 正在创建，如果没发现，则继续准备其需要的构造器参数"testC"，并将"testB"标识符放到"当前创建 bean 池"。

- Spring 容器创建"testC" bean，首先去"当前创建 bean 池"查找是否当前 bean 正在创建，如果没发现，则继续准备其需要的构造器参数"testA"，并将"testC"标识符放到"当前创建 bean 池"。
- 到此为止 Spring 容器要去创建"testA"bean，发现该 bean 标识符在"当前创建 bean 池"中，因为表示循环依赖，抛出 BeanCurrentlyInCreationException。

2. setter 循环依赖

表示通过 setter 注入方式构成的循环依赖。对于 setter 注入造成的依赖是通过 Spring 容器提前暴露刚完成构造器注入但未完成其他步骤（如 setter 注入）的 bean 来完成的，而且只能解决单例作用域的 bean 循环依赖。通过提前暴露一个单例工厂方法，从而使其他 bean 能引用到该 bean，如下代码所示：

```
addSingletonFactory(beanName, new ObjectFactory() {
    public Object getObject() throws BeansException {
        return getEarlyBeanReference(beanName, mbd, bean);
    }
});
```

具体步骤如下。

1. Spring 容器创建单例"testA"bean，首先根据无参构造器创建 bean，并暴露一个"ObjectFactory"用于返回一个提前暴露一个创建中的 bean，并将"testA"标识符放到"当前创建 bean 池"，然后进行 setter 注入"testB"。

2. Spring 容器创建单例"testB"bean，首先根据无参构造器创建 bean，并暴露一个"ObjectFactory"用于返回一个提前暴露一个创建中的 bean，并将"testB"标识符放到"当前创建 bean 池"，然后进行 setter 注入"circle"。

3. Spring 容器创建单例"testC"bean，首先根据无参构造器创建 bean，并暴露一个"ObjectFactory"用于返回一个提前暴露一个创建中的 bean，并将"testC"标识符放到"当前创建 bean 池"，然后进行 setter 注入"testA"。进行注入"testA"时由于提前暴露了"ObjectFactory"工厂，从而使用它返回提前暴露一个创建中的 bean。

4. 最后在依赖注入"testB"和"testA"，完成 setter 注入。

3. prototype 范围的依赖处理

对于"prototype"作用域 bean，Spring 容器无法完成依赖注入，因为 Spring 容器不进行缓存"prototype"作用域的 bean，因此无法提前暴露一个创建中的 bean。示例如下：

1. 创建配置文件。

```
<bean id="testA" class="com.bean.CircleA" scope="prototype">
    <property name="testB" ref="testB"/>
</bean>
<bean id="testB" class="com.bean.CircleB" scope="prototype">
    <property name="testC" ref="testC"/>
</bean>
<bean id="testC" class="com.bean.CircleC" scope="prototype">
    <property name="testA" ref="testA"/>
</bean>
```

2. 创建测试用例。

```
@Test(expected = BeanCurrentlyInCreationException.class)
public void testCircleBySetterAndPrototype () throws Throwable {
    try {
        ClassPathXmlApplicationContext ctx = new ClassPathXmlApplicationContext(
"testPrototype.xml");
        System.out.println(ctx.getBean("testA"));
    } catch (Exception e) {
        Throwable e1 = e.getCause().getCause().getCause();
        throw e1;
    }
}
```

对于"singleton"作用域 bean，可以通过"setAllowCircularReferences(false);"来禁用循环引用。

感谢互联网时代，让我可以方便地获取我想要的各种信息，当初我刚开始学习的时候，一直纠结于这里错综复杂的逻辑，幸好我看到了一篇文章解开了我心中的疑惑。在此，感谢原作者，并将原文与大家分享，帮助大家更好的理解 Spring 的依赖，大家可以从 http://www.iflym.com/index.php/code/201208280001.html 来获取原文。

5.7 创建 bean

介绍了循环依赖以及 Spring 中的循环依赖的处理方式后，我们继续 4.5 节的内容。当经历过 resolveBeforeInstantiation 方法后，程序有两个选择，如果创建了代理或者说重写了 InstantiationAwareBeanPostProcessor 的 postProcessBeforeInstantiation 方法并在方法 postProcess-BeforeInstantiation 中改变了 bean，则直接返回就可以了，否则需要进行常规 bean 的创建。而这常规 bean 的创建就是在 doCreateBean 中完成的。

```
protected Object doCreateBean(final String beanName, final RootBeanDefinition mbd, final Object[] args) {
            // Instantiate the bean.
            BeanWrapper instanceWrapper = null;
            if (mbd.isSingleton()) {
                instanceWrapper = this.factoryBeanInstanceCache.remove(beanName);
            }
            if (instanceWrapper == null) {
                //根据指定 bean 使用对应的策略创建新的实例，如：工厂方法、构造函数自动注入、简单初始化
                instanceWrapper = createBeanInstance(beanName, mbd, args);
            }
            final Object bean = (instanceWrapper != null ? instanceWrapper. getWrappedInstance() : null);
            Class beanType = (instanceWrapper != null ? instanceWrapper.getWrappedClass() : null);

            // Allow post-processors to modify the merged bean definition.
            synchronized (mbd.postProcessingLock) {
                if (!mbd.postProcessed) {
                    //应用 MergedBeanDefinitionPostProcessor
```

5.7 创建 bean

```
            applyMergedBeanDefinitionPostProcessors(mbd, beanType, beanName);
            mbd.postProcessed = true;
        }
    }

    /*
     * 是否需要提早曝光:单例&允许循环依赖&当前 bean 正在创建中,检测循环依赖
     */
    boolean earlySingletonExposure = (mbd.isSingleton() && this.allowCircularReferences &&
            isSingletonCurrentlyInCreation(beanName));
    if (earlySingletonExposure) {
        if (logger.isDebugEnabled()) {
            logger.debug("Eagerly caching bean '" + beanName +
                    "' to allow for resolving potential circular references");
        }
        //为避免后期循环依赖,可以在 bean 初始化完成前将创建实例的 ObjectFactory 加入工厂
        addSingletonFactory(beanName, new ObjectFactory() {
            public Object getObject() throws BeansException {
                //对 bean 再一次依赖引用,主要应用 SmartInstantiationAware BeanPost
                //Processor
                //其中我们熟知的 AOP 就是在这里将 advice 动态织入 bean 中,若没有则直接返回
                //bean,不做任何处理
                return getEarlyBeanReference(beanName, mbd, bean);
            }
        });
    }

    // Initialize the bean instance.
    Object exposedObject = bean;
    try {
        //对 bean 进行填充,将各个属性值注入,其中,可能存在依赖于其他 bean 的属性,则会递归初始
        //依赖 bean
        populateBean(beanName, mbd, instanceWrapper);
        if (exposedObject != null) {
            //调用初始化方法,比如 init-method
            exposedObject = initializeBean(beanName, exposedObject, mbd);
        }
    }
    catch (Throwable ex) {
        if (ex instanceof BeanCreationException && beanName.equals(((BeanCreation
Exception) ex).getBeanName()))) {
            throw (BeanCreationException) ex;
        }
        else {
            throw new BeanCreationException(mbd.getResourceDescription(), beanName,
"Initialization of bean failed", ex);
        }
    }

    if (earlySingletonExposure) {

        Object earlySingletonReference = getSingleton(beanName, false);
        //earlySingletonReference 只有在检测到有循环依赖的情况下才会不为空
        if (earlySingletonReference != null) {
```

```java
            //如果 exposedObject 没有在初始化方法中被改变，也就是没有被增强
            if (exposedObject == bean) {
                exposedObject = earlySingletonReference;
            }else if (!this.allowRawInjectionDespiteWrapping && hasDependentBean(beanName)) {
                String[] dependentBeans = getDependentBeans(beanName);
                Set<String> actualDependentBeans = new LinkedHashSet<String>(dependentBeans.length);
                for (String dependentBean : dependentBeans) {
                    //检测依赖
                    if (!removeSingletonIfCreatedForTypeCheckOnly(dependentBean)) {
                        actualDependentBeans.add(dependentBean);
                    }
                }
                /*
                 * 因为 bean 创建后其所依赖的 bean 一定是已经创建的，
                 * actualDependentBeans 不为空则表示当前 bean 创建后其依赖的 bean 却没有没
                   全部创建完，也就是说存在循环依赖
                 */
                if (!actualDependentBeans.isEmpty()) {
                    throw new BeanCurrentlyInCreationException(beanName,
                            "Bean with name '" + beanName + "' has been injected into other beans [" +
                            StringUtils.collectionToCommaDelimitedString(actualDependentBeans) +
                            "] in its raw version as part of a circular reference, " +
                            "but has eventually been " +
                            "wrapped. This means that said other beans do not use " +
                            "the final version of the " +
                            "bean. This is often the result of over-eager type " +
                            "matching - consider using " +
                            "'getBeanNamesOfType' with the 'allowEagerInit' flag " +
                            "turned off, for example.");
                }
            }
        }

        // Register bean as disposable.
        try {
            //根据 scopse 注册 bean
            registerDisposableBeanIfNecessary(beanName, bean, mbd);
        }
        catch (BeanDefinitionValidationException ex) {
            throw new BeanCreationException(mbd.getResourceDescription(), beanName, "Invalid destruction signature", ex);
        }

        return exposedObject;
    }
```

尽管日志与异常的内容非常重要，但是在阅读源码的时候似乎大部分人都会直接忽略掉。在此不深入探讨日志及异常的设计，我们看看整个函数的概要思路。

1. 如果是单例则需要首先清除缓存。
2. 实例化 bean，将 BeanDefinition 转换为 BeanWrapper。

转换是一个复杂的过程，但是我们可以尝试概括大致的功能，如下所示。

- 如果存在工厂方法则使用工厂方法进行初始化。
- 一个类有多个构造函数，每个构造函数都有不同的参数，所以需要根据参数锁定构造函数并进行初始化。
- 如果既不存在工厂方法也不存在带有参数的构造函数，则使用默认的构造函数进行 bean 的实例化。

3. MergedBeanDefinitionPostProcessor 的应用。

bean 合并后的处理，Autowired 注解正是通过此方法实现诸如类型的预解析。

4. 依赖处理。

在 Spring 中会有循环依赖的情况，例如，当 A 中含有 B 的属性，而 B 中又含有 A 的属性时就会构成一个循环依赖，此时如果 A 和 B 都是单例，那么在 Spring 中的处理方式就是当创建 B 的时候，涉及自动注入 A 的步骤时，并不是直接去再次创建 A，而是通过放入缓存中的 ObjectFactory 来创建实例，这样就解决了循环依赖的问题。

5. 属性填充。将所有属性填充至 bean 的实例中。
6. 循环依赖检查。

之前有提到过，在 Sping 中解决循环依赖只对单例有效，而对于 prototype 的 bean，Spring 没有好的解决办法，唯一要做的就是抛出异常。在这个步骤里面会检测已经加载的 bean 是否已经出现了依赖循环，并判断是否需要抛出异常。

7. 注册 DisposableBean。

如果配置了 destroy-method，这里需要注册以便于在销毁时候调用。

8. 完成创建并返回。

可以看到上面的步骤非常的繁琐，每一步骤都使用了大量的代码来完成其功能，最复杂也是最难以理解的当属循环依赖的处理，在真正进入 doCreateBean 前我们有必要先了解下循环依赖。

5.7.1 创建 bean 的实例

当我们了解了循环依赖以后就可以深入分析创建 bean 的每一个步骤了，首先我们从 createBeanInstance 开始。

```
protected BeanWrapper createBeanInstance(String beanName, RootBeanDefinition mbd,
Object[] args) {
          //解析 class
          Class beanClass = resolveBeanClass(mbd, beanName);

          if (beanClass != null && !Modifier.isPublic(beanClass.getModifiers()) && !mbd.
isNonPublicAccessAllowed()) {
```

```
            throw new BeanCreationException(mbd.getResourceDescription(), beanName,
                 "Bean class isn't public, and non-public access not allowed: " +
beanClass.getName());
    }
            //如果工厂方法不为空则使用工厂方法初始化策略
    if (mbd.getFactoryMethodName() != null)  {
        return instantiateUsingFactoryMethod(beanName, mbd, args);
    }

    // Shortcut when re-creating the same bean...
    boolean resolved = false;
    boolean autowireNecessary = false;
    if (args == null) {
        synchronized (mbd.constructorArgumentLock) {
            //一个类有多个构造函数,每个构造函数都有不同的参数,所以调用前需要先根据参数锁定构
            //造函数或对应的工厂方法
            if (mbd.resolvedConstructorOrFactoryMethod != null) {
                resolved = true;
                autowireNecessary = mbd.constructorArgumentsResolved;
            }
        }
    }
    //如果已经解析过则使用解析好的构造函数方法不需要再次锁定
    if (resolved) {
        if (autowireNecessary) {
            //构造函数自动注入
            return autowireConstructor(beanName, mbd, null, null);
        }
        else {
            //使用默认构造函数构造
            return instantiateBean(beanName, mbd);
        }
    }

    //需要根据参数解析构造函数
    Constructor[] ctors = determineConstructorsFromBeanPostProcessors(beanClass,
beanName);
    if (ctors != null ||
         mbd.getResolvedAutowireMode() == RootBeanDefinition. AUTOWIRE_
CONSTRUCTOR ||
         mbd.hasConstructorArgumentValues() || !ObjectUtils.isEmpty(args))   {
        //构造函数自动注入
        return autowireConstructor(beanName, mbd, ctors, args);
    }

    //使用默认构造函数构造
    return instantiateBean(beanName, mbd);
}
```

虽然代码中实例化的细节非常复杂,但是在 createBeanIntance 方法中我们还是可以清晰地看到实例化的逻辑的。

1. 如果在 RootBeanDefinition 中存在 factoryMethodName 属性,或者说在配置文件中配置了 factory-method,那么 Spring 会尝试使用 instantiateUsingFactoryMethod(beanName, mbd, args)方法

根据 RootBeanDefinition 中的配置生成 bean 的实例。

2. 解析构造函数并进行构造函数的实例化。因为一个 bean 对应的类中可能会有多个构造函数，而每个构造函数的参数不同，Spring 在根据参数及类型去判断最终会使用哪个构造函数进行实例化。但是，判断的过程是个比较消耗性能的步骤，所以采用缓存机制，如果已经解析过则不需要重复解析而是直接从 RootBeanDefinition 中的属性 resolvedConstructorOrFactoryMethod 缓存的值去取，否则需要再次解析，并将解析的结果添加至 RootBeanDefinition 中的属性 resolvedConstructorOrFactoryMethod 中。

1. autowireConstructor

对于实例的创建 Spring 中分成了两种情况，一种是通用的实例化，另一种是带有参数的实例化。带有参数的实例化过程相当复杂，因为存在着不确定性，所以在判断对应参数上做了大量工作。

```java
public BeanWrapper autowireConstructor(
        final String beanName, final RootBeanDefinition mbd, Constructor[] chosenCtors, final Object[] explicitArgs) {

    BeanWrapperImpl bw = new BeanWrapperImpl();
    this.beanFactory.initBeanWrapper(bw);

    Constructor constructorToUse = null;
    ArgumentsHolder argsHolderToUse = null;
    Object[] argsToUse = null;
    //explicitArgs 通过 getBean 方法传入
    //如果 getBean 方法调用的时候指定方法参数那么直接使用
    if (explicitArgs != null) {
        argsToUse = explicitArgs;
    }else {
        //如果在 getBean 方法时候没有指定则尝试从配置文件中解析
        Object[] argsToResolve = null;
        //尝试从缓存中获取
        synchronized (mbd.constructorArgumentLock) {
            constructorToUse = (Constructor) mbd.resolvedConstructorOrFactoryMethod;
            if (constructorToUse != null && mbd.constructorArgumentsResolved) {
                //从缓存中取
                argsToUse = mbd.resolvedConstructorArguments;
                if (argsToUse == null) {
                    //配置的构造函数参数
                    argsToResolve = mbd.preparedConstructorArguments;
                }
            }
        }
        //如果缓存中存在
        if (argsToResolve != null) {
            //解析参数类型，如给定方法的构造函数 A(int,int)则通过此方法后就会把配置中的
            //("1","1")转换为(1,1)
            //缓存中的值可能是原始值也可能是最终值
            argsToUse = resolvePreparedArguments(beanName, mbd, bw, constructorToUse, argsToResolve);
        }
```

```java
            }

            //没有被缓存
            if (constructorToUse == null) {
                // Need to resolve the constructor.
                boolean autowiring = (chosenCtors != null ||
                        mbd.getResolvedAutowireMode() == RootBeanDefinition.AUTOWIRE_CONSTRUCTOR);
                ConstructorArgumentValues resolvedValues = null;

                int minNrOfArgs;
                if (explicitArgs != null) {
                    minNrOfArgs = explicitArgs.length;
                }else {
                    //提取配置文件中的配置的构造函数参数
                    ConstructorArgumentValues cargs = mbd.getConstructorArgumentValues();
                    //用于承载解析后的构造函数参数的值
                    resolvedValues = new ConstructorArgumentValues();
                    //能解析到的参数个数
                    minNrOfArgs = resolveConstructorArguments(beanName, mbd, bw, cargs, resolvedValues);
                }

                // Take specified constructors, if any.
                Constructor[] candidates = chosenCtors;
                if (candidates == null) {
                    Class beanClass = mbd.getBeanClass();
                    try {
                        candidates = (mbd.isNonPublicAccessAllowed() ?
                                beanClass.getDeclaredConstructors() : beanClass.getConstructors());
                    }
                    catch (Throwable ex) {
                        throw new BeanCreationException(mbd.getResourceDescription(), beanName,
                                "Resolution of declared constructors on bean Class [" + beanClass.getName() +
                                "] from ClassLoader [" + beanClass.getClassLoader() + "] failed", ex);
                    }
                }
                //排序给定的构造函数, public 构造函数优先参数数量降序、非 public 构造函数参数数量降序
                AutowireUtils.sortConstructors(candidates);

                int minTypeDiffWeight = Integer.MAX_VALUE;
                Set<Constructor> ambiguousConstructors = null;
                List<Exception> causes = null;

                for (int i = 0; i < candidates.length; i++) {
                    Constructor<?> candidate = candidates[i];
                    Class[] paramTypes = candidate.getParameterTypes();

                    if (constructorToUse != null && argsToUse.length > paramTypes.length) {
                        //如果已经找到选用的构造函数或者需要的参数个数小于当前的构造函数参数个数则终止,
                        //因为已经按照参数个数降序排列
```

5.7 创建 bean

```java
                            break;
                    }
                    if (paramTypes.length < minNrOfArgs) {
                        //参数个数不相等
                        continue;
                    }

                    ArgumentsHolder argsHolder;
                    if (resolvedValues != null) {
                        //有参数则根据值构造对应参数类型的参数
                        try {
                            String[] paramNames = null;
                            if (constructorPropertiesAnnotationAvailable) {
                                //注释上获取参数名称
                                paramNames = ConstructorPropertiesChecker.evaluateAnnotation(candidate, paramTypes.length);
                            }
                            if (paramNames == null) {
                                //获取参数名称探索器
                                ParameterNameDiscoverer pnd = this.beanFactory.getParameterNameDiscoverer();
                                if (pnd != null) {
                                    //获取指定构造函数的参数名称
                                    paramNames = pnd.getParameterNames(candidate);
                                }
                            }
                            //根据名称和数据类型创建参数持有者
                            argsHolder = createArgumentArray(
                                    beanName, mbd, resolvedValues, bw, paramTypes, paramNames, candidate, autowiring);
                        }
                        catch (UnsatisfiedDependencyException ex) {
                            if (this.beanFactory.logger.isTraceEnabled()) {
                                this.beanFactory.logger.trace(
                                        "Ignoring constructor [" + candidate + "] of bean '" + beanName + "': " + ex);
                            }
                            if (i == candidates.length - 1 && constructorToUse == null) {
                                if (causes != null) {
                                    for (Exception cause : causes) {
                                        this.beanFactory.onSuppressedException(cause);
                                    }
                                }
                                throw ex;
                            }
                            else {
                                // Swallow and try next constructor.
                                if (causes == null) {
                                    causes = new LinkedList<Exception>();
                                }
                                causes.add(ex);
                                continue;
                            }
                        }
                    }
```

```java
            }else {
                if (paramTypes.length != explicitArgs.length) {
                    continue;
                }
                //构造函数没有参数的情况
                argsHolder = new ArgumentsHolder(explicitArgs);
            }

            //探测是否有不确定性的构造函数存在,例如不同构造函数的参数为父子关系
            int typeDiffWeight = (mbd.isLenientConstructorResolution() ?
                    argsHolder.getTypeDifferenceWeight(paramTypes) : argsHolder.getAssignabilityWeight(paramTypes));
            //如果它代表着当前最接近的匹配则选择作为构造函数
            if (typeDiffWeight < minTypeDiffWeight) {
                constructorToUse = candidate;
                argsHolderToUse = argsHolder;
                argsToUse = argsHolder.arguments;
                minTypeDiffWeight = typeDiffWeight;
                ambiguousConstructors = null;
            }else if (constructorToUse != null && typeDiffWeight == minTypeDiffWeight) {
                if (ambiguousConstructors == null) {
                    ambiguousConstructors = new LinkedHashSet<Constructor>();
                    ambiguousConstructors.add(constructorToUse);
                }
                ambiguousConstructors.add(candidate);
            }
        }

        if (constructorToUse == null) {
            throw new BeanCreationException(mbd.getResourceDescription(), beanName,
                    "Could not resolve matching constructor " +
                    "(hint: specify index/type/name arguments for simple parameters to avoid type ambiguities)");
        }else if (ambiguousConstructors != null && !mbd.isLenientConstructorResolution()) {
            throw new BeanCreationException(mbd.getResourceDescription(), beanName,
                    "Ambiguous constructor matches found in bean '" + beanName + "' " +
                    "(hint: specify index/type/name arguments for simple parameters to avoid type ambiguities): " +
                    ambiguousConstructors);
        }

        if (explicitArgs == null) {
            //将解析的构造函数加入缓存
            argsHolderToUse.storeCache(mbd, constructorToUse);
        }
    }

    try {
        Object beanInstance;

        if (System.getSecurityManager() != null) {
            final Constructor ctorToUse = constructorToUse;
```

```
                final Object[] argumentsToUse = argsToUse;
                beanInstance = AccessController.doPrivileged(new PrivilegedAction< Object>() {
                    public Object run() {
                        return beanFactory.getInstantiationStrategy().instantiate(
                                mbd, beanName, beanFactory, ctorToUse, argumentsToUse);
                    }
                }, beanFactory.getAccessControlContext());
            }
            else {
                beanInstance = this.beanFactory.getInstantiationStrategy().instantiate(
                        mbd, beanName, this.beanFactory, constructorToUse, argsToUse);
            }
            //将构建的实例加入 BeanWrapper 中
            bw.setWrappedInstance(beanInstance);
            return bw;
        }
        catch (Throwable ex) {
            throw new BeanCreationException(mbd.getResourceDescription(), beanName, "Instantiation of bean failed", ex);
        }
    }
```

逻辑很复杂，函数代码量很大，不知道你是否坚持读完了整个函数并理解了整个功能呢？这里要先吐个槽，作者觉得这个函数的写法完全不符合 Spring 的一贯风格，如果你一直跟随作者的分析思路到这里，相信你或多或少对 Spring 的编码风格有所了解，Spring 的一贯做法是将复杂的逻辑分解，分成 N 个小函数的嵌套，每一层都是对下一层逻辑的总结及概要，这样使得每一层的逻辑会变得简单容易理解。在上面的函数中，包含着很多的逻辑实现，作者觉得至少应该将逻辑封装在不同函数中而使得在 autowireConstructor 中的逻辑清晰明了。

我们总览一下整个函数，其实现的功能考虑了以下几个方面。

1. 构造函数参数的确定。

- 根据 explicitArgs 参数判断。

如果传入的参数 explicitArgs 不为空，那边可以直接确定参数，因为 explicitArgs 参数是在调用 Bean 的时候用户指定的，在 BeanFactory 类中存在这样的方法：

```
Object getBean(String name, Object... args) throws BeansException;
```

在获取 bean 的时候，用户不但可以指定 bean 的名称还可以指定 bean 所对应类的构造函数或者工厂方法的方法参数，主要用于静态工厂方法的调用，而这里是需要给定完全匹配的参数的，所以，便可以判断，如果传入参数 explicitArgs 不为空，则可以确定构造函数参数就是它。

- 缓存中获取。

除此之外，确定参数的办法如果之前已经分析过，也就是说构造函数参数已经记录在缓存中，那么便可以直接拿来使用。而且，这里要提到的是，在缓存中缓存的可能是参数的最终类型也可能是参数的初始类型，例如：构造函数参数要求的是 int 类型，但是原始的参数值可能是 String 类型的 "1"，那么即使在缓存中得到了参数，也需要经过类型转换器的过滤以确保参数类型与对应的构造函数参数类型完全对应。

- 配置文件获取。

如果不能根据传入的参数 explicitArgs 确定构造函数的参数也无法在缓存中得到相关信息，那么只能开始新一轮的分析了。

分析从获取配置文件中配置的构造函数信息开始，经过之前的分析，我们知道，Spring 中配置文件中的信息经过转换都会通过 BeanDefinition 实例承载，也就是参数 mbd 中包含，那么可以通过调用 mbd.getConstructorArgumentValues() 来获取配置的构造函数信息。有了配置中的信息便可以获取对应的参数值信息了，获取参数值的信息包括直接指定值，如：直接指定构造函数中某个值为原始类型 String 类型，或者是一个对其他 bean 的引用，而这一处理委托给 resolveConstructorArguments 方法，并返回能解析到的参数的个数。

2. 构造函数的确定。

经过了第一步后已经确定了构造函数的参数，接下来的任务就是根据构造函数参数在所有构造函数中锁定对应的构造函数，而匹配的方法就是根据参数个数匹配，所以在匹配之前需要先对构造函数按照 public 构造函数优先参数数量降序、非 public 构造函数参数数量降序。这样可以在遍历的情况下迅速判断排在后面的构造函数参数个数是否符合条件。

由于在配置文件中并不是唯一限制使用参数位置索引的方式去创建，同样还支持指定参数名称进行设定参数值的情况，如<constructor-arg name="aa">，那么这种情况就需要首先确定构造函数中的参数名称。

获取参数名称可以有两种方式，一种是通过注解的方式直接获取，另一种就是使用 Spring 中提供的工具类 ParameterNameDiscoverer 来获取。构造函数、参数名称、参数类型、参数值都确定后就可以锁定构造函数以及转换对应的参数类型了。

3. 根据确定的构造函数转换对应的参数类型。

主要是使用 Spring 中提供的类型转换器或者用户提供的自定义类型转换器进行转换。

4. 构造函数不确定性的验证。

当然，有时候即使构造函数、参数名称、参数类型、参数值都确定后也不一定会直接锁定构造函数，不同构造函数的参数为父子关系，所以 Spring 在最后又做了一次验证。

5. 根据实例化策略以及得到的构造函数及构造函数参数实例化 Bean。后面章节中将进行讲解。

2. instantiateBean

经历了带有参数的构造函数的实例构造，相信你会非常轻松愉快地理解不带参数的构造函数的实例化过程。

```
    protected BeanWrapper instantiateBean(final String beanName, final RootBean
Definition mbd) {
        try {
            Object beanInstance;
            final BeanFactory parent = this;
            if (System.getSecurityManager() != null) {
```

```
                    beanInstance = AccessController.doPrivileged(new PrivilegedAction
<Object>() {
                        public Object run() {
                            return getInstantiationStrategy().instantiate(mbd, beanName, parent);
                        }
                    }, getAccessControlContext());
                }
                else {
                    beanInstance = getInstantiationStrategy().instantiate(mbd, beanName, parent);
                }
                BeanWrapper bw = new BeanWrapperImpl(beanInstance);
                initBeanWrapper(bw);
                return bw;
            }
            catch (Throwable ex) {
                throw new BeanCreationException(mbd.getResourceDescription(), beanName, "
Instantiation of bean failed", ex);
            }
        }
```

你会发现，此方法并没有什么实质性的逻辑，带有参数的实例构造中，Spring 把精力都放在了构造函数以及参数的匹配上，所以如果没有参数的话那将是非常简单的一件事，直接调用实例化策略进行实例化就可以了。

3. 实例化策略

实例化过程中反复提到过实例化策略，那这又是做什么用的呢？其实，经过前面的分析，我们已经得到了足以实例化的所有相关信息，完全可以使用最简单的反射方法直接反射来构造实例对象，但是 Spring 却并没有这么做。

SimpleInstantiationStrategy.java
```
    public Object instantiate(RootBeanDefinition beanDefinition, String beanName,
BeanFactory owner) {
        //如果有需要覆盖或者动态替换的方法则当然需要使用 cglib 进行动态代理,因为可以在创建代理的同时
        //将动态方法织入类中
        //但是如果没有需要动态改变得方法，为了方便直接反射就可以了
        if (beanDefinition.getMethodOverrides().isEmpty()) {
            Constructor<?> constructorToUse;
            synchronized (beanDefinition.constructorArgumentLock) {
                constructorToUse = (Constructor<?>)beanDefinition.resolvedConstructor
OrFactoryMethod;
                if (constructorToUse == null) {
                    final Class clazz = beanDefinition.getBeanClass();
                    if (clazz.isInterface()) {
                        throw new BeanInstantiationException(clazz, "Specified class
is an interface");
                    }
                    try {
                        if (System.getSecurityManager() != null) {
                            constructorToUse = AccessController.doPrivileged(new
PrivilegedExceptionAction<Constructor>() {
```

```
                                public Constructor run() throws Exception {
                                    return clazz.getDeclaredConstructor((Class[]) null);
                                }
                            });
                        }
                        else {
                            constructorToUse = clazz.getDeclaredConstructor((Class[]) null);
                        }
                        beanDefinition.resolvedConstructorOrFactoryMethod = constructorToUse;
                    }
                    catch (Exception ex) {
                        throw new BeanInstantiationException(clazz, "No default
constructor found", ex);
                    }
                }
            }
            return BeanUtils.instantiateClass(constructorToUse);
        }else {
            // Must generate CGLIB subclass.
            return instantiateWithMethodInjection(beanDefinition, beanName, owner);
        }
    }
}
CglibSubclassingInstantiationStrategy.java
public Object instantiate(Constructor ctor, Object[] args) {
        Enhancer enhancer = new Enhancer();
        enhancer.setSuperclass(this.beanDefinition.getBeanClass());
        enhancer.setCallbackFilter(new CallbackFilterImpl());
        enhancer.setCallbacks(new Callback[] {
                NoOp.INSTANCE,
                new LookupOverrideMethodInterceptor(),
                new ReplaceOverrideMethodInterceptor()
        });

        return (ctor == null) ?
                enhancer.create() :
                enhancer.create(ctor.getParameterTypes(), args);
}
```

看了上面两个函数后似乎我们已经感受到了 Spring 的良苦用心以及为了能更方便地使用 Spring 而做了大量的工作。程序中，首先判断如果 beanDefinition.getMethodOverrides()为空也就是用户没有使用 replace 或者 lookup 的配置方法，那么直接使用反射的方式，简单快捷，但是如果使用了这两个特性，在直接使用反射的方式创建实例就不妥了，因为需要将这两个配置提供的功能切入进去，所以就必须要使用动态代理的方式将包含两个特性所对应的逻辑的拦截增强器设置进去，这样才可以保证在调用方法的时候会被相应的拦截器增强，返回值为包含拦截器的代理实例。

对于拦截器的处理方法非常简单，不再详细介绍，如果读者有兴趣，可以仔细研读第 7 章中关于 AOP 的介绍，对动态代理方面的知识会有更详细地介绍。

5.7.2 记录创建 bean 的 ObjectFactory

在 doCreate 函数中有这样一段代码：

```
boolean earlySingletonExposure = (mbd.isSingleton() && this.allowCircularReferences &&
        isSingletonCurrentlyInCreation(beanName));
if (earlySingletonExposure) {
    if (logger.isDebugEnabled()) {
        logger.debug("Eagerly caching bean '" + beanName +
            "' to allow for resolving potential circular references");
    }
    //为避免后期循环依赖，可以在 bean 初始化完成前创建实例的 ObjectFactory 加入工厂
    addSingletonFactory(beanName, new ObjectFactory() {
        public Object getObject() throws BeansException {
            //对 bean 再一次依赖引用，主要应用 SmartInstantiationAware BeanPost Processor,
            //其中我们熟知的 AOP 就是在这里将 advice 动态织入 bean 中，若没有则直接返回
            //bean, 不做任何处理
            return getEarlyBeanReference(beanName, mbd, bean);
        }
    });
}
```

这段代码不是很复杂，但是很多人不是太理解这段代码的作用，而且，这段代码仅从此函数中去理解也很难弄懂其中的含义，我们需要从全局的角度去思考 Spring 的依赖解决办法。

- earlySingletonExposure：从字面的意思理解就是提早曝光的单例，我们暂不定义它的学名叫什么，我们感兴趣的是有哪些条件影响这个值。
- mbd.isSingleton()：没有太多可以解释的，此 RootBeanDefinition 代表的是否是单例。
- this.allowCircularReferences：是否允许循环依赖，很抱歉，并没有找到在配置文件中如何配置，但是在 AbstractRefreshableApplicationContext 中提供了设置函数，可以通过硬编码的方式进行设置或者可以通过自定义命名空间进行配置，其中硬编码的方式代码如下。
    ```
    ClassPathXmlApplicationContext bf = new ClassPathXmlApplicationContext ("aspectTest.xml");
        bf.setAllowBeanDefinitionOverriding(false);
    ```
- isSingletonCurrentlyInCreation(beanName)：该 bean 是否在创建中。在 Spring 中，会有个专门的属性默认为 DefaultSingletonBeanRegistry 的 singletonsCurrentlyInCreation 来记录 bean 的加载状态，在 bean 开始创建前会将 beanName 记录在属性中，在 bean 创建结束后会将 beanName 从属性中移除。那么我们跟随代码一路走来可是对这个属性的记录并没有多少印象，这个状态是在哪里记录的呢？不同 scope 的记录位置并不一样，我们以 singleton 为例，在 singleton 下记录属性的函数是在 DefaultSingletonBeanRegistry 类的 public Object getSingleton(String beanName, ObjectFactory singletonFactory) 函数的 beforeSingletonCreation(beanName) 和 afterSingletonCreation(beanName) 中，在这两段函数中分别 this.singletonsCurrentlyInCreation.add(beanName) 与 this.singletonsCurrentlyIn-Creation.remove(beanName) 来进行状态的记录与移除。

经过以上分析我们了解变量 earlySingletonExposure 是否是单例、是否允许循环依赖、是否对应的 bean 正在创建的条件的综合。当这 3 个条件都满足时会执行 addSingletonFactory 操作，那么加入 SingletonFactory 的作用是什么呢？又是在什么时候调用呢？

我们还是以简单的 AB 循环依赖为例,类 A 中含有属性类 B,而类 B 中又会含有属性类 A,那么初始化 beanA 的过程如图 5-3 所示。

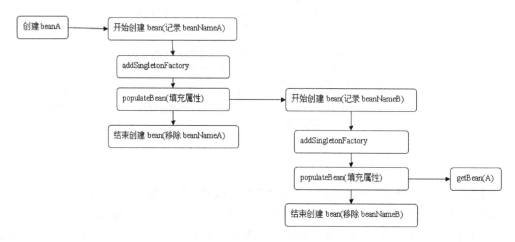

图 5-3 处理循环依赖

图 5-3 中展示了创建 beanA 的流程,图中我们看到,在创建 A 的时候首先会记录类 A 所对应的 beanName,并将 beanA 的创建工厂加入缓存中,而在对 A 的属性填充也就是调用 populate 方法的时候又会再一次的对 B 进行递归创建。同样的,因为在 B 中同样存在 A 属性,因此在实例化 B 的的 populate 方法中又会再次地初始化 B,也就是图形的最后,调用 getBean(A)。关键是在这里,有心的同学可以去找找这个代码的实现方式,我们之前已经讲过,在这个函数中并不是直接去实例化 A,而是先去检测缓存中是否有已经创建好的对应的 bean,或者是否已经创建好的 ObjectFactory,而此时对于 A 的 ObjectFactory 我们早已经创建,所以便不会再去向后执行,而是直接调用 ObjectFactory 去创建 A。这里最关键的是 ObjectFactory 的实现。

```
addSingletonFactory(beanName, new ObjectFactory() {
            public Object getObject() throws BeansException {
                //对bean再一次依赖引用,主要应用SmartInstantiationAware BeanPost Processor,
                //其中我们熟知的 AOP 就是在这里将 advice 动态织入 bean 中,若没有则直接返回
                //bean, 不做任何处理
                return getEarlyBeanReference(beanName, mbd, bean);
            }
        });
```

其中 getEarlyBeanReference 的代码如下:

```
protected Object getEarlyBeanReference(String beanName, RootBeanDefinition mbd, Object bean) {
        Object exposedObject = bean;
        if (bean != null && !mbd.isSynthetic() && hasInstantiationAwareBeanPostProcessors()) {
            for (BeanPostProcessor bp : getBeanPostProcessors()) {
                if (bp instanceof SmartInstantiationAwareBeanPostProcessor) {
                    SmartInstantiationAwareBeanPostProcessor ibp = (SmartInstantiationAwareBeanPostProcessor) bp;
```

```
                    exposedObject = ibp.getEarlyBeanReference(exposedObject, beanName);
                    if (exposedObject == null) {
                        return exposedObject;
                    }
                }
            }
        }
        return exposedObject;
}
```

在 getEarlyBeanReference 函数中并没有太多的逻辑处理，或者说除了后处理器的调用外没有别的处理工作，根据以上分析，基本可以理清 Spring 处理循环依赖的解决办法，在 B 中创建依赖 A 时通过 ObjectFactory 提供的实例化方法来中断 A 中的属性填充，使 B 中持有的 A 仅仅是刚刚初始化并没有填充任何属性的 A，而这正初始化 A 的步骤还是在最开始创建 A 的时候进行的，但是因为 A 与 B 中的 A 所表示的属性地址是一样的，所以在 A 中创建好的属性填充自然可以通过 B 中的 A 获取，这样就解决了循环依赖的问题。

5.7.3 属性注入

在了解循环依赖的时候，我们曾经反复提到了 populateBean 这个函数，也多少了解了这个函数的主要功能就是属性填充，那么究竟是如何实现填充的呢？

```
protected void populateBean(String beanName, AbstractBeanDefinition mbd, BeanWrapper bw) {
        PropertyValues pvs = mbd.getPropertyValues();

        if (bw == null) {
            if (!pvs.isEmpty()) {
                throw new BeanCreationException(
                    mbd.getResourceDescription(), beanName, "Cannot apply property values to null instance");
            }
            else {
                //没有可填充的属性
                return;
            }
        }


        //给 InstantiationAwareBeanPostProcessors 最后一次机会在属性设置前来改变 bean
        //如：可以用来支持属性注入的类型
        boolean continueWithPropertyPopulation = true;

        if (!mbd.isSynthetic() && hasInstantiationAwareBeanPostProcessors()) {
            for (BeanPostProcessor bp : getBeanPostProcessors()) {
                if (bp instanceof InstantiationAwareBeanPostProcessor) {
                    InstantiationAwareBeanPostProcessor ibp = (InstantiationAwareBean PostProcessor) bp;
                    //返回值为是否继续填充 bean
                    if (!ibp.postProcessAfterInstantiation(bw.getWrappedInstance(), beanName)) {
                        continueWithPropertyPopulation = false;
```

```
                    break;
                }
            }
        }
        //如果后处理器发出停止填充命令则终止后续的执行
        if (!continueWithPropertyPopulation) {
            return;
        }

        if (mbd.getResolvedAutowireMode() == RootBeanDefinition.AUTOWIRE_BY_NAME ||
                mbd.getResolvedAutowireMode() == RootBeanDefinition.AUTOWIRE_BY_TYPE) {
            MutablePropertyValues newPvs = new MutablePropertyValues(pvs);

            // Add property values based on autowire by name if applicable.
            //根据名称自动注入
            if (mbd.getResolvedAutowireMode() == RootBeanDefinition.AUTOWIRE_BY_NAME) {
                autowireByName(beanName, mbd, bw, newPvs);
            }

            // Add property values based on autowire by type if applicable.
            //根据类型自动注入
            if (mbd.getResolvedAutowireMode() == RootBeanDefinition.AUTOWIRE_BY_TYPE) {
                autowireByType(beanName, mbd, bw, newPvs);
            }

            pvs = newPvs;
        }

        //后处理器已经初始化
        boolean hasInstAwareBpps = hasInstantiationAwareBeanPostProcessors();
        //需要依赖检查
        boolean needsDepCheck = (mbd.getDependencyCheck() != RootBeanDefinition.DEPENDENCY_CHECK_NONE);

        if (hasInstAwareBpps || needsDepCheck) {
            PropertyDescriptor[] filteredPds = filterPropertyDescriptors ForDependency Check(bw);
            if (hasInstAwareBpps) {
                for (BeanPostProcessor bp : getBeanPostProcessors()) {
                    if (bp instanceof InstantiationAwareBeanPostProcessor) {
                        InstantiationAwareBeanPostProcessor ibp = (InstantiationAwareBean PostProcessor) bp;
                        //对所有需要依赖检查的属性进行后处理
                        pvs = ibp.postProcessPropertyValues(pvs, filteredPds, bw. getWrappedInstance(), beanName);
                        if (pvs == null) {
                            return;
                        }
                    }
                }
            }

            if (needsDepCheck) {
```

```
            //依赖检查，对应depends-on属性，3.0已经弃用此属性
            checkDependencies(beanName, mbd, filteredPds, pvs);
        }
    }
    //将属性应用到bean中
    applyPropertyValues(beanName, mbd, bw, pvs);
}
```

在populateBean函数中提供了这样的处理流程。

1. InstantiationAwareBeanPostProcessor处理器的postProcessAfterInstantiation函数的应用，此函数可以控制程序是否继续进行属性填充。

2. 根据注入类型（byName/byType），提取依赖的bean，并统一存入PropertyValues中。

3. 应用InstantiationAwareBeanPostProcessor处理器的postProcessPropertyValues方法，对属性获取完毕填充前对属性的再次处理，典型应用是RequiredAnnotationBeanPostProcessor类中对属性的验证。

4. 将所有PropertyValues中的属性填充至BeanWrapper中。

在上面的步骤中有几个地方是我们比较感兴趣的，它们分别是依赖注入（autowire-ByName/autowireByType）以及属性填充，那么，接下来进一步分析这几个功能的实现细节。

1. autowireByName

上文提到根据注入类型（byName/byType），提取依赖的bean，并统一存入PropertyValues中，那么我们首先了解下byName功能是如何实现的。

```
protected void autowireByName(
            String beanName, AbstractBeanDefinition mbd, BeanWrapper bw, MutablePropertyValues pvs) {
        
        //寻找bw中需要依赖注入的属性
        String[] propertyNames = unsatisfiedNonSimpleProperties(mbd, bw);
        for (String propertyName : propertyNames) {
            if (containsBean(propertyName)) {
                //递归初始化相关的bean
                Object bean = getBean(propertyName);
                pvs.add(propertyName, bean);
                //注册依赖
                registerDependentBean(propertyName, beanName);
                if (logger.isDebugEnabled()) {
                    logger.debug("Added autowiring by name from bean name '" + beanName +
                            "' via property '" + propertyName + "' to bean named '" + propertyName + "'");
                }
            }
            else {
                if (logger.isTraceEnabled()) {
                    logger.trace("Not autowiring property '" + propertyName + "' of bean '" + beanName +
                            "' by name: no matching bean found");
                }
            }
```

                }
            }
        }

如果读者之前了解了 autowire 的使用方法，相信理解这个函数的功能不会太困难，无非是在传入的参数 pvs 中找出已经加载的 bean，并递归实例化，进而加入到 pvs 中。

2. autowireByType

autowireByType 与 autowireByName 对于我们理解与使用来说复杂程度都很相似，但是其实现功能的复杂度却完全不一样。

```java
protected void autowireByType(
        String beanName, AbstractBeanDefinition mbd, BeanWrapper bw, MutablePropertyValues pvs) {

            TypeConverter converter = getCustomTypeConverter();
            if (converter == null) {
                converter = bw;
            }

            Set<String> autowiredBeanNames = new LinkedHashSet<String>(4);
            //寻找bw中需要依赖注入的属性
            String[] propertyNames = unsatisfiedNonSimpleProperties(mbd, bw);
            for (String propertyName : propertyNames) {
                try {
                    PropertyDescriptor pd = bw.getPropertyDescriptor(propertyName);
                    // Don't try autowiring by type for type Object: never makes sense,
                    // even if it technically is a unsatisfied, non-simple property.
                    if (!Object.class.equals(pd.getPropertyType())) {
                        //探测指定属性的set方法
                        MethodParameter methodParam = BeanUtils.getWriteMethodParameter(pd);
                        boolean eager = !PriorityOrdered.class.isAssignableFrom(bw.getWrappedClass());
                        DependencyDescriptor desc = new AutowireByTypeDependencyDescriptor(methodParam, eager);
                        //解析指定beanName的属性所匹配的值，并把解析到的属性名称存储在
                        //autowiredBeanNames 中，当属性存在多个封装bean时如：
                        //@Autowired private List<A> aList; 将会找到所有匹配A类型的bean并将
                        //其注入
                        Object autowiredArgument = resolveDependency(desc, beanName, autowiredBeanNames, converter);
                        if (autowiredArgument != null) {
                            pvs.add(propertyName, autowiredArgument);
                        }
                        for (String autowiredBeanName : autowiredBeanNames) {
                            //注册依赖
                            registerDependentBean(autowiredBeanName, beanName);
                            if (logger.isDebugEnabled()) {
                                logger.debug("Autowiring by type from bean name '" + beanName + "' via property '" +
                                        propertyName + "' to bean named '" + autowiredBeanName + "'");
```

```
                    }
                }
                autowiredBeanNames.clear();
            }
        }
        catch (BeansException ex) {
            throw new UnsatisfiedDependencyException(mbd.getResourceDescription(),
beanName, propertyName, ex);
        }
    }
}
```

实现根据名称自动匹配的第一步就是寻找 bw 中需要依赖注入的属性,同样对于根据类型自动匹配的实现来讲第一步也是寻找 bw 中需要依赖注入的属性,然后遍历这些属性并寻找类型匹配的 bean,其中最复杂的就是寻找类型匹配的 bean。同时,Spring 中提供了对集合的类型注入的支持,如使用注解的方式:

```
@Autowired
    private List<Test> tests;
```

Spring 将会把所有与 Test 匹配的类型找出来并注入到 tests 属性中,正是由于这一因素,所以在 autowireByType 函数中,新建了局部遍历 autowiredBeanNames,用于存储所有依赖的 bean,如果只是对非集合类的属性注入来说,此属性并无用处。

对于寻找类型匹配的逻辑实现封装在了 resolveDependency 函数中。

DefaultListableBeanFactory.java
```
public Object resolveDependency(DependencyDescriptor descriptor, String beanName,
        Set<String> autowiredBeanNames, TypeConverter typeConverter) throws BeansException {

            descriptor.initParameterNameDiscovery(getParameterNameDiscoverer());
            if (descriptor.getDependencyType().equals(ObjectFactory.class)) {
                //ObjectFactory 类注入的特殊处理
                return new DependencyObjectFactory(descriptor, beanName);
            }
            else if (descriptor.getDependencyType().equals(javaxInjectProviderClass)) {
                //javaxInjectProviderClass 类注入的特殊处理
                return new DependencyProviderFactory().createDependencyProvider (descriptor, beanName);
            }
            else {
                //通用处理逻辑
                return doResolveDependency(descriptor, descriptor.getDependencyType(),
beanName, autowiredBeanNames, typeConverter);
            }
    }

    protected Object doResolveDependency(DependencyDescriptor descriptor, Class<?> type,
String beanName,
            Set<String> autowiredBeanNames, TypeConverter typeConverter) throws BeansException {
```

```java
        /*
         * 用于支持 Spring 中新增的注解@Value
         */
        Object value = getAutowireCandidateResolver().getSuggestedValue(descriptor);
        if (value != null) {
            if (value instanceof String) {
                String strVal = resolveEmbeddedValue((String) value);
                BeanDefinition bd = (beanName != null && containsBean(beanName) ? getMergedBeanDefinition(beanName) : null);
                value = evaluateBeanDefinitionString(strVal, bd);
            }
            TypeConverter converter = (typeConverter != null ? typeConverter : getTypeConverter());
            return converter.convertIfNecessary(value, type);
        }

        //如果解析器没有成功解析,则需要考虑各种情况
        //属性是数组类型
        if (type.isArray()) {

            Class<?> componentType = type.getComponentType();
            //根据属性类型找到beanFacotry中所有类型的匹配bean,
            //返回值的构成为：key=匹配的beanName,value=beanName对应的实例化后的bean(通过
            //getBean(beanName)返回)
            Map<String, Object> matchingBeans = findAutowireCandidates(beanName, componentType, descriptor);
            if (matchingBeans.isEmpty()) {
                //如果autowire的require属性为true而找到的匹配项却为空则只能抛出异常
                if (descriptor.isRequired()) {
                    raiseNoSuchBeanDefinitionException(componentType, "array of " + componentType.getName(), descriptor);
                }
                return null;
            }
            if (autowiredBeanNames != null) {
                autowiredBeanNames.addAll(matchingBeans.keySet());
            }
            TypeConverter converter = (typeConverter != null ? typeConverter : getTypeConverter());
            //通过转换器将bean的值转换为对应的type类型
            return converter.convertIfNecessary(matchingBeans.values(), type);
        }
        //属性是Collection类型
        else if (Collection.class.isAssignableFrom(type) && type.isInterface()) {
            Class<?> elementType = descriptor.getCollectionType();
            if (elementType == null) {
                if (descriptor.isRequired()) {
                    throw new FatalBeanException("No element type declared for collection [" + type.getName() + "]");
                }
                return null;
            }
            Map<String, Object> matchingBeans = findAutowireCandidates(beanName, elementType, descriptor);
```

```
                if (matchingBeans.isEmpty()) {
                    if (descriptor.isRequired()) {
                        raiseNoSuchBeanDefinitionException(elementType, "collection of "
+ elementType.getName(), descriptor);
                    }
                    return null;
                }
                if (autowiredBeanNames != null) {
                    autowiredBeanNames.addAll(matchingBeans.keySet());
                }
                TypeConverter converter = (typeConverter != null ? typeConverter :
getTypeConverter());
                return converter.convertIfNecessary(matchingBeans.values(), type);
            }
            //属性是 Map 类型
            else if (Map.class.isAssignableFrom(type) && type.isInterface()) {
                Class<?> keyType = descriptor.getMapKeyType();
                if (keyType == null || !String.class.isAssignableFrom(keyType)) {
                    if (descriptor.isRequired()) {
                        throw new FatalBeanException("Key type [" + keyType + "] of
map [" + type.getName() +
                                "] must be assignable to [java.lang.String]");
                    }
                    return null;
                }
                Class<?> valueType = descriptor.getMapValueType();
                if (valueType == null) {
                    if (descriptor.isRequired()) {
                        throw new FatalBeanException("No value type declared for map [" +
type.getName() + "]");
                    }
                    return null;
                }
                Map<String, Object> matchingBeans = findAutowireCandidates(beanName,
valueType, descriptor);
                if (matchingBeans.isEmpty()) {
                    if (descriptor.isRequired()) {
                        raiseNoSuchBeanDefinitionException(valueType, "map with value type
" + valueType.getName(), descriptor);
                    }
                    return null;
                }
                if (autowiredBeanNames != null) {
                    autowiredBeanNames.addAll(matchingBeans.keySet());
                }
                return matchingBeans;
            }else {
                Map<String, Object> matchingBeans = findAutowireCandidates(beanName, type,
descriptor);
                if (matchingBeans.isEmpty()) {
                    if (descriptor.isRequired()) {
                        raiseNoSuchBeanDefinitionException(type, "", descriptor);
                    }
                    return null;
```

```
            }
            if (matchingBeans.size() > 1) {
                String primaryBeanName = determinePrimaryCandidate(matchingBeans, descriptor);
                if (primaryBeanName == null) {
                    throw new NoSuchBeanDefinitionException(type, "expected single matching bean but found " +
                            matchingBeans.size() + ": " + matchingBeans.keySet());
                }
                if (autowiredBeanNames != null) {
                    autowiredBeanNames.add(primaryBeanName);
                }
                return matchingBeans.get(primaryBeanName);
            }
            //已经可以确定只有一个匹配项
            Map.Entry<String, Object> entry = matchingBeans.entrySet().iterator().next();
            if (autowiredBeanNames != null) {
                autowiredBeanNames.add(entry.getKey());
            }
            return entry.getValue();
        }
    }
```

寻找类型的匹配执行顺序时，首先尝试使用解析器进行解析，如果解析器没有成功解析，那么可能是使用默认的解析器没有做任何处理，或者是使用了自定义的解析器，但是对于集合等类型来说并不在解析范围之内，所以再次对不同类型进行不同情况的处理，虽说对于不同类型处理方式不一致，但是大致的思路还是很相似的，所以函数中只对数组类型进行了详细地注释。

3. applyPropertyValues

程序运行到这里，已经完成了对所有注入属性的获取，但是获取的属性是以PropertyValues形式存在的，还并没有应用到已经实例化的bean中，这一工作是在applyPropertyValues中。

```
protected void applyPropertyValues(String beanName, BeanDefinition mbd, BeanWrapper bw, PropertyValues pvs) {
    if (pvs == null || pvs.isEmpty()) {
        return;
    }

    MutablePropertyValues mpvs = null;
    List<PropertyValue> original;

    if (System.getSecurityManager()!= null) {
        if (bw instanceof BeanWrapperImpl) {
            ((BeanWrapperImpl) bw).setSecurityContext(getAccessControlContext());
        }
    }

    if (pvs instanceof MutablePropertyValues) {
        mpvs = (MutablePropertyValues) pvs;
        //如果mpvs中的值已经被转换为对应的类型那么可以直接设置到beanwapper中
```

5.7 创建 bean

```java
                    if (mpvs.isConverted()) {
                        // Shortcut: use the pre-converted values as-is.
                        try {
                            bw.setPropertyValues(mpvs);
                            return;
                        }
                        catch (BeansException ex) {
                            throw new BeanCreationException(
                                    mbd.getResourceDescription(), beanName, "Error setting property values", ex);
                        }
                    }
                    original = mpvs.getPropertyValueList();
                }else {
                    //如果pvs并不是使用MutablePropertyValues封装的类型,那么直接使用原始的属性获取方法
                    original = Arrays.asList(pvs.getPropertyValues());
                }

                TypeConverter converter = getCustomTypeConverter();
                if (converter == null) {
                    converter = bw;
                }
                //获取对应的解析器
                BeanDefinitionValueResolver valueResolver = new BeanDefinitionValueResolver(this, beanName, mbd, converter);

                // Create a deep copy, resolving any references for values.
                List<PropertyValue> deepCopy = new ArrayList<PropertyValue>(original.size());
                boolean resolveNecessary = false;
                //遍历属性,将属性转换为对应类的对应属性的类型
                for (PropertyValue pv : original) {
                    if (pv.isConverted()) {
                        deepCopy.add(pv);
                    }else {
                        String propertyName = pv.getName();
                        Object originalValue = pv.getValue();
                        Object resolvedValue = valueResolver.resolveValueIfNecessary(pv, originalValue);
                        Object convertedValue = resolvedValue;
                        boolean convertible = bw.isWritableProperty(propertyName) &&
                                !PropertyAccessorUtils.isNestedOrIndexedProperty (propertyName);
                        if (convertible) {
                            convertedValue = convertForProperty(resolvedValue, propertyName, bw, converter);
                        }
                        if (resolvedValue == originalValue) {
                            if (convertible) {
                                pv.setConvertedValue(convertedValue);
                            }
                            deepCopy.add(pv);
                        }
                        else if (convertible && originalValue instanceof TypedStringValue &&
                                !((TypedStringValue) originalValue).isDynamic() &&
                                !(convertedValue instanceof Collection || ObjectUtils.isArray(convertedValue))) {
```

```
                pv.setConvertedValue(convertedValue);
                deepCopy.add(pv);
            }
            else {
                resolveNecessary = true;
                deepCopy.add(new PropertyValue(pv, convertedValue));
            }
        }
    }
    if (mpvs != null && !resolveNecessary) {
        mpvs.setConverted();
    }

    try {
        bw.setPropertyValues(new MutablePropertyValues(deepCopy));
    }
    catch (BeansException ex) {
        throw new BeanCreationException(
            mbd.getResourceDescription(), beanName, "Error setting property values", ex);
    }
}
```

5.7.4 初始化 bean

大家应该记得在 bean 配置时 bean 中有一个 init-method 的属性,这个属性的作用是在 bean 实例化前调用 init-method 指定的方法来根据用户业务进行相应的实例化。我们现在就已经进入这个方法了,首先看一下这个方法的执行位置,Spring 中程序已经执行过 bean 的实例化,并且进行了属性的填充,而就在这时将会调用用户设定的初始化方法。

```
protected Object initializeBean(final String beanName, final Object bean, RootBean Definition mbd) {
            if (System.getSecurityManager() != null) {
                AccessController.doPrivileged(new PrivilegedAction<Object>() {
                    public Object run() {
                        invokeAwareMethods(beanName, bean);
                        return null;
                    }
                }, getAccessControlContext());
            }
            else {
                //对特殊的 bean 处理:Aware、BeanClassLoaderAware、BeanFactoryAware
                invokeAwareMethods(beanName, bean);
            }

            Object wrappedBean = bean;
            if (mbd == null || !mbd.isSynthetic()) {
                //应用后处理器
                wrappedBean = applyBeanPostProcessorsBeforeInitialization(wrappedBean, beanName);
            }

            try {
```

```java
        //激活用户自定义的init方法
        invokeInitMethods(beanName, wrappedBean, mbd);
    }
    catch (Throwable ex) {
        throw new BeanCreationException(
                (mbd != null ? mbd.getResourceDescription() : null),
                beanName, "Invocation of init method failed", ex);
    }

    if (mbd == null || !mbd.isSynthetic()) {
        //后处理器应用
        wrappedBean = applyBeanPostProcessorsAfterInitialization(wrappedBean, beanName);
    }
    return wrappedBean;
}
```

虽然说此函数的主要目的是进行客户设定的初始化方法的调用，但是除此之外还有些其他必要的工作。

1. 激活 Aware 方法

在分析其原理之前，我们先了解一下 Aware 的使用。Spring 中提供一些 Aware 相关接口，比如 BeanFactoryAware、ApplicationContextAware、ResourceLoaderAware、ServletContextAware 等，实现这些 Aware 接口的 bean 在被初始之后，可以取得一些相对应的资源，例如实现 BeanFactoryAware 的 bean 在初始后，Spring 容器将会注入 BeanFactory 的实例，而实现 ApplicationContextAware 的 bean，在 bean 被初始后，将会被注入 ApplicationContext 的实例等。我们首先通过示例方法来了解一下 Aware 的使用。

1. 定义普通 bean。

```java
public class Hello {
    public void say() {
        System.out.println("hello");
    }
}
```

2. 定义 BeanFactoryAware 类型的 bean。

```java
public class Test implements BeanFactoryAware {
    private BeanFactory beanFactory;

    // 声明bean的时候Spring会自动注入BeanFactory
    @Override
    public void setBeanFactory(BeanFactory beanFactory) throws BeansException {
        this.beanFactory = beanFactory;
    }

    public void testAware() {
        // 通过hello这个bean id从beanFactory获取实例
        Hello hello = (Hello) beanFactory.getBean("hello");
        hello.say();
    }
}
```

3. 使用 main 方法测试。
```java
public static void main(String[] s) {
    ApplicationContext ctx = new ClassPathXmlApplicationContext("applicationContext.xml");
    Test test = (Test) ctx.getBean("test");
    test.testAware();
}
```
运行测试类，控制台输出：
```
hello
```
按照上面的方法我们可以获取到 Spring 中 BeanFactory，并且可以根据 BeanFactory 获取所有 bean，以及进行相关设置。当然还有其他 Aware 的使用方法都大同小异，看一下 Spring 的实现方式，相信读者便会使用了。
```java
private void invokeAwareMethods(final String beanName, final Object bean) {
    if (bean instanceof Aware) {
        if (bean instanceof BeanNameAware) {
            ((BeanNameAware) bean).setBeanName(beanName);
        }
        if (bean instanceof BeanClassLoaderAware) {
            ((BeanClassLoaderAware) bean).setBeanClassLoader(getBeanClassLoader());
        }
        if (bean instanceof BeanFactoryAware) {
            ((BeanFactoryAware) bean).setBeanFactory(AbstractAutowireCapableBeanFactory.this);
        }
    }
}
```
代码简单得已经没有什么好说的了。读者可以自己尝试使用别的 Aware，都比较简单。

2. 处理器的应用

BeanPostProcessor 相信大家都不陌生，这是 Spring 中开放式架构中一个必不可少的亮点，给用户充足的权限去更改或者扩展 Spring，而除了 BeanPostProcessor 外还有很多其他的 PostProcessor，当然大部分都是以此为基础，继承自 BeanPostProcessor。BeanPostProcessor 的使用位置就是这里，在调用客户自定义初始化方法前以及调用自定义初始化方法后分别会调用 BeanPostProcessor 的 postProcessBeforeInitialization 和 postProcessAfterInitialization 方法，使用户可以根据自己的业务需求进行响应的处理。
```java
public Object applyBeanPostProcessorsBeforeInitialization(Object existingBean, String beanName)
        throws BeansException {
    Object result = existingBean;
    for (BeanPostProcessor beanProcessor : getBeanPostProcessors()) {
        result = beanProcessor.postProcessBeforeInitialization(result, beanName);
        if (result == null) {
            return result;
        }
    }
    return result;
```

```
    }
    public Object applyBeanPostProcessorsAfterInitialization(Object existingBean, String beanName)
            throws BeansException {
        Object result = existingBean;
        for (BeanPostProcessor beanProcessor : getBeanPostProcessors()) {
            result = beanProcessor.postProcessAfterInitialization(result, beanName);
            if (result == null) {
                return result;
            }
        }
        return result;
    }
```

3. 激活自定义的 init 方法

客户定制的初始化方法除了我们熟知的使用配置 init-method 外，还有使自定义的 bean 实现 InitializingBean 接口，并在 afterPropertiesSet 中实现自己的初始化业务逻辑。

init-method 与 afterPropertiesSet 都是在初始化 bean 时执行，执行顺序是 afterPropertiesSet 先执行，而 init-method 后执行。

在 invokeInitMethods 方法中就实现了这两个步骤的初始化方法调用。

```
protected void invokeInitMethods(String beanName, final Object bean, RootBeanDefinition mbd)
        throws Throwable {
    //首先会检查是否是 InitializingBean，如果是的话需要调用 afterPropertiesSet 方法
    boolean isInitializingBean = (bean instanceof InitializingBean);
    if (isInitializingBean && (mbd == null || !mbd.isExternallyManagedInitMethod("afterPropertiesSet"))) {
        if (logger.isDebugEnabled()) {
            logger.debug("Invoking afterPropertiesSet() on bean with name '" + beanName + "'");
        }
        if (System.getSecurityManager() != null) {
            try {
                AccessController.doPrivileged(new PrivilegedExceptionAction<Object>() {
                    public Object run() throws Exception {
                        ((InitializingBean) bean).afterPropertiesSet();
                        return null;
                    }
                }, getAccessControlContext());
            }
            catch (PrivilegedActionException pae) {
                throw pae.getException();
            }
        }else {
            //属性初始化后的处理
            ((InitializingBean) bean).afterPropertiesSet();
        }
    }
```

```java
            if (mbd != null) {
                String initMethodName = mbd.getInitMethodName();
                if (initMethodName != null && !(isInitializingBean && "afterPropertiesSet".equals(initMethodName)) &&
                        !mbd.isExternallyManagedInitMethod(initMethodName)) {
                    //调用自定义初始化方法
                    invokeCustomInitMethod(beanName, bean, mbd);
                }
            }
        }
    }
```

5.7.5 注册 DisposableBean

Spring 中不但提供了对于初始化方法的扩展入口，同样也提供了销毁方法的扩展入口，对于销毁方法的扩展，除了我们熟知的配置属性 destroy-method 方法外，用户还可以注册后处理器 DestructionAwareBeanPostProcessor 来统一处理 bean 的销毁方法，代码如下：

```java
protected void registerDisposableBeanIfNecessary(String beanName, Object bean, RootBeanDefinition mbd) {
    AccessControlContext acc = (System.getSecurityManager() != null ? getAccessControlContext() : null);
    if (!mbd.isPrototype() && requiresDestruction(bean, mbd)) {
        if (mbd.isSingleton()) {
            /*
             * 单例模式下注册需要销毁的 bean, 此方法中会处理实现 DisposableBean 的 bean,
             * 并且对所有的 bean 使用 DestructionAwareBeanPostProcessors 处理
             * DisposableBean DestructionAwareBeanPostProcessors,
             */
            registerDisposableBean(beanName,
                    new DisposableBeanAdapter(bean, beanName, mbd, getBeanPostProcessors(), acc));
        }else {
            /*
             * 自定义 scope 的处理
             */
            Scope scope = this.scopes.get(mbd.getScope());
            if (scope == null) {
                throw new IllegalStateException("No Scope registered for scope '" + mbd.getScope() + "'");
            }
            scope.registerDestructionCallback(beanName,
                    new DisposableBeanAdapter(bean, beanName, mbd, getBeanPostProcessors(), acc));
        }
    }
}
```

第 6 章 容器的功能扩展

经过前面几章的分析,相信大家已经对 Spring 中的容器功能有了简单的了解,在前面的章节中我们一直以 BeanFacotry 接口以及它的默认实现类 XmlBeanFactory 为例进行分析,但是,Spring 中还提供了另一个接口 ApplicationContext,用于扩展 BeanFacotry 中现有的功能。

ApplicationContext 和 BeanFacotry 两者都是用于加载 Bean 的,但是相比之下,ApplicationContext 提供了更多的扩展功能,简单一点说:ApplicationContext 包含 BeanFactory 的所有功能。通常建议比 BeanFactory 优先,除非在一些限制的场合,比如字节长度对内存有很大的影响时(Applet)。绝大多数"典型的"企业应用和系统,ApplicationContext 就是你需要使用的。

那么究竟 ApplicationContext 比 BeanFactory 多出了哪些功能呢?还需要我们进一步的探索。首先我们来看看使用两个不同的类去加载配置文件在写法上的不同。

- 使用 BeanFactory 方式加载 XML。
```
BeanFactory    bf = new XmlBeanFactory(new ClassPathResource("beanFactoryTest.xml"));
```
- 使用 ApplicationContext 方式加载 XML。
```
ApplicationContext bf = new ClassPathXmlApplicationContext("beanFactoryTest.xml");
```
同样,我们还是以 ClassPathXmlApplicationContext 作为切入点,开始对整体功能进行分析。

```
public ClassPathXmlApplicationContext(String configLocation) throws BeansException {
      this(new String[] {configLocation}, true, null);
}
public ClassPathXmlApplicationContext(String[] configLocations, boolean refresh,
ApplicationContext parent) throws BeansException {

      super(parent);
      setConfigLocations(configLocations);
      if (refresh) {
          refresh();
      }
}
```

设置路径是必不可少的步骤,ClassPathXmlApplicationContext 中可以将配置文件路径以数组的方式传入,ClassPathXmlApplicationContext 可以对数组进行解析并进行加载。而对于解析及功能实现都在 refresh()中实现。

6.1 设置配置路径

在 ClassPathXmlApplicationContext 中支持多个配置文件以数组方式同时传入：

```
public void setConfigLocations(String[] locations) {
    if (locations != null) {
        Assert.noNullElements(locations, "Config locations must not be null");
        this.configLocations = new String[locations.length];
        for (int i = 0; i < locations.length; i++) {
            //解析给定路径
            this.configLocations[i] = resolvePath(locations[i]).trim();
        }
    }
    else {
        this.configLocations = null;
    }
}
```

此函数主要用于解析给定的路径数组，当然，如果数组中包含特殊符号，如${var}，那么在 resolvePath 中会搜寻匹配的系统变量并替换。

6.2 扩展功能

设置了路径之后，便可以根据路径做配置文件的解析以及各种功能的实现了。可以说 refresh 函数中包含了几乎 ApplicationContext 中提供的全部功能，而且此函数中逻辑非常清晰明了，使我们很容易分析对应的层次及逻辑。

```
public void refresh() throws BeansException, IllegalStateException {
    synchronized (this.startupShutdownMonitor) {
        //准备刷新的上下文环境
        prepareRefresh();

        // Tell the subclass to refresh the internal bean factory.
        //初始化 BeanFactory，并进行 XML 文件读取
        ConfigurableListableBeanFactory beanFactory = obtainFreshBeanFactory();

        // Prepare the bean factory for use in this context.
        //对 BeanFactory 进行各种功能填充
        prepareBeanFactory(beanFactory);

        try {
            // Allows post-processing of the bean factory in context subclasses.
            // 子类覆盖方法做额外的处理
            postProcessBeanFactory(beanFactory);

            //激活各种 BeanFactory 处理器
            invokeBeanFactoryPostProcessors(beanFactory);
```

```
            // 注册拦截 Bean 创建的 Bean 处理器，这里只是注册，真正的调用是在 getBean 时候
            registerBeanPostProcessors(beanFactory);

            // 为上下文初始化 Message 源，即不同语言的消息体 ，国际化处理
            initMessageSource();

            // Initialize event multicaster for this context.
            // 初始化应用消息广播器，并放入 "applicationEventMulticaster" bean 中
            initApplicationEventMulticaster();

            // Initialize other special beans in specific context subclasses.
            // 留给子类来初始化其它的 Bean
            onRefresh();

            // Check for listener beans and register them.
            // 在所有注册的 bean 中查找 Listener bean, 注册到消息广播器中
            registerListeners();

            // Instantiate all remaining (non-lazy-init) singletons.
            // 初始化剩下的单实例（非惰性的）
            finishBeanFactoryInitialization(beanFactory);

            // Last step: publish corresponding event.
            // 完成刷新过程，通知生命周期处理器 lifecycleProcessor 刷新过程，同时发出
            //ContextRefreshEvent 通知别人
            finishRefresh();
        }

        catch (BeansException ex) {
            // Destroy already created singletons to avoid dangling resources.
            destroyBeans();

            // Reset 'active' flag.
            cancelRefresh(ex);

            // Propagate exception to caller.
            throw ex;
        }
    }
}
```

下面概括一下 ClassPathXmlApplicationContext 初始化的步骤，并从中解释一下它为我们提供的功能。

1. 初始化前的准备工作，例如对系统属性或者环境变量进行准备及验证。

在某种情况下项目的使用需要读取某些系统变量，而这个变量的设置很可能会影响着系统的正确性，那么 ClassPathXmlApplicationContext 为我们提供的这个准备函数就显得非常必要，它可以在 Spring 启动的时候提前对必需的变量进行存在性验证。

2. 初始化 BeanFactory，并进行 XML 文件读取。

之前有提到 ClassPathXmlApplicationContext 包含着 BeanFactory 所提供的一切特征，那么在这一步骤中将会复用 BeanFactory 中的配置文件读取解析及其他功能，这一步之后，

ClassPathXmlApplicationContext 实际上就已经包含了 BeanFactory 所提供的功能，也就是可以进行 bean 的提取等基础操作了。

3. 对 BeanFactory 进行各种功能填充。

@Qualifier 与@Autowired 应该是大家非常熟悉的注解，那么这两个注解正是在这一步骤中增加的支持。

4. 子类覆盖方法做额外的处理。

Spring 之所以强大，为世人所推崇，除了它功能上为大家提供了便例外，还有一方面是它的完美架构，开放式的架构让使用它的程序员很容易根据业务需要扩展已经存在的功能。这种开放式的设计在 Spring 中随处可见，例如在本例中就提供了一个空的函数实现 postProcess-BeanFactory 来方便程序员在业务上做进一步扩展。

5. 激活各种 BeanFactory 处理器。
6. 注册拦截 bean 创建的 bean 处理器，这里只是注册，真正的调用是在 getBean 时候。
7. 为上下文初始化 Message 源，即对不同语言的消息体进行国际化处理。
8. 初始化应用消息广播器，并放入 "applicationEventMulticaster" bean 中。
9. 留给子类来初始化其他的 bean。
10. 在所有注册的 bean 中查找 listener bean，注册到消息广播器中。
11. 初始化剩下的单实例（非惰性的）。
12. 完成刷新过程，通知生命周期处理器 lifecycleProcessor 刷新过程，同时发出 Context-RefreshEvent 通知别人。

6.3 环境准备

prepareRefresh 函数主要是做些准备工作，例如对系统属性及环境变量的初始化及验证。

```
protected void prepareRefresh() {
    this.startupDate = System.currentTimeMillis();

    synchronized (this.activeMonitor) {
        this.active = true;
    }

    if (logger.isInfoEnabled()) {
        logger.info("Refreshing " + this);
    }

    //留给子类覆盖
    initPropertySources();
    //验证需要的属性文件是否都已经放入环境中
    getEnvironment().validateRequiredProperties();
}
```

网上有人说其实这个函数没什么用，因为最后两句代码才是最为关键的，但是却没有什么逻辑处理，initPropertySources 是空的，没有任何逻辑，而 getEnvironment().validateRequiredProperties

也因为没有需要验证的属性而没有做任何处理。其实这都是因为没有彻底理解才会这么说，这个函数如果用好了作用还是挺大的。那么，该怎么用呢？我们先探索下各个函数的作用。

1. initPropertySources 正符合 Spring 的开放式结构设计，给用户最大扩展 Spring 的能力。用户可以根据自身的需要重写 initPropertySources 方法，并在方法中进行个性化的属性处理及设置。

2. validateRequiredProperties 则是对属性进行验证，那么如何验证呢？我们举个融合两句代码的小例子来帮助大家理解。

假如现在有这样一个需求，工程在运行过程中用到的某个设置（例如 VAR）是从系统环境变量中取得的，而如果用户没有在系统环境变量中配置这个参数，那么工程可能不会工作。这一要求可能会有各种各样的解决办法，当然，在 Spring 中可以这样做，你可以直接修改 Spring 的源码，例如修改 ClassPathXmlApplicationContext。当然，最好的办法还是对源码进行扩展，我们可以自定义类：

```java
public class MyClassPathXmlApplicationContext extends ClassPathXmlApplicationContext{
    public MyClassPathXmlApplicationContext(String... configLocations ){
        super(configLocations);
    }

    protected void initPropertySources() {
    //添加验证要求
        getEnvironment().setRequiredProperties("VAR");
    }
}
```

我们自定义了继承自 ClassPathXmlApplicationContext 的 MyClassPathXmlApplicationContext，并重写了 initPropertySources 方法，在方法中添加了我们的个性化需求，那么在验证的时候也就是程序走到 getEnvironment().validateRequiredProperties()代码的时候，如果系统并没有检测到对应 VAR 的环境变量，那么将抛出异常。当然我们还需要在使用的时候替换掉原有的 ClassPathXmlApplicationContext：

```java
public static void main(String[] args) {
        ApplicationContext bf = new MyClassPathXmlApplicationContext ("test/customtag/test.xml");
        User user=(User) bf.getBean("testbean");
}
```

6.4 加载 BeanFactory

obtainFreshBeanFactory 方法从字面理解是获取 BeanFactory。之前有说过，ApplicationContext 是对 BeanFactory 的功能上的扩展，不但包含了 BeanFactory 的全部功能更在其基础上添加了大量的扩展应用，那么 obtainFreshBeanFactory 正是实现 BeanFactory 的地方，也就是经过了这个函数后 ApplicationContext 就已经拥有了 BeanFactory 的全部功能。

```java
protected ConfigurableListableBeanFactory obtainFreshBeanFactory() {
        //初始化BeanFactory，并进行XML文件读取,并将得到的BeanFacotry记录在当前实体的属性中
```

```
        refreshBeanFactory();
//返回当前实体的beanFactory属性
        ConfigurableListableBeanFactory beanFactory = getBeanFactory();
        if (logger.isDebugEnabled()) {
            logger.debug("Bean factory for " + getDisplayName() + ": " + beanFactory);
        }
        return beanFactory;
    }
```

方法中将核心实现委托给了 refreshBeanFactory：

```
@Override
protected final void refreshBeanFactory() throws BeansException {
        if (hasBeanFactory()) {
            destroyBeans();
            closeBeanFactory();
        }
        try {
            //创建 DefaultListableBeanFactory
            DefaultListableBeanFactory beanFactory = createBeanFactory();

            //为了序列化指定id,如果需要的话,让这个BeanFactory从id反序列化到BeanFactory对象
            beanFactory.setSerializationId(getId());
            //定制beanFactory,设置相关属性,包括是否允许覆盖同名称的不同定义的对象以及循环依赖以及
            //设置@Autowired 和 @Qualifier 注解解析器 QualifierAnnotationAutowire-
            //CandidateResolver
            customizeBeanFactory(beanFactory);
            //初始化 DodumentReader,并进行 XML 文件读取及解析
            loadBeanDefinitions(beanFactory);
            synchronized (this.beanFactoryMonitor) {
                this.beanFactory = beanFactory;
            }
        }
        catch (IOException ex) {
            throw new ApplicationContextException("I/O error parsing bean definition source for " + getDisplayName(), ex);
        }
    }
```

我们详细分析上面的每个步骤。

1. 创建 DefaultListableBeanFactory。

在介绍 BeanFactory 的时候，不知道读者是否还有印象，声明方式为：BeanFactory bf = new XmlBeanFactory("beanFactoryTest.xml")，其中的 XmlBeanFactory 继承自 DefaultListableBean-Factory，并提供了 XmlBeanDefinitionReader 类型的 reader 属性，也就是说 DefaultListableBean-Factory 是容器的基础。必须首先要实例化，那么在这里就是实例化 DefaultListableBeanFactory 的步骤。

2. 指定序列化 ID。
3. 定制 BeanFactory。
4. 加载 BeanDefinition。
5. 使用全局变量记录 BeanFactory 类实例。

因为 DefaultListableBeanFactory 类型的变量 beanFactory 是函数内的局部变量，所以要使用全局变量记录解析结果。

6.4.1 定制 BeanFactory

这里已经开始了对 BeanFactory 的扩展，在基本容器的基础上，增加了是否允许覆盖是否允许扩展的设置并提供了注解@Qualifier 和@Autowired 的支持。

```
protected void customizeBeanFactory(DefaultListableBeanFactory beanFactory) {
    //如果属性 allowBeanDefinitionOverriding 不为空，设置给 beanFactory 对象相应属性，
    //此属性的含义：是否允许覆盖同名称的不同定义的对象
    if (this.allowBeanDefinitionOverriding != null) {
        beanFactory.setAllowBeanDefinitionOverriding(this.allowBeanDefinitionOverriding);
    }
    //如果属性 allowCircularReferences 不为空，设置给 beanFactory 对象相应属性，
    //此属性的含义：是否允许 bean 之间存在循环依赖
    if (this.allowCircularReferences != null) {
        beanFactory.setAllowCircularReferences(this.allowCircularReferences);
    }
    //用于@Qualifier 和@Autowired
    beanFactory.setAutowireCandidateResolver(new QualifierAnnotationAutowireCandidateResolver());
}
```

对于允许覆盖和允许依赖的设置这里只是判断了是否为空，如果不为空要进行设置，但是并没有看到在哪里进行设置，究竟这个设置是在哪里进行设置的呢？还是那句话，使用子类覆盖方法，例如：

```
public class MyClassPathXmlApplicationContext extends ClassPathXmlApplicationContext{
    ... ...
    protected void  customizeBeanFactory(DefaultListableBeanFactory beanFactory) {
        super.setAllowBeanDefinitionOverriding(false);
        super.setAllowCircularReferences(false);
        super.customizeBeanFactory(beanFactory);
    }
}
```

设置完后相信大家已经对于这两个属性的使用有所了解，或者可以回到前面的章节进行再一次查看。对于定制 BeanFactory，Spring 还提供了另外一个重要的扩展，就是设置 AutowireCandidateResolver，在 bean 加载部分中讲解创建 Bean 时，如果采用 autowireByType 方式注入，那么默认会使用 Spring 提供的 SimpleAutowireCandidateResolver，而对于默认的实现并没有过多的逻辑处理。在这里，Spring 使用了 QualifierAnnotationAutowireCandidateResolver，设置了这个解析器后 Spring 就可以支持注解方式的注入了。

在讲解根据类型自定注入的时候，我们说过解析 autowire 类型时首先会调用方法：

```
Object value = getAutowireCandidateResolver().getSuggestedValue(descriptor);
```

因此我们知道，在 QualifierAnnotationAutowireCandidateResolver 中一定会提供了解析 Qualifier 与 Autowire 注解的方法。

QualifierAnnotationAutowireCandidateResolver.java
```
public Object getSuggestedValue(DependencyDescriptor descriptor) {
    Object value = findValue(descriptor.getAnnotations());
    if (value == null) {
        MethodParameter methodParam = descriptor.getMethodParameter();
        if (methodParam != null) {
            value = findValue(methodParam.getMethodAnnotations());
        }
    }
    return value;
}
```

6.4.2 加载 BeanDefinition

在第一步中提到了将 ClassPathXmlApplicationContext 与 XmlBeanFactory 创建的对比，在实现配置文件的加载功能中除了我们在第一步中已经初始化的 DefaultListableBeanFactory 外，还需要 XmlBeanDefinitionReader 来读取 XML，那么在这个步骤中首先要做的就是初始化 XmlBeanDefinitionReader。

```
@Override
    protected void loadBeanDefinitions(DefaultListableBeanFactory beanFactory) throws BeansException, IOException {
        //为指定beanFactory创建XmlBeanDefinitionReader
        XmlBeanDefinitionReader beanDefinitionReader = new XmlBeanDefinitionReader(beanFactory);

        //对beanDefinitionReader进行环境变量的设置
        beanDefinitionReader.setEnvironment(this.getEnvironment());
        beanDefinitionReader.setResourceLoader(this);
        beanDefinitionReader.setEntityResolver(new ResourceEntityResolver(this));

        //对BeanDefinitionReader进行设置，可以覆盖
        initBeanDefinitionReader(beanDefinitionReader);

        loadBeanDefinitions(beanDefinitionReader);
    }
```

在初始化了 DefaultListableBeanFactory 和 XmlBeanDefinitionReader 后就可以进行配置文件的读取了。

```
    protected void loadBeanDefinitions(XmlBeanDefinitionReader reader) throws BeansException, IOException {
        Resource[] configResources = getConfigResources();
        if (configResources != null) {
            reader.loadBeanDefinitions(configResources);
        }
        String[] configLocations = getConfigLocations();
        if (configLocations != null) {
            reader.loadBeanDefinitions(configLocations);
        }
    }
```

使用 XmlBeanDefinitionReader 的 loadBeanDefinitions 方法进行配置文件的加载机注册相信大家已经不陌生，这完全就是开始 BeanFactory 的套路。因为在 XmlBeanDefinitionReader 中已经将之前初始化的 DefaultListableBeanFactory 注册进去了，所以 XmlBeanDefinitionReader 所读取的 BeanDefinitionHolder 都会注册到 DefaultListableBeanFactory 中，也就是经过此步骤，类型 DefaultListableBeanFactory 的变量 beanFactory 已经包含了所有解析好的配置。

6.5 功能扩展

进入函数 prepareBeanFactory 前，Spring 已经完成了对配置的解析，而 ApplicationContext 在功能上的扩展也由此展开。

```
protected void prepareBeanFactory(ConfigurableListableBeanFactory beanFactory) {
    //设置 beanFactory 的 classLoader 为当前 context 的 classLoader
    beanFactory.setBeanClassLoader(getClassLoader());

    //设置 beanFactory 的表达式语言处理器，Spring3 增加了表达式语言的支持，
    //默认可以使用#{bean.xxx}的形式来调用相关属性值。
    beanFactory.setBeanExpressionResolver(new StandardBeanExpressionResolver());

    //为 beanFactory 增加了一个默认的 propertyEditor，这个主要是对 bean 的属性等设置管理的一
    //个工具
    beanFactory.addPropertyEditorRegistrar(new ResourceEditorRegistrar(this, getEnvironment()));

    /*
     * 添加 BeanPostProcessor,
     */
    beanFactory.addBeanPostProcessor(new ApplicationContextAwareProcessor(this));

    //设置了几个忽略自动装配的接口
    beanFactory.ignoreDependencyInterface(ResourceLoaderAware.class);
    beanFactory.ignoreDependencyInterface(ApplicationEventPublisherAware.class);
    beanFactory.ignoreDependencyInterface(MessageSourceAware.class);
    beanFactory.ignoreDependencyInterface(ApplicationContextAware.class);
    beanFactory.ignoreDependencyInterface(EnvironmentAware.class);

    //设置了几个自动装配的特殊规则
    beanFactory.registerResolvableDependency(BeanFactory.class, beanFactory);
    beanFactory.registerResolvableDependency(ResourceLoader.class, this);
    beanFactory.registerResolvableDependency(ApplicationEventPublisher.class, this);
    beanFactory.registerResolvableDependency(ApplicationContext.class, this);

    //增加对 AspectJ 的支持
    if (beanFactory.containsBean(LOAD_TIME_WEAVER_BEAN_NAME)) {
        beanFactory.addBeanPostProcessor(new LoadTimeWeaverAwareProcessor(beanFactory));
        // Set a temporary ClassLoader for type matching.
        beanFactory.setTempClassLoader(new ContextTypeMatchClassLoader (beanFactory.getBeanClassLoader()));
    }
```

```
        //添加默认的系统环境bean
        if (!beanFactory.containsLocalBean(ENVIRONMENT_BEAN_NAME)) {
            beanFactory.registerSingleton(ENVIRONMENT_BEAN_NAME, getEnvironment());
        }
        if (!beanFactory.containsLocalBean(SYSTEM_PROPERTIES_BEAN_NAME)) {
            beanFactory.registerSingleton(SYSTEM_PROPERTIES_BEAN_NAME, getEnvironment().getSystemProperties());
        }
        if (!beanFactory.containsLocalBean(SYSTEM_ENVIRONMENT_BEAN_NAME)) {
            beanFactory.registerSingleton(SYSTEM_ENVIRONMENT_BEAN_NAME, getEnvironment().getSystemEnvironment());
        }
    }
```

上面函数中主要进行了几个方面的扩展。

- 增加对 SpEL 语言的支持。
- 增加对属性编辑器的支持。
- 增加对一些内置类，比如 EnvironmentAware、MessageSourceAware 的信息注入。
- 设置了依赖功能可忽略的接口。
- 注册一些固定依赖的属性。
- 增加 AspectJ 的支持（会在第 7 章中进行详细的讲解）。
- 将相关环境变量及属性注册以单例模式注册。

可能读者不是很理解每个步骤的具体含义，接下来我们会对各个步骤进行详细地分析。

6.5.1 增加 SpEL 语言的支持

Spring 表达式语言全称为 Spring Expression Language，缩写为 SpEL，类似于 Struts 2x 中使用的 OGNL 表达式语言，能在运行时构建复杂表达式、存取对象图属性、对象方法调用等，并且能与 Spring 功能完美整合，比如能用来配置 bean 定义。SpEL 是单独模块，只依赖于 core 模块，不依赖于其他模块，可以单独使用。

SpEL 使用#{...}作为定界符，所有在大框号中的字符都将被认为是 SpEL，使用格式如下：
```
<bean id="saxophone" value="com.xxx.xxx.Xxx"/>
<bean >
    <property name="instrument" value="#{saxophone}"/>
<bean/>
```
相当于：
```
<bean id="saxophone" value="com.xxx.xxx.Xxx"/>
<bean >
    <property name="instrument" ref="saxophone"/>
<bean/>
```
当然，上面只是列举了其中最简单的使用方式，SpEL 功能非常强大，使用好可以大大提高开发效率，这里只为唤起读者的记忆来帮助我们理解源码，有兴趣的读者可以进一步深入研究。

在源码中通过代码 beanFactory.setBeanExpressionResolver(new StandardBeanExpressionResolver())注册语言解析器，就可以对 SpEL 进行解析了，那么在注册解析器后 Spring 又是在

什么时候调用这个解析器进行解析呢？

之前我们讲解过 Spring 在 bean 进行初始化的时候会有属性填充的一步，而在这一步中 Spring 会调用 AbstractAutowireCapableBeanFactory 类的 applyPropertyValues 函数来完成功能。就在这个函数中，会通过构造 BeanDefinitionValueResolver 类型实例 valueResolver 来进行属性值的解析。同时，也是在这个步骤中一般通过 AbstractBeanFactory 中的 evaluateBeanDefinitionString 方法去完成 SpEL 的解析。

```
protected Object evaluateBeanDefinitionString(String value, BeanDefinition beanDefinition) {
        if (this.beanExpressionResolver == null) {
            return value;
        }
        Scope scope = (beanDefinition != null ? getRegisteredScope(beanDefinition.getScope()
) : null);
        return this.beanExpressionResolver.evaluate(value, new BeanExpressionContext(this,
 scope));
    }
```

当调用这个方法时会判断是否存在语言解析器，如果存在则调用语言解析器的方法进行解析，解析的过程是在 Spring 的 expression 的包内，这里不做过多解释。我们通过查看对 evaluateBeanDefinitionString 方法的调用层次可以看出，应用语言解析器的调用主要是在解析依赖注入 bean 的时候，以及在完成 bean 的初始化和属性获取后进行属性填充的时候。

6.5.2 增加属性注册编辑器

在 Spring DI 注入的时候可以把普通属性注入进来，但是像 Date 类型就无法被识别。例如：

```
public class UserManager {
    private Date dataValue;

    public Date getDataValue() {
        return dataValue;
    }

    public void setDataValue(Date dataValue) {
        this.dataValue = dataValue;
    }

    public String toString(){
        return "dataValue: " + dataValue;
    }
}
```

上面代码中，需要对日期型属性进行注入：

```
<bean id="userManager" class="com.test.UserManager">
    <property name="dataValue">
        <value>2013-03-15</value>
    </property>
</bean>
```

测试代码：

```
@Test
public void testDate(){
```

```java
ApplicationContext ctx = new ClassPathXmlApplicationContext("beans.xml");
UserManager userManager = (UserManager)ctx.getBean("userManager");
System.out.println(userManager);
}
```

如果直接这样使用，程序则会报异常，类型转换不成功。因为在 UserManager 中的 dataValue 属性是 Date 类型的，而在 XML 中配置的却是 String 类型的，所以当然会报异常。

Spring 针对此问题提供了两种解决办法。

1. 使用自定义属性编辑器

使用自定义属性编辑器，通过继承 PropertyEditorSupport，重写 setAsText 方法，具体步骤如下。

1. 编写自定义的属性编辑器。

```java
public class DatePropertyEditor extends PropertyEditorSupport {
    private String format = "yyyy-MM-dd";
    public void setFormat(String format) {
        this.format = format;
    }
    public void setAsText(String arg0) throws IllegalArgumentException {
        System.out.println("arg0: " + arg0);
        SimpleDateFormat sdf = new SimpleDateFormat(format);
        try {
            Date d = sdf.parse(arg0);
            this.setValue(d);
        } catch (ParseException e) {
            e.printStackTrace();
        }
    }
}
```

2. 将自定义属性编辑器注册到 Spring 中。

```xml
<!-- 自定义属性编辑器 -->
<bean class="org.Springframework.beans.factory.config.CustomEditorConfigurer">
    <property name="customEditors">
            <map>
                <entry key="java.util.Date">
                    <bean class="com.test.DatePropertyEditor">
                        <property name="format" value="yyyy-MM-dd"/>
                    </bean>
                </entry>
            </map>
    </property>
</bean>
```

在配置文件中引入类型为 org.Springframework.beans.factory.config.CustomEditorConfigurer 的 bean，并在属性 customEditors 中加入自定义的属性编辑器，其中 key 为属性编辑器所对应的类型。通过这样的配置，当 Spring 在注入 bean 的属性时一旦遇到了 java.util.Date 类型的属性会自动调用自定义的 DatePropertyEditor 解析器进行解析，并用解析结果代替配置属性进行注入。

2. 注册 Spring 自带的属性编辑器 CustomDateEditor

通过注册 Spring 自带的属性编辑器 CustomDateEditor，具体步骤如下。
1. 定义属性编辑器。
```
public class DatePropertyEditorRegistrar implements PropertyEditorRegistrar{
    public void registerCustomEditors(PropertyEditorRegistry registry) {
        registry.registerCustomEditor(Date.class, new CustomDateEditor(new SimpleDateFormat ("yyyy-MM-dd"),true));
    }
}
```
2. 注册到 Spring 中。
```
<!-- 注册 Spring 自带编辑器 -->
<bean class="org.Springframework.beans.factory.config.CustomEditorConfigurer">
    <property name="propertyEditorRegistrars">
        <list>
            <bean class="com.test.DatePropertyEditorRegistrar"></bean>
        </list>
    </property>
</bean>
```

通过在配置文件中将自定义的 DatePropertyEditorRegistrar 注册进入 org.Springframework.beans.factory.config.CustomEditorConfigurer 的 propertyEditorRegistrars 属性中，可以具有与方法 1 同样的效果。

我们了解了自定义属性编辑器的使用，但是，似乎这与本节中围绕的核心代码 beanFactory.addPropertyEditorRegistrar(new ResourceEditorRegistrar(this, getEnvironment()))并无联系，因为在注册自定义属性编辑器的时候使用的是 PropertyEditorRegistry 的 registerCustomEditor 方法，而这里使用的是 ConfigurableListableBeanFactory 的 addPropertyEditorRegistrar 方法。我们不妨深入探索一下 ResourceEditorRegistrar 的内部实现，在 ResourceEditorRegistrar 中，我们最关心的方法是 registerCustomEditors。

```
public void registerCustomEditors(PropertyEditorRegistry registry) {
    ResourceEditor baseEditor = new ResourceEditor(this.resourceLoader, this.propertyResolver);
    doRegisterEditor(registry, Resource.class, baseEditor);
    doRegisterEditor(registry, ContextResource.class, baseEditor);
    doRegisterEditor(registry, InputStream.class, new InputStreamEditor(baseEditor));
    doRegisterEditor(registry, InputSource.class, new InputSourceEditor(baseEditor));
    doRegisterEditor(registry, File.class, new FileEditor(baseEditor));
    doRegisterEditor(registry, URL.class, new URLEditor(baseEditor));

    ClassLoader classLoader = this.resourceLoader.getClassLoader();
    doRegisterEditor(registry, URI.class, new URIEditor(classLoader));
    doRegisterEditor(registry, Class.class, new ClassEditor(classLoader));
    doRegisterEditor(registry, Class[].class, new ClassArrayEditor(classLoader));

    if (this.resourceLoader instanceof ResourcePatternResolver) {
        doRegisterEditor(registry, Resource[].class,
            new ResourceArrayPropertyEditor((ResourcePatternResolver) this.resourceLoader, this.propertyResolver));
```

```
            }
        }
        private void doRegisterEditor(PropertyEditorRegistry registry, Class<?> requiredType,
PropertyEditor editor) {
            if (registry instanceof PropertyEditorRegistrySupport) {
                ((PropertyEditorRegistrySupport) registry).overrideDefaultEditor (requiredType,
editor);
            }
            else {
                registry.registerCustomEditor(requiredType, editor);
            }
        }
```

在 doRegisterEditor 函数中，可以看到在之前提到的自定义属性中使用的关键代码：registry.registerCustomEditor(requiredType, editor)，回过头来看 ResourceEditorRegistrar 类的 registerCustomEditors 方法的核心功能，其实无非是注册了一系列的常用类型的属性编辑器，例如，代码 doRegisterEditor(registry, Class.class, new ClassEditor(classLoader))实现的功能就是注册 Class 类对应的属性编辑器。那么，注册后，一旦某个实体 bean 中存在一些 Class 类型的属性，那么 Spring 会调用 ClassEditor 将配置中定义的 String 类型转换为 Class 类型并进行赋值。

分析到这里，我们不禁有个疑问，虽说 ResourceEditorRegistrar 类的 registerCustomEditors 方法实现了批量注册的功能，但是 beanFactory.addPropertyEditorRegistrar(new ResourceEditorRegistrar(this, getEnvironment()))仅仅是注册了 ResourceEditorRegistrar 实例，却并没有调用 ResourceEditorRegistrar 的 registerCustomEditors 方法进行注册，那么到底是在什么时候进行注册的呢？进一步查看 ResourceEditorRegistrar 的 registerCustomEditors 方法的调用层次结构，如图 6-1 所示。

```
● registerCustomEditors(PropertyEditorRegistry) : void - org.springframework.beans.support.ResourceEditorRegistrar
    ◇ registerCustomEditors(PropertyEditorRegistry) : void - org.springframework.beans.factory.support.AbstractBeanFactory
```

图 6-1　ResourceEditorRegistrar 的 registerCustomEditors 方法的调用层次结构

发现在 AbstractBeanFactory 中的 registerCustomEditors 方法中被调用过，继续查看 AbstractBeanFactory 中的 registerCustomEditors 方法的调用层次结构，如图 6-2 所示。

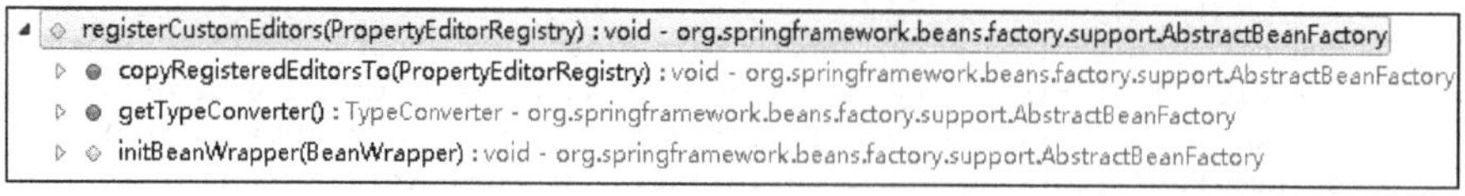

```
▲ ● registerCustomEditors(PropertyEditorRegistry) : void - org.springframework.beans.factory.support.AbstractBeanFactory
    ▷ ● copyRegisteredEditorsTo(PropertyEditorRegistry) : void - org.springframework.beans.factory.support.AbstractBeanFactory
    ▷ ● getTypeConverter() : TypeConverter - org.springframework.beans.factory.support.AbstractBeanFactory
    ▷ ● initBeanWrapper(BeanWrapper) : void - org.springframework.beans.factory.support.AbstractBeanFactory
```

图 6-2　AbstractBeanFactory 中的 registerCustomEditors 方法的调用层次结构

其中我们看到一个方法是我们熟悉的，就是 AbstractBeanFactory 类中的 initBeanWrapper 方法，这是在 bean 初始化时使用的一个方法，之前已经使用过大量的篇幅进行讲解，主要是在将 BeanDefinition 转换为 BeanWrapper 后用于对属性的填充。到此，逻辑已经明了，在 bean 的初始化后会调用 ResourceEditorRegistrar 的 registerCustomEditors 方法进行批量的通用属性编辑器注册。注册后，在属性填充的环节便可以直接让 Spring 使用这些编辑器进行属性的解析了。

6.5 功能扩展

既然提到了 BeanWrapper，这里也有必要强调下，Spring 中用于封装 bean 的是 BeanWrapper 类型，而它又间接继承了 PropertyEditorRegistry 类型，也就是我们之前反复看到的方法参数 PropertyEditorRegistry registry，其实大部分情况下都是 BeanWrapper，对于 BeanWrapper 在 Spring 中的默认实现是 BeanWrapperImpl，而 BeanWrapperImpl 除了实现 BeanWrapper 接口外还继承了 PropertyEditorRegistrySupport，在 PropertyEditorRegistrySupport 中有这样一个方法：

```java
private void createDefaultEditors() {
    this.defaultEditors = new HashMap<Class<?>, PropertyEditor>(64);

    // Simple editors, without parameterization capabilities.
    // The JDK does not contain a default editor for any of these target types.
    this.defaultEditors.put(Charset.class, new CharsetEditor());
    this.defaultEditors.put(Class.class, new ClassEditor());
    this.defaultEditors.put(Class[].class, new ClassArrayEditor());
    this.defaultEditors.put(Currency.class, new CurrencyEditor());
    this.defaultEditors.put(File.class, new FileEditor());
    this.defaultEditors.put(InputStream.class, new InputStreamEditor());
    this.defaultEditors.put(InputSource.class, new InputSourceEditor());
    this.defaultEditors.put(Locale.class, new LocaleEditor());
    this.defaultEditors.put(Pattern.class, new PatternEditor());
    this.defaultEditors.put(Properties.class, new PropertiesEditor());
    this.defaultEditors.put(Resource[].class, new ResourceArrayPropertyEditor());
    this.defaultEditors.put(TimeZone.class, new TimeZoneEditor());
    this.defaultEditors.put(URI.class, new URIEditor());
    this.defaultEditors.put(URL.class, new URLEditor());
    this.defaultEditors.put(UUID.class, new UUIDEditor());

    // Default instances of collection editors.
    // Can be overridden by registering custom instances of those as custom editors.
    this.defaultEditors.put(Collection.class, new CustomCollectionEditor(Collection.class));
    this.defaultEditors.put(Set.class, new CustomCollectionEditor(Set.class));
    this.defaultEditors.put(SortedSet.class, new CustomCollectionEditor(SortedSet.class));
    this.defaultEditors.put(List.class, new CustomCollectionEditor(List.class));
    this.defaultEditors.put(SortedMap.class, new CustomMapEditor(SortedMap.class));

    // Default editors for primitive arrays.
    this.defaultEditors.put(byte[].class, new ByteArrayPropertyEditor());
    this.defaultEditors.put(char[].class, new CharArrayPropertyEditor());

    // The JDK does not contain a default editor for char!
    this.defaultEditors.put(char.class, new CharacterEditor(false));
    this.defaultEditors.put(Character.class, new CharacterEditor(true));

    // Spring's CustomBooleanEditor accepts more flag values than the JDK's default editor.
    this.defaultEditors.put(boolean.class, new CustomBooleanEditor(false));
    this.defaultEditors.put(Boolean.class, new CustomBooleanEditor(true));

    // The JDK does not contain default editors for number wrapper types!
    // Override JDK primitive number editors with our own CustomNumberEditor.
    this.defaultEditors.put(byte.class, new CustomNumberEditor(Byte.class, false));
    this.defaultEditors.put(Byte.class, new CustomNumberEditor(Byte.class, true));
    this.defaultEditors.put(short.class, new CustomNumberEditor(Short.class, false));
```

```
            this.defaultEditors.put(Short.class, new CustomNumberEditor(Short.class, true));
            this.defaultEditors.put(int.class, new CustomNumberEditor(Integer.class, false));
            this.defaultEditors.put(Integer.class, new CustomNumberEditor(Integer.class, true));
            this.defaultEditors.put(long.class, new CustomNumberEditor(Long.class, false));
            this.defaultEditors.put(Long.class, new CustomNumberEditor(Long.class, true));
            this.defaultEditors.put(float.class, new CustomNumberEditor(Float.class, false));
            this.defaultEditors.put(Float.class, new CustomNumberEditor(Float.class, true));
            this.defaultEditors.put(double.class, new CustomNumberEditor(Double.class, false));
            this.defaultEditors.put(Double.class, new CustomNumberEditor(Double.class, true));
            this.defaultEditors.put(BigDecimal.class, new CustomNumberEditor(BigDecimal.class, true));
            this.defaultEditors.put(BigInteger.class, new CustomNumberEditor(BigInteger.class, true));

            // Only register config value editors if explicitly requested.
            if (this.configValueEditorsActive) {
                StringArrayPropertyEditor sae = new StringArrayPropertyEditor();
                this.defaultEditors.put(String[].class, sae);
                this.defaultEditors.put(short[].class, sae);
                this.defaultEditors.put(int[].class, sae);
                this.defaultEditors.put(long[].class, sae);
            }
        }
```

具体的调用方法我们就不去深究了，但是至少通过这个方法我们已经知道了在 Spring 中定义了上面一系列常用的属性编辑器使我们可以方便地进行配置。如果我们定义的 bean 中的某个属性的类型不在上面的常用配置中的话，才需要我们进行个性化属性编辑器的注册。

6.5.3 添加 ApplicationContextAwareProcessor 处理器

了解了属性编辑器的使用后，接下来我们继续通过 AbstractApplicationContext 的 prepareBeanFactory 方法的主线来进行函数跟踪。对于 beanFactory.addBeanPostProcessor(new ApplicationContextAwareProcessor(this))其实主要目的就是注册个 BneaPostProcessor，而真正的逻辑还是在 ApplicationContextAwareProcessor 中。

ApplicationContextAwareProcessor 实现 BeanPostProcessor 接口，我们回顾下之前讲过的内容，在 bean 实例化的时候，也就是 Spring 激活 bean 的 init-method 的前后，会调用 BeanPostProcessor 的 postProcessBeforeInitialization 方法和 postProcessAfterInitialization 方法。同样，对于 ApplicationContextAwareProcessor 我们也关心这两个方法。

对于 postProcessAfterInitialization 方法，在 ApplicationContextAwareProcessor 中并没有做过多逻辑处理。

```
public Object postProcessAfterInitialization(Object bean, String beanName) {
    return bean;
}
```

那么，我们重点看一下 postProcessBeforeInitialization 方法。

```
public Object postProcessBeforeInitialization(final Object bean, String beanName) throws BeansException {
```

6.5 功能扩展

```java
            AccessControlContext acc = null;

            if (System.getSecurityManager() != null &&
                    (bean instanceof EnvironmentAware || bean instanceof EmbeddedValueResolverAware ||
                            bean instanceof ResourceLoaderAware || bean instanceof ApplicationEventPublisherAware ||
                            bean instanceof MessageSourceAware || bean instanceof ApplicationContextAware)) {
                acc = this.applicationContext.getBeanFactory().getAccessControlContext();
            }

            if (acc != null) {
                AccessController.doPrivileged(new PrivilegedAction<Object>() {
                    public Object run() {
                        invokeAwareInterfaces(bean);
                        return null;
                    }
                }, acc);
            }
            else {
                invokeAwareInterfaces(bean);
            }

            return bean;
    }

    private void invokeAwareInterfaces(Object bean) {
            if (bean instanceof Aware) {
                if (bean instanceof EnvironmentAware) {
                    ((EnvironmentAware) bean).setEnvironment(this.applicationContext.getEnvironment());
                }
                if (bean instanceof EmbeddedValueResolverAware) {
                    ((EmbeddedValueResolverAware) bean).setEmbeddedValueResolver(
                            new EmbeddedValueResolver(this.applicationContext. getBeanFactory()));
                }
                if (bean instanceof ResourceLoaderAware) {
                    ((ResourceLoaderAware) bean).setResourceLoader(this.applicationContext);
                }
                if (bean instanceof ApplicationEventPublisherAware) {
                    ((ApplicationEventPublisherAware) bean).setApplicationEventPublisher(this.applicationContext);
                }
                if (bean instanceof MessageSourceAware) {
                    ((MessageSourceAware) bean).setMessageSource(this.applicationContext);
                }
                if (bean instanceof ApplicationContextAware) {
                    ((ApplicationContextAware) bean).setApplicationContext(this.applicationContext);
                }
            }
    }
```

postProcessBeforeInitialization 方法中调用了 invokeAwareInterfaces。从 invokeAwareInterfaces 方法中，我们或许已经或多或少了解了 Spring 的用意，实现这些 Aware 接口的 bean 在被初始化之后，可以取得一些对应的资源。

6.5.4 设置忽略依赖

当 Spring 将 ApplicationContextAwareProcessor 注册后，那么在 invokeAwareInterfaces 方法中间接调用的 Aware 类已经不是普通的 bean 了，如 ResourceLoaderAware、ApplicationEventPublisherAware 等，那么当然需要在 Spring 做 bean 的依赖注入的时候忽略它们。而 ignoreDependencyInterface 的作用正是在此。

```
//设置了几个忽略自动装配的接口
        beanFactory.ignoreDependencyInterface(ResourceLoaderAware.class);
        beanFactory.ignoreDependencyInterface(ApplicationEventPublisherAware.class);
        beanFactory.ignoreDependencyInterface(MessageSourceAware.class);
        beanFactory.ignoreDependencyInterface(ApplicationContextAware.class);
        beanFactory.ignoreDependencyInterface(EnvironmentAware.class);
```

6.5.5 注册依赖

Spring 中有了忽略依赖的功能，当然也必不可少地会有注册依赖的功能。

```
beanFactory.registerResolvableDependency(BeanFactory.class, beanFactory);
        beanFactory.registerResolvableDependency(ResourceLoader.class, this);
        beanFactory.registerResolvableDependency(ApplicationEventPublisher.class, this);
        beanFactory.registerResolvableDependency(ApplicationContext.class, this);
```

当注册了依赖解析后，例如当注册了对 BeanFactory.class 的解析依赖后，当 bean 的属性注入的时候，一旦检测到属性为 BeanFactory 类型便会将 beanFactory 的实例注入进去。

6.6 BeanFactory 的后处理

BeanFacotry 作为 Spring 中容器功能的基础，用于存放所有已经加载的 bean，为了保证程序上的高可扩展性，Spring 针对 BeanFactory 做了大量的扩展，比如我们熟知的 PostProcessor 等都是在这里实现的。

6.6.1 激活注册的 BeanFactoryPostProcessor

正式开始介绍之前我们先了解下 BeanFactoryPostProcessor 的用法。

BeanFactoryPostProcessor 接口跟 BeanPostProcessor 类似，可以对 bean 的定义（配置元数据）进行处理。也就是说，Spring IoC 容器允许 BeanFactoryPostProcessor 在容器实际实例化任何其他的 bean 之前读取配置元数据，并有可能修改它。如果你愿意，你可以配置多个 BeanFactoryPostProcessor。你还能通过设置"order"属性来控制 BeanFactoryPostProcessor 的执行次序（仅当 BeanFactoryPostProcessor 实现了 Ordered 接口时你才可以设置此属性，因此在实现 BeanFactoryPostProcessor 时，就应当考虑实现 Ordered 接口）。请参考 BeanFactoryPostProcessor 和 Ordered 接口的 JavaDoc 以获取更详细的信息。

如果你想改变实际的 bean 实例（例如从配置元数据创建的对象），那么你最好使用 BeanPostProcessor。同样地，BeanFactoryPostProcessor 的作用域范围是容器级的。它只和你所使用的容器有关。如果你在容器中定义一个 BeanFactoryPostProcessor，它仅仅对此容器中的 bean 进行后置处理。BeanFactoryPostProcessor 不会对定义在另一个容器中的 bean 进行后置处理，即使这两个容器都是在同一层次上。在 Spring 中存在对于 BeanFactoryPostProcessor 的典型应用，比如 PropertyPlaceholderConfigurer。

1. BeanFactoryPostProcessor 的典型应用：PropertyPlaceholderConfigurer

有时候，阅读 Spring 的 Bean 描述文件时，你也许会遇到类似如下的一些配置：

```
<bean id="message" class="distConfig.HelloMessage">
    <property name="mes">
       <value>${bean.message}</value>
    </property>
</bean>
```

其中竟然出现了变量引用：${bean.message}。这就是 Spring 的分散配置，可以在另外的配置文件中为 bean.message 指定值。如在 bean.property 配置如下定义：

```
bean.message=Hi,can you find me?
```

当访问名为 message 的 bean 时，mes 属性就会被置为字符串"Hi,can you find me?"，但 Spring 框架是怎么知道存在这样的配置文件呢？这就要靠 PropertyPlaceholderConfigurer 这个类的 bean：

```
<bean id="mesHandler" class="org.Springframework.beans.factory.config. Property Placeholder Configurer">
        <property name="locations">
           <list>
              <value>config/bean.properties</value>
           </list>
        </property>
    </bean>
```

在这个 bean 中指定了配置文件为 config/bean.properties。到这里似乎找到问题的答案了，但是其实还有个问题。这个"mesHandler"只不过是 Spring 框架管理的一个 bean，并没有被别的 bean 或者对象引用，Spring 的 beanFactory 是怎么知道要从这个 bean 中获取配置信息的呢？

查看层级结构可以看出 PropertyPlaceholderConfigurer 这个类间接继承了 BeanFactory-PostProcessor 接口。这是一个很特别的接口，当 Spring 加载任何实现了这个接口的 bean 的配置时，都会在 bean 工厂载入所有 bean 的配置之后执行 postProcessBeanFactory 方法。在 PropertyResourceConfigurer 类中实现了 postProcessBeanFactory 方法，在方法中先后调用了 mergeProperties、convertProperties、processProperties 这 3 个方法，分别得到配置，将得到的配置转换为合适的类型，最后将配置内容告知 BeanFactory。

正是通过实现 BeanFactoryPostProcessor 接口,BeanFactory 会在实例化任何 bean 之前获得配置信息,从而能够正确解析 bean 描述文件中的变量引用。

2. 使用自定义 BeanFactoryPostProcessor

我们以实现一个 BeanFactoryPostProcessor,去除潜在的"流氓"属性值的功能来展示自定义 BeanFactoryPostProcessor 的创建及使用,例如 bean 定义中留下 bollocks 这样的字眼。

配置文件 BeanFactory.xml

```xml
<?xml version="1.0" encoding="UTF-8"?>
<beans xmlns="http://www.Springframework.org/schema/beans"
xmlns:xsi="http://www.w3.org/2001/XMLSchema-instance"
xsi:schemaLocation="http://www.Springframework.org/schema/beans
 http://www.Springframework.org/schema/beans/Spring-beans.xsd
">
<bean id="bfpp" class="com.Spring.ch04.ObscenityRemovingBeanFactoryPostProcessor">
    <property name="obscenties">
        <set>
            <value>bollocks</value>
            <value>winky</value>
            <value>bum</value>
            <value>Microsoft</value>
        </set>
    </property>

</bean>
<bean id="simpleBean" class="com.Spring.ch04.SimplePostProcessor">
    <property name="connectionString" value="bollocks"/>
    <property name="password" value="imaginecup"/>
    <property name="username" value="Microsoft"/>

</bean>

</beans>
```

ObscenityRemovingBeanFactoryPostProcessor.java

```java
public class ObscenityRemovingBeanFactoryPostProcessor implements
        BeanFactoryPostProcessor {
    private Set<String> obscenties;
    public ObscenityRemovingBeanFactoryPostProcessor(){
        this.obscenties=new HashSet<String>();
    }
    public void postProcessBeanFactory(
            ConfigurableListableBeanFactory beanFactory) throws BeansException {
        String[] beanNames=beanFactory.getBeanDefinitionNames();
        for(String beanName:beanNames){
            BeanDefinition bd=beanFactory.getBeanDefinition(beanName);
            StringValueResolver valueResover=new StringValueResolver() {
                public String resolveStringValue(String strVal) {
                    if(isObscene(strVal)) return "*****";
```

6.6 BeanFactory 的后处理

```
                return strVal;
            }
        };
        BeanDefinitionVisitor visitor=new BeanDefinitionVisitor(valueResover);
        visitor.visitBeanDefinition(bd);
    }
}
public boolean isObscene(Object value){
    String potentialObscenity=value.toString().toUpperCase();
    return this.obscenties.contains(potentialObscenity);
}
public void setObscenties(Set<String> obscenties) {
    this.obscenties.clear();
    for(String obscenity:obscenties){
        this.obscenties.add(obscenity.toUpperCase());
    }
}
```
}

执行类：
```
public class PropertyConfigurerDemo {
    public static void main(String[] args) {

        ConfigurableListableBeanFactory bf=new XmlBeanFactory(new ClassPathResource ("/META-INF/BeanFactory.xml"));


        BeanFactoryPostProcessor bfpp=(BeanFactoryPostProcessor)bf.getBean("bfpp");
        bfpp.postProcessBeanFactory(bf);
        System.out.println(bf.getBean("simpleBean"));

    }
}
```

输出结果：
```
SimplePostProcessor{connectionString=*****,username=*****,password=imaginecup
```

通过 ObscenityRemovingBeanFactoryPostProcessor Spring 很好地实现了屏蔽掉 obscenties 定义的不应该展示的属性。

3. 激活 BeanFactoryPostProcessor

了解了 BeanFactoryPostProcessor 的用法后便可以深入研究 BeanFactoryPostProcessor 的调用过程了。

```
protected void invokeBeanFactoryPostProcessors(ConfigurableListableBeanFactory beanFactory) {
        // Invoke BeanDefinitionRegistryPostProcessors first, if any.
        Set<String> processedBeans = new HashSet<String>();
        //对BeanDefinitionRegistry类型的处理
        if (beanFactory instanceof BeanDefinitionRegistry) {
            BeanDefinitionRegistry registry = (BeanDefinitionRegistry) beanFactory;
```

```java
                    List<BeanFactoryPostProcessor> regularPostProcessors = new LinkedList
<BeanFactoryPostProcessor>();
                    /**
                     * BeanDefinitionRegistryPostProcessor
                     */
                    List<BeanDefinitionRegistryPostProcessor> registryPostProcessors = new
LinkedList<BeanDefinitionRegistryPostProcessor>();
                    /*
                     * 硬编码注册的后处理器
                     */
                    for (BeanFactoryPostProcessor postProcessor : getBeanFactoryPostProcessors()) {
                        if (postProcessor instanceof BeanDefinitionRegistryPostProcessor) {
                            BeanDefinitionRegistryPostProcessor registryPostProcessor =(Bean
DefinitionRegistryPostProcessor) postProcessor;
                            //对于BeanDefinitionRegistryPostProcessor类型,在BeanFactoryPostProcessor的
基础上还有自己定义的方法,需要先调用
                            registryPostProcessor.postProcessBeanDefinitionRegistry(registry);
                            registryPostProcessors.add(registryPostProcessor);
                        }else {
                            //记录常规BeanFactoryPostProcessor
                            regularPostProcessors.add(postProcessor);
                        }
                    }
                    /*
                     * 配置注册的后处理器
                     */
                    Map<String, BeanDefinitionRegistryPostProcessor> beanMap = beanFactory.
getBeansOfType(BeanDefinitionRegistryPostProcessor.class, true, false);
                    List<BeanDefinitionRegistryPostProcessor> registryPostProcessorBeans =
new ArrayList<BeanDefinitionRegistryPostProcessor>(beanMap.values());
                    OrderComparator.sort(registryPostProcessorBeans);
                    for (BeanDefinitionRegistryPostProcessor postProcessor : registryPostProcessorBeans) {
                        //BeanDefinitionRegistryPostProcessor的特殊处理
                        postProcessor.postProcessBeanDefinitionRegistry(registry);
                    }

                    //激活postProcessBeanFactory方法,之前激活的是postProcessBeanDefinitionRegistry
                    //硬编码设置的BeanDefinitionRegistryPostProcessor
                    invokeBeanFactoryPostProcessors(registryPostProcessors, beanFactory);
                    //配置的BeanDefinitionRegistryPostProcessor
                    invokeBeanFactoryPostProcessors(registryPostProcessorBeans, beanFactory);
                    //常规BeanFactoryPostProcessor
                    invokeBeanFactoryPostProcessors(regularPostProcessors, beanFactory);
                    processedBeans.addAll(beanMap.keySet());
                }
                else {
                    // Invoke factory processors registered with the context instance.
                    invokeBeanFactoryPostProcessors(getBeanFactoryPostProcessors(), beanFactory);
                }

                //对于配置中读取的BeanFactoryPostProcessor的处理
                String[] postProcessorNames = beanFactory.getBeanNamesForType(BeanFactoryPost
Processor.class, true, false);

                List<BeanFactoryPostProcessor> priorityOrderedPostProcessors = new ArrayList
<BeanFactoryPostProcessor>();
```

```java
            List<String> orderedPostProcessorNames = new ArrayList<String>();
            List<String> nonOrderedPostProcessorNames = new ArrayList<String>();
            //对后处理器进行分类
            for (String ppName : postProcessorNames) {
                if (processedBeans.contains(ppName)) {
                    //已经处理过
                }else if (isTypeMatch(ppName, PriorityOrdered.class)) {
                    priorityOrderedPostProcessors.add(beanFactory.getBean(ppName, BeanFactoryPostProcessor.class));
                }else if (isTypeMatch(ppName, Ordered.class)) {
                    orderedPostProcessorNames.add(ppName);
                }else {
                    nonOrderedPostProcessorNames.add(ppName);
                }
            }

            //按照优先级进行排序
            OrderComparator.sort(priorityOrderedPostProcessors);
            invokeBeanFactoryPostProcessors(priorityOrderedPostProcessors, beanFactory);

            // Next, invoke the BeanFactoryPostProcessors that implement Ordered.
            List<BeanFactoryPostProcessor> orderedPostProcessors = new ArrayList<BeanFactoryPostProcessor>();
            for (String postProcessorName : orderedPostProcessorNames) {
                orderedPostProcessors.add(getBean(postProcessorName, BeanFactoryPostProcessor.class));
            }
            //按照 order 排序
            OrderComparator.sort(orderedPostProcessors);
            invokeBeanFactoryPostProcessors(orderedPostProcessors, beanFactory);


            //无序，直接调用
            List<BeanFactoryPostProcessor> nonOrderedPostProcessors = new ArrayList<BeanFactoryPostProcessor>();
            for (String postProcessorName : nonOrderedPostProcessorNames) {
                nonOrderedPostProcessors.add(getBean(postProcessorName, BeanFactoryPostProcessor.class));
            }
            invokeBeanFactoryPostProcessors(nonOrderedPostProcessors, beanFactory);
    }
```

从上面的方法中我们看到，对于 BeanFactoryPostProcessor 的处理主要分两种情况进行，一个是对于 BeanDefinitionRegistry 类的特殊处理，另一种是对普通的 BeanFactoryPostProcessor 进行处理。而对于每种情况都需要考虑硬编码注入注册的后处理器以及通过配置注入的后处理器。

对于 BeanDefinitionRegistry 类型的处理类的处理主要包括以下内容。

1. 对于硬编码注册的后处理器的处理，主要是通过 AbstractApplicationContext 中的添加处理器方法 addBeanFactoryPostProcessor 进行添加。

```java
public void addBeanFactoryPostProcessor(BeanFactoryPostProcessor beanFactoryPostProcessor) {
        this.beanFactoryPostProcessors.add(beanFactoryPostProcessor);
}
```

添加后的后处理器会存放在 beanFactoryPostProcessors 中，而在处理 BeanFactoryPostProcessor 时候会首先检测 beanFactoryPostProcessors 是否有数据。当然，BeanDefinitionRegistryPostProcessor 继承自 BeanFactoryPostProcessor，不但有 BeanFactoryPostProcessor 的特性，同时还有自己定义的个性化方法，也需要在此调用。所以，这里需要从 beanFactoryPostProcessors 中挑出 BeanDefinitionRegistryPostProcessor 的后处理器，并进行其 postProcessBeanDefinitionRegistry 方法的激活。

2．记录后处理器主要使用了 3 个 List 完成。

- registryPostProcessors：记录通过硬编码方式注册的 BeanDefinitionRegistryPostProcessor 类型的处理器。
- regularPostProcessors：记录通过硬编码方式注册的 BeanFactoryPostProcessor 类型的处理器。
- registryPostProcessorBeans：记录通过配置方式注册的 BeanDefinitionRegistryPostProcessor 类型的处理器。

3．对以上所记录的 List 中的后处理器进行统一调用 BeanFactoryPostProcessor 的 postProcessBeanFactory 方法。

4．对 beanFactoryPostProcessors 中非 BeanDefinitionRegistryPostProcessor 类型的后处理器进行统一的 BeanFactoryPostProcessor 的 postProcessBeanFactory 方法调用。

5．普通 beanFactory 处理。

BeanDefinitionRegistryPostProcessor 只对 BeanDefinitionRegistry 类型的 ConfigurableListableBeanFactory 有效，所以如果判断所示的 beanFactory 并不是 BeanDefinitionRegistry，那么便可以忽略 BeanDefinitionRegistryPostProcessor，而直接处理 BeanFactoryPostProcessor，当然获取的方式与上面的获取类似。

这里需要提到的是，对于硬编码方式手动添加的后处理器是不需要做任何排序的，但是在配置文件中读取的处理器，Sping 并不保证读取的顺序。所以，为了保证用户的调用顺序的要求，Spring 对于后处理器的调用支持按照 PriorityOrdered 或者 Ordered 的顺序调用。

6.6.2 注册 BeanPostProcessor

上文中提到了 BeanFacotoryPostProcessors 的调用，现在我们来探索下 BeanPostProcessor，但是这里并不是调用，而是注册。真正的调用其实是在 bean 的实例化阶段进行的。这是一个很重要的步骤，也是很多功能 BeanFactory 不支持的重要原因。Spring 中大部分功能都是通过后处理器的方式进行扩展的，这是 Spring 框架的一个特性，但是在 BeanFactory 中其实并没有实现后处理器的自动注册，所以在调用的时候如果没有进行手动注册其实是不能使用的。但是在 ApplicationContext 中却添加了自动注册功能，如自定义这样一个后处理器：

```
public class MyInstantiationAwareBeanPostProcessor implements  InstantiationAwareBean
PostProcessor{
```

6.6 BeanFactory 的后处理

```java
public Object postProcessBeforeInitialization(Object bean, String beanName)
    throws BeansException {
    System.out.println("====");
    return null;
}
    ... ...
}
```

在配置文件中添加配置：

```xml
<bean class="processors.MyInstantiationAwareBeanPostProcessor"/>
```

那么使用 BeanFactory 方式进行 Spring 的 bean 的加载时是不会有任何改变的，但是使用 ApplicationContext 方式获取 bean 的时候会在获取每个 bean 时打印出"===="，而这个特性就是在 registerBeanPostProcessors 方法中完成的。

我们继续探索 registerBeanPostProcessors 的方法实现。

```java
protected void registerBeanPostProcessors(ConfigurableListableBeanFactory beanFactory) {
    String[] postProcessorNames = beanFactory.getBeanNamesForType(BeanPostProcessor.class, true, false);

    /*
     * BeanPostProcessorChecker 是一个普通的信息打印，可能会有些情况，
     * 当 Spring 的配置中的后处理器还没有被注册就已经开始了 bean 的初始化时
     * 便会打印出 BeanPostProcessorChecker 中设定的信息
     */
    int beanProcessorTargetCount = beanFactory.getBeanPostProcessorCount() + 1 + postProcessorNames.length;
    beanFactory.addBeanPostProcessor(new BeanPostProcessorChecker(beanFactory, beanProcessorTargetCount));

    //使用 PriorityOrdered 保证顺序
    List<BeanPostProcessor> priorityOrderedPostProcessors = new ArrayList<BeanPostProcessor>();
    //MergedBeanDefinitionPostProcessor
    List<BeanPostProcessor> internalPostProcessors = new ArrayList<BeanPostProcessor>();
    //使用 Ordered 保证顺序
    List<String> orderedPostProcessorNames = new ArrayList<String>();
    //无序 BeanPostProcessor
    List<String> nonOrderedPostProcessorNames = new ArrayList<String>();

    for (String ppName : postProcessorNames) {
        if (isTypeMatch(ppName, PriorityOrdered.class)) {
            BeanPostProcessor pp = beanFactory.getBean(ppName, BeanPostProcessor.class);
            priorityOrderedPostProcessors.add(pp);
            if (pp instanceof MergedBeanDefinitionPostProcessor) {
                internalPostProcessors.add(pp);
            }
        }else if (isTypeMatch(ppName, Ordered.class)) {
            orderedPostProcessorNames.add(ppName);
        }else {
            nonOrderedPostProcessorNames.add(ppName);
```

```
            }
        }

        //第 1 步，注册所有实现 PriorityOrdered 的 BeanPostProcessor
        OrderComparator.sort(priorityOrderedPostProcessors);
        registerBeanPostProcessors(beanFactory, priorityOrderedPostProcessors);

        //第 2 步，注册所有实现 Ordered 的 BeanPostProcessor
        List<BeanPostProcessor> orderedPostProcessors = new ArrayList<BeanPostProcessor>();
        for (String ppName : orderedPostProcessorNames) {
            BeanPostProcessor pp = beanFactory.getBean(ppName, BeanPostProcessor.class);
            orderedPostProcessors.add(pp);
            if (pp instanceof MergedBeanDefinitionPostProcessor) {
                internalPostProcessors.add(pp);
            }
        }
        OrderComparator.sort(orderedPostProcessors);
        registerBeanPostProcessors(beanFactory, orderedPostProcessors);

        //第 3 步，注册所有无序的 BeanPostProcessor
        List<BeanPostProcessor> nonOrderedPostProcessors = new ArrayList<BeanPostProcessor>();
        for (String ppName : nonOrderedPostProcessorNames) {
            BeanPostProcessor pp = beanFactory.getBean(ppName, BeanPostProcessor.class);
            nonOrderedPostProcessors.add(pp);
            if (pp instanceof MergedBeanDefinitionPostProcessor) {
                internalPostProcessors.add(pp);
            }
        }
        registerBeanPostProcessors(beanFactory, nonOrderedPostProcessors);

        //第 4 步，注册所有 MergedBeanDefinitionPostProcessor 类型的 BeanPostProcessor，并非
        //重复注册，
        //在 beanFactory.addBeanPostProcessor 中会先移除已经存在的 BeanPostProcessor
        OrderComparator.sort(internalPostProcessors);
        registerBeanPostProcessors(beanFactory, internalPostProcessors);

        //添加 ApplicationListener 探测器
        beanFactory.addBeanPostProcessor(new ApplicationListenerDetector());
}
```

配合源码以及注释，在 registerBeanPostProcessors 方法中所做的逻辑相信大家已经很清楚了，我们再做一下总结。

首先我们会发现，对于 BeanPostProcessor 的处理与 BeanFactoryPostProcessor 的处理极为相似，但是似乎又有些不一样的地方。经过反复的对比发现，对于 BeanFactoryPostProcessor 的处理要区分两种情况，一种方式是通过硬编码方式的处理，另一种是通过配置文件方式的处理。那么为什么在 BeanPostProcessor 的处理中只考虑了配置文件的方式而不考虑硬编码的方式呢？提出这个问题，还是因为读者没有完全理解两者实现的功能。对于 BeanFactoryPostProcessor 的处理，不但要实现注册功能，而且还要实现对后处理器的激活操作，所以需要载入配置中的定义，并进行激活；而对于 BeanPostProcessor 并不需要马上调用，再说，硬编码的方式实现的

功能是将后处理器提取并调用，这里并不需要调用，当然不需要考虑硬编码的方式了，这里的功能只需要将配置文件的 BeanPostProcessor 提取出来并注册进入 beanFactory 就可以了。

对于 beanFactory 的注册，也不是直接注册就可以的。在 Spring 中支持对于 BeanPostProcessor 的排序，比如根据 PriorityOrdered 进行排序、根据 Ordered 进行排序或者无序，而 Spring 在 BeanPostProcessor 的激活顺序的时候也会考虑对于顺序的问题而先进行排序。

这里可能有个地方读者不是很理解，对于 internalPostProcessors 中存储的后处理器也就是 MergedBeanDefinitionPostProcessor 类型的处理器，在代码中似乎是被重复调用了，如：

```
for (String ppName : postProcessorNames) {
            if (isTypeMatch(ppName, PriorityOrdered.class)) {
                BeanPostProcessor pp=beanFactory.getBean(ppName, BeanPostProcessor.class);
                priorityOrderedPostProcessors.add(pp);
                if (pp instanceof MergedBeanDefinitionPostProcessor) {
                    internalPostProcessors.add(pp);
                }
            }else if (isTypeMatch(ppName, Ordered.class)) {
                orderedPostProcessorNames.add(ppName);
            }else {
                nonOrderedPostProcessorNames.add(ppName);
            }
}
```

其实不是，我们可以看看对于 registerBeanPostProcessors 方法的实现方式。

```
private void registerBeanPostProcessors(
            ConfigurableListableBeanFactory beanFactory, List<BeanPostProcessor> postProcessors) {
        for (BeanPostProcessor postProcessor : postProcessors) {
            beanFactory.addBeanPostProcessor(postProcessor);
        }
}
    public void addBeanPostProcessor(BeanPostProcessor beanPostProcessor) {
        Assert.notNull(beanPostProcessor, "BeanPostProcessor must not be null");
        this.beanPostProcessors.remove(beanPostProcessor);
        this.beanPostProcessors.add(beanPostProcessor);
        if (beanPostProcessor instanceof InstantiationAwareBeanPostProcessor) {
            this.hasInstantiationAwareBeanPostProcessors = true;
        }
        if (beanPostProcessor instanceof DestructionAwareBeanPostProcessor) {
            this.hasDestructionAwareBeanPostProcessors = true;
        }
}
```

可以看到，在 registerBeanPostProcessors 方法的实现中其实已经确保了 beanPostProcessor 的唯一性，个人猜想，之所以选择在 registerBeanPostProcessors 中没有进行重复移除操作或许是为了保持分类的效果，使逻辑更为清晰吧。

6.6.3　初始化消息资源

在进行这段函数的解析之前，我们同样先来回顾 Spring 国际化的使用方法。

假设我们正在开发一个支持多国语言的 Web 应用程序，要求系统能够根据客户端的系统的语言类型返回对应的界面：英文的操作系统返回英文界面，而中文的操作系统则返回中文界面——这便是典型的 i18n 国际化问题。对于有国际化要求的应用系统，我们不能简单地采用硬编码的方式编写用户界面信息、报错信息等内容，而必须为这些需要国际化的信息进行特殊处理。简单来说，就是为每种语言提供一套相应的资源文件，并以规范化命名的方式保存在特定的目录中，由系统自动根据客户端语言选择适合的资源文件。

"国际化信息"也称为"本地化信息"，一般需要两个条件才可以确定一个特定类型的本地化信息，它们分别是"语言类型"和"国家/地区的类型"。如中文本地化信息既有中国大陆地区的中文，又有中国台湾地区、中国香港地区的中文，还有新加坡地区的中文。Java 通过 java.util.Locale 类表示一个本地化对象，它允许通过语言参数和国家/地区参数创建一个确定的本地化对象。

java.util.Locale 是表示语言和国家/地区信息的本地化类，它是创建国际化应用的基础。下面给出几个创建本地化对象的示例：

```
//① 带有语言和国家/地区信息的本地化对象
Locale locale1 = new Locale("zh","CN");

//② 只有语言信息的本地化对象
Locale locale2 = new Locale("zh");

//③ 等同于 Locale("zh","CN")
Locale locale3 = Locale.CHINA;

//④ 等同于 Locale("zh")
Locale locale4 = Locale.CHINESE;

//⑤ 获取本地系统默认的本地化对象
Locale locale 5= Locale.getDefault();
```

JDK 的 java.util 包中提供了几个支持本地化的格式化操作工具类：NumberFormat、DateFormat、MessageFormat，而在 Spring 中的国际化资源操作也无非是对于这些类的封装操作，我们仅仅介绍下 MessageFormat 的用法以帮助大家回顾：

```
//①信息格式化串
String pattern1 = "{0}，你好! 你于{1}在工商银行存入{2} 元。";
String pattern2 = "At {1,time,short} On{1,date,long} {0} paid {2,number, currency}.";

//②用于动态替换占位符的参数
Object[] params = {"John", new GregorianCalendar().getTime(),1.0E3};

//③使用默认本地化对象格式化信息
String msg1 = MessageFormat.format(pattern1,params);

//④使用指定的本地化对象格式化信息
MessageFormat mf = new MessageFormat(pattern2,Locale.US);
String msg2 = mf.format(params);
System.out.println(msg1);
System.out.println(msg2);
```

Spring 定义了访问国际化信息的 MessageSource 接口，并提供了几个易用的实现类。MessageSource 分别被 HierarchicalMessageSource 和 ApplicationContext 接口扩展，这里我们主

要看一下 HierarchicalMessageSource 接口的几个实现类，如图 6-3 所示。

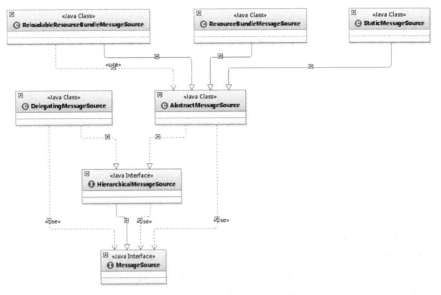

图 6-3　MessageSource 类图结构

HierarchicalMessageSource 接口最重要的两个实现类是 ResourceBundleMessageSource 和 ReloadableResourceBundleMessageSource。它们基于 Java 的 ResourceBundle 基础类实现，允许仅通过资源名加载国际化资源。ReloadableResourceBundleMessageSource 提供了定时刷新功能，允许在不重启系统的情况下，更新资源的信息。StaticMessageSource 主要用于程序测试，它允许通过编程的方式提供国际化信息。而 DelegatingMessageSource 是为方便操作父 MessageSource 而提供的代理类。仅仅举例 ResourceBundleMessageSource 的实现方式。

1．定义资源文件

- messages.properties（默认：英文），内容仅一句，如下：
  ```
  test=test
  ```
- messages_zh_CN.properties（简体中文）：
  ```
  test=测试
  ```

然后 cmd，打开命令行窗口，输入 native2ascii -encoding gbk C:\messages_zh_CN.properties C:\messages_zh_CN_tem.properties，并将 C:\messages_zh_CN_tem.properties 中的内容替换到 messages_zh_CN.properties 中，这样 messages_zh_CN.properties 文件就存放的是转码后的内容了，比较简单。

2．定义配置文件

```
<bean id="messageSource"    class="org.Springframework.context.support.ResourceBundleMessageSource">
    <property name="basenames">
```

```
        <list>
          <value>test/messages</value>
        </list>
      </property>
</bean>
```

其中,这个 Bean 的 ID 必须命名为 messageSource,否则会抛出 NoSuchMessageException 异常。

3. 使用 ApplicationContext 访问国际化信息

```
String[] configs = {"applicationContext.xml"};
ApplicationContext ctx = new ClassPathXmlApplicationContext(configs);
//①直接通过容器访问国际化信息
Object[] params = {"John", new GregorianCalendar().getTime()};

String str1 = ctx.getMessage("test",params,Locale.US);
String str2 = ctx.getMessage("test",params,Locale.CHINA);
System.out.println(str1);
System.out.println(str2);
```

了解了 Spring 国际化的使用后便可以进行源码的分析了。

在 initMessageSource 中的方法主要功能是提取配置中定义的 messageSource,并将其记录在 Spring 的容器中,也就是 AbstractApplicationContext 中。当然,如果用户未设置资源文件的话,Spring 中也提供了默认的配置 DelegatingMessageSource。

在 initMessageSource 中获取自定义资源文件的方式为 beanFactory.getBean(MESSAGE_SOURCE_BEAN_NAME, MessageSource.class),在这里 Spring 使用了硬编码的方式硬性规定了子定义资源文件必须为 message,否则便会获取不到自定义资源配置,这也是为什么之前提到 Bean 的 id 如果部位 message 会抛出异常。

```
protected void initMessageSource() {
            ConfigurableListableBeanFactory beanFactory = getBeanFactory();
            if (beanFactory.containsLocalBean(MESSAGE_SOURCE_BEAN_NAME)) {
                //如果在配置中已经配置了 messageSource,那么将 messageSource 提取并记录在
                //this.messageSource 中
                this.messageSource = beanFactory.getBean(MESSAGE_SOURCE_BEAN_NAME, MessageSource.class);
                // Make MessageSource aware of parent MessageSource.
                if (this.parent != null && this.messageSource instanceof HierarchicalMessageSource) {
                    HierarchicalMessageSource hms = (HierarchicalMessageSource) this.messageSource;
                    if (hms.getParentMessageSource() == null) {
                        hms.setParentMessageSource(getInternalParentMessageSource());
                    }
                }
                if (logger.isDebugEnabled()) {
                    logger.debug("Using MessageSource [" + this.messageSource + "]");
                }
            }else {
                //如果用户并没有定义配置文件,那么使用临时的 DelegatingMessageSource 以便于作为调用
                //getMessage 方法的返回
                DelegatingMessageSource dms = new DelegatingMessageSource();
```

```
            dms.setParentMessageSource(getInternalParentMessageSource());
            this.messageSource = dms;
            beanFactory.registerSingleton(MESSAGE_SOURCE_BEAN_NAME, this.messageSource);
            if (logger.isDebugEnabled()) {
                logger.debug("Unable to locate MessageSource with name '" + MESSAGE_SOURCE_BEAN_NAME +
                    "': using default [" + this.messageSource + "]");
            }
        }
    }
```

通过读取并将自定义资源文件配置记录在容器中，那么就可以在获取资源文件的时候直接使用了，例如，在 AbstractApplicationContext 中的获取资源文件属性的方法：

```
public String getMessage(String code, Object args[], Locale locale) throws NoSuchMessage Exception {
            return getMessageSource().getMessage(code, args, locale);
}
```

其中的 getMessageSource() 方法正是获取了之前定义的自定义资源配置。

6.6.4 初始化 ApplicationEventMulticaster

在讲解 Spring 的时间传播器之前，我们还是先来看一下 Spring 的事件监听的简单用法。

1. 定义监听事件

```
 public class TestEvent extends ApplicationEvent {

public String msg;

public TestEvent (Object source) {
    super(source);
}

public TestEvent (Object source, String msg) {
    super(source);
    this.msg = msg;
}

public void print(){
    System.out.println(msg);
}

}
```

2. 定义监听器

```
 public class TestListener implements ApplicationListener {

public void onApplicationEvent(ApplicationEvent  event) {
    if(event instanceof TestEvent){
        TestEvent testEvent = (TestEvent)event;
        testEvent .print();
```

        }
    }
}

3. 添加配置文件

```
<bean id="testListener" class="com.test.event.TestListener "/>
```

4. 测试

```
public class Test {
    public static void main(String[] args) {
        ApplicationContext context = new ClassPathXmlApplicationContext ("classpath:applicationContext.xml");

        TestEvent event = new TestEvent ("hello","msg");
        context.publishEvent(event);
    }
}
```

当程序运行时，Spring 会将发出的 TestEvent 事件转给我们自定义的 TestListener 进行进一步处理。

或许很多人一下子会反映出设计模式中的观察者模式，这确实是个典型的应用，可以在比较关心的事件结束后及时处理。那么我们看看 ApplicationEventMulticaster 是如何被初始化的，以确保功能的正确运行。

initApplicationEventMulticaster 的方式比较简单，无非考虑两种情况。

- 如果用户自定义了事件广播器，那么使用用户自定义的事件广播器。
- 如果用户没有自定义事件广播器，那么使用默认的 ApplicationEventMulticaster。

```
protected void initApplicationEventMulticaster() {
    ConfigurableListableBeanFactory beanFactory = getBeanFactory();
    if (beanFactory.containsLocalBean(APPLICATION_EVENT_MULTICASTER_BEAN_NAME)) {
        this.applicationEventMulticaster =
                beanFactory.getBean(APPLICATION_EVENT_MULTICASTER_BEAN_NAME, ApplicationEventMulticaster.class);
        if (logger.isDebugEnabled()) {
            logger.debug("Using ApplicationEventMulticaster [" + this.applicationEventMulticaster + "]");
        }
    }else {
        this.applicationEventMulticaster = new **SimpleApplicationEventMulticaster**(beanFactory);
        beanFactory.registerSingleton(APPLICATION_EVENT_MULTICASTER_BEAN_NAME, this.applicationEventMulticaster);
        if (logger.isDebugEnabled()) {
            logger.debug("Unable to locate ApplicationEventMulticaster with name '" +
                    APPLICATION_EVENT_MULTICASTER_BEAN_NAME +
                    "': using default [" + this.applicationEventMulticaster + "]");
        }
    }
}
```

按照之前介绍的顺序及逻辑，我们推断，作为广播器，一定是用于存放监听器并在合适的时候调用监听器，那么我们不妨进入默认的广播器实现 SimpleApplicationEventMulticaster 来一探究竟。

其中的一段代码是我们感兴趣的。

```java
public void multicastEvent(final ApplicationEvent event) {
    for (final ApplicationListener listener : getApplicationListeners(event)) {
        Executor executor = getTaskExecutor();
        if (executor != null) {
            executor.execute(new Runnable() {
                @SuppressWarnings("unchecked")
                public void run() {
                    listener.onApplicationEvent(event);
                }
            });
        }
        else {
            listener.onApplicationEvent(event);
        }
    }
}
```

可以推断，当产生 Spring 事件的时候会默认使用 SimpleApplicationEventMulticaster 的 multicastEvent 来广播事件，遍历所有监听器，并使用监听器中的 onApplicationEvent 方法来进行监听器的处理。而对于每个监听器来说其实都可以获取到产生的事件，但是是否进行处理则由事件监听器来决定。

6.6.5 注册监听器

之前在介绍 Spring 的广播器时反复提到了事件监听器，那么在 Spring 注册监听器的时候又做了哪些逻辑操作呢？

```java
protected void registerListeners() {
            //硬编码方式注册的监听器处理
            for (ApplicationListener<?> listener : getApplicationListeners()) {
                getApplicationEventMulticaster().addApplicationListener(listener);
            }
            //配置文件注册的监听器处理
            String[] listenerBeanNames = getBeanNamesForType(ApplicationListener.class, true, false);
            for (String lisName : listenerBeanNames) {
                getApplicationEventMulticaster().addApplicationListenerBean(lisName);
            }
    }
```

6.7 初始化非延迟加载单例

完成 BeanFactory 的初始化工作，其中包括 ConversionService 的设置、配置冻结以及非延迟加载的 bean 的初始化工作。

```
protected void finishBeanFactoryInitialization(ConfigurableListableBeanFactory beanFactory) {
        // Initialize conversion service for this context.
        if (beanFactory.containsBean(CONVERSION_SERVICE_BEAN_NAME) &&
                beanFactory.isTypeMatch(CONVERSION_SERVICE_BEAN_NAME, ConversionService.class)) {
            beanFactory.setConversionService(
                    beanFactory.getBean(CONVERSION_SERVICE_BEAN_NAME, ConversionService.class));
        }

        // Initialize LoadTimeWeaverAware beans early to allow for registering their
        //transformers early.
        String[] weaverAwareNames = beanFactory.getBeanNamesForType (LoadTimeWeaverAware.class, false, false);
        for (String weaverAwareName : weaverAwareNames) {
            getBean(weaverAwareName);
        }

        // Stop using the temporary ClassLoader for type matching.
        beanFactory.setTempClassLoader(null);

        //冻结所有的 bean 定义，说明注册的 bean 定义将不被修改或任何进一步的处理
        beanFactory.freezeConfiguration();

        // Instantiate all remaining (non-lazy-init) singletons.
        //初始化剩下的单实例（非惰性的）
        beanFactory.preInstantiateSingletons();
}
```

首先我们来了解一下 ConversionService 类所提供的作用。

1. ConversionService 的设置

之前我们提到过使用自定义类型转换器从 String 转换为 Date 的方式，那么，在 Spring 中还提供了另一种转换方式：使用 Converter。同样，我们使用一个简单的示例来了解下 Converter 的使用方式。

1. 定义转换器。

```
public class String2DateConverter implements Converter<String, Date> {

    @Override
    public Date convert(String arg0) {
        try {
            return DateUtils.parseDate(arg0,
                    new String[] { "yyyy-MM-dd HH:mm:ss" });
        } catch (ParseException e) {
            return null;
        }
    }
}
```

2. 注册。
```xml
<bean id="conversionService"
    class="org.Springframework.context.support.ConversionServiceFactoryBean">
    <property name="converters">
        <list>
            <bean class="String2DateConverter" />
        </list>
    </property>
</bean>
```
3. 测试。

这样便可以使用 Converter 为我们提供的功能了，下面我们通过一个简便的方法来对此直接测试。
```java
public void testStringToPhoneNumberConvert() {
    DefaultConversionService conversionService = new DefaultConversionService();
    conversionService.addConverter(new StringToPhoneNumberConverter());

    String phoneNumberStr = "010-12345678";
    PhoneNumberModel phoneNumber = conversionService.convert(phoneNumberStr,
PhoneNumber Model.class);

    Assert.assertEquals("010", phoneNumber.getAreaCode());
}
```
通过以上的功能我们看到了 Converter 以及 ConversionService 提供的便利功能，其中的配置就是在当前函数中被初始化的。

2. 冻结配置

冻结所有的 bean 定义，说明注册的 bean 定义将不被修改或进行任何进一步的处理。
```java
public void freezeConfiguration() {
        this.configurationFrozen = true;
        synchronized (this.beanDefinitionMap) {
            this.frozenBeanDefinitionNames = StringUtils.toStringArray(this.beanDefinitionNames);
        }
}
```

3. 初始化非延迟加载

ApplicationContext 实现的默认行为就是在启动时将所有单例 bean 提前进行实例化。提前实例化意味着作为初始化过程的一部分，ApplicationContext 实例会创建并配置所有的单例 bean。通常情况下这是一件好事，因为这样在配置中的任何错误就会即刻被发现（否则的话可能要花几个小时甚至几天）。而这个实例化的过程就是在 finishBeanFactoryInitialization 中完成的。
```java
public void preInstantiateSingletons() throws BeansException {
        if (this.logger.isInfoEnabled()) {
            this.logger.info("Pre-instantiating singletons in " + this);
        }
```

```
        List<String> beanNames;
        synchronized (this.beanDefinitionMap) {
            // Iterate over a copy to allow for init methods which in turn register
            //new bean definitions.
            // While this may not be part of the regular factory bootstrap, it does
            //otherwise work fine.
            beanNames = new ArrayList<String>(this.beanDefinitionNames);
        }
        for (String beanName : beanNames) {
            RootBeanDefinition bd = getMergedLocalBeanDefinition(beanName);
            if (!bd.isAbstract() && bd.isSingleton() && !bd.isLazyInit()) {
                if (isFactoryBean(beanName)) {
                    final FactoryBean<?> factory = (FactoryBean<?>) getBean(FACTORY_
BEAN_PREFIX + beanName);
                    boolean isEagerInit;
                    if (System.getSecurityManager() != null && factory instanceof
SmartFactoryBean) {
                        isEagerInit = AccessController.doPrivileged(new PrivilegedAction <
Boolean>() {
                            public Boolean run() {
                                return ((SmartFactoryBean<?>) factory).isEagerInit();
                            }
                        }, getAccessControlContext());
                    }
                    else {
                        isEagerInit = (factory instanceof SmartFactoryBean &&
                                ((SmartFactoryBean<?>) factory).isEagerInit());
                    }
                    if (isEagerInit) {
                        getBean(beanName);
                    }
                }
                else {
                    getBean(beanName);
                }
            }
        }
    }
```

6.8 finishRefresh

在 Spring 中还提供了 Lifecycle 接口，Lifecycle 中包含 start/stop 方法，实现此接口后 Spring 会保证在启动的时候调用其 start 方法开始生命周期，并在 Spring 关闭的时候调用 stop 方法来结束生命周期，通常用来配置后台程序，在启动后一直运行（如对 MQ 进行轮询等）。而 ApplicationContext 的初始化最后正是保证了这一功能的实现。

```
protected void finishRefresh() {

        initLifecycleProcessor();

        // Propagate refresh to lifecycle processor first.
```

```
            getLifecycleProcessor().onRefresh();

            // Publish the final event.
            publishEvent(new ContextRefreshedEvent(this));
    }
```

1. initLifecycleProcessor

当 ApplicationContext 启动或停止时，它会通过 LifecycleProcessor 来与所有声明的 bean 的周期做状态更新，而在 LifecycleProcessor 的使用前首先需要初始化。

```
    protected void initLifecycleProcessor() {
            ConfigurableListableBeanFactory beanFactory = getBeanFactory();
            if (beanFactory.containsLocalBean(LIFECYCLE_PROCESSOR_BEAN_NAME)) {
                this.lifecycleProcessor =
                        beanFactory.getBean(LIFECYCLE_PROCESSOR_BEAN_NAME, Lifecycle-
Processor.class);
                if (logger.isDebugEnabled()) {
                    logger.debug("Using LifecycleProcessor [" + this.lifecycleProcessor + "]");
                }
            }else {
                DefaultLifecycleProcessor defaultProcessor = new DefaultLifecycleProcessor();
                defaultProcessor.setBeanFactory(beanFactory);
                this.lifecycleProcessor = defaultProcessor;
                beanFactory.registerSingleton(LIFECYCLE_PROCESSOR_BEAN_NAME, this.
lifecycleProcessor);
                if (logger.isDebugEnabled()) {
                    logger.debug("Unable to locate LifecycleProcessor with name '" +
                            LIFECYCLE_PROCESSOR_BEAN_NAME +
                            "': using default [" + this.lifecycleProcessor + "]");
                }
            }
    }
```

2. onRefresh

启动所有实现了 Lifecycle 接口的 bean。

```
    public void onRefresh() {
            startBeans(true);
            this.running = true;
    }

    private void startBeans(boolean autoStartupOnly) {
            Map<String, Lifecycle> lifecycleBeans = getLifecycleBeans();
            Map<Integer, LifecycleGroup> phases = new HashMap<Integer, LifecycleGroup>();
            for (Map.Entry<String, ? extends Lifecycle> entry : lifecycleBeans.entrySet()) {
                Lifecycle bean = entry.getValue();
                if (!autoStartupOnly || (bean instanceof SmartLifecycle && ((SmartLifecycle)
bean).isAutoStartup())) {
                    int phase = getPhase(bean);
                    LifecycleGroup group = phases.get(phase);
                    if (group == null) {
```

```
                    group = new LifecycleGroup(phase, this.timeoutPerShutdownPhase,
lifecycleBeans, autoStartupOnly);
                    phases.put(phase, group);
                }
                group.add(entry.getKey(), bean);
            }
        }
        if (phases.size() > 0) {
            List<Integer> keys = new ArrayList<Integer>(phases.keySet());
            Collections.sort(keys);
            for (Integer key : keys) {
                phases.get(key).start();
            }
        }
    }
}
```

3. publishEvent

当完成 ApplicationContext 初始化的时候,要通过 Spring 中的事件发布机制来发出 Context-RefreshedEvent 事件,以保证对应的监听器可以做进一步的逻辑处理。

```
public void publishEvent(ApplicationEvent event) {
        Assert.notNull(event, "Event must not be null");
        if (logger.isTraceEnabled()) {
            logger.trace("Publishing event in " + getDisplayName() + ": " + event);
        }
        getApplicationEventMulticaster().multicastEvent(event);
        if (this.parent != null) {
            this.parent.publishEvent(event);
        }
    }
```

第 7 章 AOP

我们知道，使用**面向对象编程**（OOP）有一些弊端，当需要为多个不具有继承关系的对象引入同一个公共行为时，例如日志、安全检测等，我们只有在每个对象里引用公共行为，这样程序中就产生了大量的重复代码，程序就不便于维护了，所以就有了一个对面向对象编程的补充，即**面向方面编程**（AOP），AOP 所关注的方向是横向的，不同于 OOP 的纵向。

Spring 中提供了 AOP 的实现，但是在低版本 Spring 中定义一个切面是比较麻烦的，需要实现特定的接口，并进行一些较为复杂的配置。低版本 Spring AOP 的配置是被批评最多的地方。Spring 听取了这方面的批评声音，并下决心彻底改变这一现状。在 Spring 2.0 中，Spring AOP 已经焕然一新，你可以使用@AspectJ 注解非常容易地定义一个切面，不需要实现任何的接口。

Spring 2.0 采用@AspectJ 注解对 POJO 进行标注，从而定义一个包含切点信息和增强横切逻辑的切面。Spring 2.0 可以将这个切面织入到匹配的目标 Bean 中。@AspectJ 注解使用 AspectJ 切点表达式语法进行切点定义，可以通过切点函数、运算符、通配符等高级功能进行切点定义，拥有强大的连接点描述能力。我们先来直观地浏览一下 Spring 中的 AOP 实现。

7.1 动态 AOP 使用示例

1. 创建用于拦截的 bean

在实际工作中，此 bean 可能是满足业务需要的核心逻辑，例如 test 方法中可能会封装着某个核心业务，但是，如果我们想在 test 前后加入日志来跟踪调试，如果直接修改源码并不符合面向对象的设计方法，而且随意改动原有代码也会造成一定的风险，还好接下来的 Spring 帮我们做到了这一点。

```
public class TestBean{

    private String testStr = "testStr";

    public String getTestStr() {
```

```
        return testStr;
    }

    public void setTestStr(String testStr) {
        this.testStr = testStr;
    }

    public void test(){
        System.out.println("test");
    }
}
```

2. 创建 Advisor

Spring 中摒弃了最原始的繁杂配置方式而采用 @AspectJ 注解对 POJO 进行标注，使 AOP 的工作大大简化，例如，在 AspectJTest 类中，我们要做的就是在所有类的 test 方法执行前在控制台中打印 beforeTest，而在所有类的 test 方法执行后打印 afterTest，同时又使用环绕的方式在所有类的方法执行前后再次分别打印 before1 和 after1。

```
@Aspect
public class AspectJTest {

    @Pointcut("execution(* *.test(..))")
    public void test(){

    }

    @Before("test()")
    public void beforeTest(){
        System.out.println("beforeTest");
    }

    @After("test()")
    public void afterTest(){
        System.out.println("afterTest");
    }

    @Around("test()")
    public Object arountTest(ProceedingJoinPoint p){
        System.out.println("before1");
        Object o=null;
        try {
            o = p.proceed();
        } catch (Throwable e) {
            e.printStackTrace();
        }
        System.out.println("after1");
        return o;
    }
}
```

3. 创建配置文件

XML 是 Spring 的基础。尽管 Spring 一再简化配置，并且大有使用注解取代 XML 配置之势,但是无论如何,至少现在 XML 还是 Spring 的基础。要在 Spring 中开启 AOP 功能,还需要在配置文件中作如下声明:

```xml
<?xml version="1.0" encoding="UTF-8"?>
<beans xmlns="http://www.Springframework.org/schema/beans"
       xmlns:xsi="http://www.w3.org/2001/XMLSchema-instance"
       xmlns:aop="http://www.Springframework.org/schema/aop"
       xmlns:context="http://www.Springframework.org/schema/context"
       xsi:schemaLocation="http://www.Springframework.org/schema/beans
           http://www.Springframework.org/schema/beans/Spring-beans-3.0.xsd

           http://www.Springframework.org/schema/aop
           http://www.Springframework.org/schema/aop/Spring-aop-3.0.xsd
           http://www.Springframework.org/schema/context
           http://www.Springframework.org/schema/context/Spring- context- 3.0.xsd
       ">
    <aop:aspectj-autoproxy />

    <bean id="test" class="test.TestBean"/>
    <bean class="test.AspectJTest"/>
</beans>
```

4. 测试

经过以上步骤后,便可以验证 Spring 的 AOP 为我们提供的神奇效果了。

```
public static void main(String[] args) {
    ApplicationContext bf = new ClassPathXmlApplicationContext("aspectTest.xml");
    TestBean bean=(TestBean) bf.getBean("test");
    bean.test();
}
```

不出意外,我们会看到控制台中打印了如下代码:

```
beforeTest
before1
test
afterTest
after1
```

Spring 实现了对所有类的 test 方法进行增强,使辅助功能可以独立于核心业务之外,方便与程序的扩展和解耦。

那么,Spring 究竟是如何实现 AOP 的呢?首先我们知道,Spring 是否支持注解的 AOP 是由一个配置文件控制的,也就是<aop:aspectj-autoproxy />,当在配置文件中声明了这句配置的时候,Spring 就会支持注解的 AOP,那么我们的分析就从这句注解开始。

7.2 动态 AOP 自定义标签

之前讲过 Spring 中的自定义注解,如果声明了自定义的注解,那么就一定会在程序中的某个地方注册了对应的解析器。我们搜索整个代码,尝试找到注册的地方,全局搜索后我们发现了在 AopNamespaceHandler 中对应着这样一段函数:

```
public void init() {
    // In 2.0 XSD as well as in 2.1 XSD.
    registerBeanDefinitionParser("config", new ConfigBeanDefinitionParser());
    registerBeanDefinitionParser("aspectj-autoproxy", new AspectJAutoProxyBeanDefinitionParser());
    registerBeanDefinitionDecorator("scoped-proxy", new ScopedProxyBeanDefinitionDecorator());

    // Only in 2.0 XSD: moved to context namespace as of 2.1
    registerBeanDefinitionParser("Spring-configured", new SpringConfiguredBeanDefinitionParser());
}
```

此处不再对 Spring 中的自定义注解方式进行讨论。有兴趣的读者可以回顾之前的内容。

我们可以得知，在解析配置文件的时候，一旦遇到 aspectj-autoproxy 注解时就会使用解析器 AspectJAutoProxyBeanDefinitionParser 进行解析，那么我们来看一看 AspectJAutoProxyBeanDefinitionParser 的内部实现。

7.2.1 注册 AnnotationAwareAspectJAutoProxyCreator

所有解析器，因为是对 BeanDefinitionParser 接口的统一实现，入口都是从 parse 函数开始的，AspectJAutoProxyBeanDefinitionParser 的 parse 函数如下：

```
public BeanDefinition parse(Element element, ParserContext parserContext) {
    //注册 AnnotationAwareAspectJAutoProxyCreator
    AopNamespaceUtils.registerAspectJAnnotationAutoProxyCreatorIfNecessary(parserContext, element);
    //对于注解中子类的处理
    extendBeanDefinition(element, parserContext);
    return null;
}
```

其中 registerAspectJAnnotationAutoProxyCreatorIfNecessary 函数是我们比较关心的，也是关键逻辑的实现。

```
/**
 * 注册 AnnotationAwareAspectJAutoProxyCreator
 * @param parserContext
 * @param sourceElement
 */
public static void registerAspectJAnnotationAutoProxyCreatorIfNecessary(
        ParserContext parserContext, Element sourceElement) {
    //注册或升级 AutoProxyCreator 定义 beanName 为 org.Springframework.aop.config.
    //internalAutoProxyCreator 的 BeanDefinition

    BeanDefinition beanDefinition = AopConfigUtils.registerAspectJAnnotationAutoProxyCreatorIfNecessary(
            parserContext.getRegistry(), parserContext.extractSource(sourceElement));

    //对于 proxy-target-class 以及 expose-proxy 属性的处理
    useClassProxyingIfNecessary(parserContext.getRegistry(), sourceElement);

    //注册组件并通知，便于监听器做进一步处理
    //其中 beanDefinition 的 className 为 AnnotationAwareAspectJAutoProxyCreator
    registerComponentIfNecessary(beanDefinition, parserContext);
}
```

7.2 动态 AOP 自定义标签

在 registerAspectJAnnotationAutoProxyCreatorIfNecessary 方法中主要完成了 3 件事情，基本上每行代码就是一个完整的逻辑。

1. 注册或者升级 AnnotationAwareAspectJAutoProxyCreator

对于 AOP 的实现，基本上都是靠 AnnotationAwareAspectJAutoProxyCreator 去完成，它可以根据@Point 注解定义的切点来自动代理相匹配的 bean。但是为了配置简便，Spring 使用了自定义配置来帮助我们自动注册 AnnotationAwareAspectJAutoProxyCreator，其注册过程就是在这里实现的。

```java
public static BeanDefinition registerAspectJAnnotationAutoProxyCreatorIfNecessary
(BeanDefinitionRegistry registry, Object source) {
        return registerOrEscalateApcAsRequired(AnnotationAwareAspectJAutoProxyCreator.
class, registry, source);
    }

    private static BeanDefinition registerOrEscalateApcAsRequired(Class cls, BeanDefinition
Registry registry, Object source) {
            Assert.notNull(registry, "BeanDefinitionRegistry must not be null");
//如果已经存在了自动代理创建器且存在的自动代理创建器与现在的不一致，那么需要根据优先级来判断到底需要使用哪个
            if (registry.containsBeanDefinition(AUTO_PROXY_CREATOR_BEAN_NAME)) {
                //AUTO_PROXY_CREATOR_BEAN_NAME =
                //          "org.Springframework.aop.config.internalAutoProxyCreator";
                BeanDefinition apcDefinition = registry.getBeanDefinition(AUTO_PROXY_
CREATOR_BEAN_NAME);

                if (!cls.getName().equals(apcDefinition.getBeanClassName())) {
                    int currentPriority = findPriorityForClass(apcDefinition.getBean
ClassName());
                    int requiredPriority = findPriorityForClass(cls);
                    if (currentPriority < requiredPriority) {
                        //改变 bean 最重要的就是改变 bean 所对应的 className 属性
                        apcDefinition.setBeanClassName(cls.getName());
                    }
                }
                //如果已经存在自动代理创建器并且与将要创建的一致，那么无须再次创建
                return null;
            }
            RootBeanDefinition beanDefinition = new RootBeanDefinition(cls);
            beanDefinition.setSource(source);
            beanDefinition.getPropertyValues().add("order", Ordered.HIGHEST_PRECEDENCE);
            beanDefinition.setRole(BeanDefinition.ROLE_INFRASTRUCTURE);
            //AUTO_PROXY_CREATOR_BEAN_NAME =
            //          "org.Springframework.aop.config.internalAutoProxyCreator";
            registry.registerBeanDefinition(AUTO_PROXY_CREATOR_BEAN_NAME, beanDefinition);
            return beanDefinition;
    }
```

以上代码中实现了自动注册 AnnotationAwareAspectJAutoProxyCreator 类的功能，同时这里还涉及了一个优先级的问题，如果已经存在了自动代理创建器，而且存在的自动代理创建器与现在的不一致，那么需要根据优先级来判断到底需要使用哪个。

2. 处理 proxy-target-class 以及 expose-proxy 属性

useClassProxyingIfNecessary 实现了 proxy-target-class 属性以及 expose-proxy 属性的处理。

```
private static void useClassProxyingIfNecessary(BeanDefinitionRegistry registry,
Element sourceElement) {
            if (sourceElement != null) {
                //对于 proxy-target-class 属性的处理
                boolean proxyTargetClass = Boolean.valueOf(sourceElement.getAttribute
(PROXY_TARGET_CLASS_ATTRIBUTE));
                if (proxyTargetClass) {
                    AopConfigUtils.forceAutoProxyCreatorToUseClassProxying(registry);
                }
                //对于 expose-proxy 属性的处理
                boolean exposeProxy = Boolean.valueOf(sourceElement.getAttribute (EXPOSE_
PROXY_ATTRIBUTE));
                if (exposeProxy) {
                    AopConfigUtils.forceAutoProxyCreatorToExposeProxy(registry);
                }
            }
    }

    //强制使用的过程其实也是一个属性设置的过程
    public static void forceAutoProxyCreatorToUseClassProxying(BeanDefinitionRegistry
registry) {
            if (registry.containsBeanDefinition(AUTO_PROXY_CREATOR_BEAN_NAME)) {
                BeanDefinition definition = registry.getBeanDefinition(AUTO_PROXY_CREATOR
_BEAN_NAME);
                definition.getPropertyValues().add("proxyTargetClass", Boolean.TRUE);
            }
    }

    static void forceAutoProxyCreatorToExposeProxy(BeanDefinitionRegistry registry) {
            if (registry.containsBeanDefinition(AUTO_PROXY_CREATOR_BEAN_NAME)) {
                BeanDefinition definition = registry.getBeanDefinition(AUTO_PROXY_CREATOR
_BEAN_NAME);
                definition.getPropertyValues().add("exposeProxy", Boolean.TRUE);
            }
    }
```

- proxy-target-class：Spring AOP 部分使用 JDK 动态代理或者 CGLIB 来为目标对象创建代理（建议尽量使用 JDK 的动态代理）。如果被代理的目标对象实现了至少一个接口，则会使用 JDK 动态代理。所有该目标类型实现的接口都将被代理。若该目标对象没有实现任何接口，则创建一个 CGLIB 代理。如果你希望强制使用 CGLIB 代理（例如希望代理目标对象的所有方法，而不只是实现自接口的方法），那也可以。但是需要考虑以下两个问题。

  ◆ 无法通知（advise）Final 方法，因为它们不能被覆写。

  ◆ 你需要将 CGLIB 二进制发行包放在 classpath 下面。

与之相比，JDK 本身就提供了动态代理，强制使用 CGLIB 代理需要将 <aop:config> 的

proxy-target-class 属性设为 true：

```
<aop:config proxy-target-class="true"> ... </aop:config>
```

当需要使用 CGLIB 代理和@AspectJ 自动代理支持，可以按照以下方式设置 <aop:aspectj-autoproxy>的 proxy-target-class 属性：

```
<aop:aspectj-autoproxy proxy-target-class="true"/>
```

而实际使用的过程中才会发现细节问题的差别，*The devil is in the details*。

- JDK 动态代理：其代理对象必须是某个接口的实现，它是通过在运行期间创建一个接口的实现类来完成对目标对象的代理。
- CGLIB 代理：实现原理类似于 JDK 动态代理，只是它在运行期间生成的代理对象是针对目标类扩展的子类。CGLIB 是高效的代码生成包，底层是依靠 ASM（开源的 Java 字节码编辑类库）操作字节码实现的，性能比 JDK 强。
- expose-proxy：有时候目标对象内部的自我调用将无法实施切面中的增强，如下示例：

```
public interface AService {
    public void a();
    public void b();
}

@Service()
public class AServiceImpl1 implements AService{
    @Transactional(propagation = Propagation.REQUIRED)
    public void a() {
        this.b();
    }
    @Transactional(propagation = Propagation.REQUIRES_NEW)
    public void b() {
    }
}
```

此处的 this 指向目标对象，因此调用 this.b()将不会执行 b 事务切面，即不会执行事务增强，因此 b 方法的事务定义"@Transactional(propagation = Propagation.REQUIRES_NEW)"将不会实施，为了解决这个问题，我们可以这样做：

```
<aop:aspectj-autoproxy expose-proxy="true"/>
```

然后将以上代码中的"this.b();"修改为"((AService) AopContext.currentProxy()).b();"即可。通过以上的修改便可以完成对 a 和 b 方法的同时增强。

最后注册组件并通知，便于监听器做进一步处理，这里就不再一一赘述了。

7.3 创建 AOP 代理

上文中讲解了通过自定义配置完成了对 AnnotationAwareAspectJAutoProxyCreator 类型的自动注册，那么这个类到底做了什么工作来完成 AOP 的操作呢？首先我们看看 Annotation-AwareAspectJAutoProxyCreator 类的层次结构，如图 7-1 所示。

第 7 章　AOP

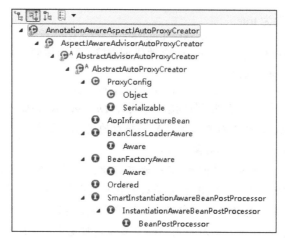

图 7-1　AnnotationAwareAspectJAutoProxyCreator 类的层次结构图

在类的层级中，我们看到 AnnotationAwareAspectJAutoProxyCreator 实现了 BeanPostProcessor 接口，而实现 BeanPostProcessor 后，当 Spring 加载这个 Bean 时会在实例化前调用其 postProcess-AfterInitialization 方法，而我们对于 AOP 逻辑的分析也由此开始。

在父类 AbstractAutoProxyCreator 的 postProcessAfterInitialization 中代码如下：

```
public Object postProcessAfterInitialization(Object bean, String beanName) throws BeansException {
            if (bean != null) {
                //根据给定的bean的class和name构建出一个key,格式:beanClassName_beanName
                Object cacheKey = getCacheKey(bean.getClass(), beanName);
                if (!this.earlyProxyReferences.contains(cacheKey)) {
                    //如果它适合被代理,则需要封装指定bean
                    return wrapIfNecessary(bean, beanName, cacheKey);
                }
            }
            return bean;
    }

    protected Object wrapIfNecessary(Object bean, String beanName, Object cacheKey) {
            //如果已经处理过
            if (this.targetSourcedBeans.contains(beanName)) {
                return bean;
            }
            //无须增强
            if (this.nonAdvisedBeans.contains(cacheKey)) {
                return bean;
            }
            //给定的bean类是否代表一个基础设施类,基础设施类不应代理,或者配置了指定bean不需要自动代理
            if (isInfrastructureClass(bean.getClass()) || shouldSkip(bean.getClass(), beanName)) {
                this.nonAdvisedBeans.add(cacheKey);
                return bean;
            }

            //如果存在增强方法则创建代理
```

```
                Object[] specificInterceptors = getAdvicesAndAdvisorsForBean(bean.getClass(),
beanName, null);
                //如果获取到了增强则需要针对增强创建代理
                if (specificInterceptors != DO_NOT_PROXY) {
                    this.advisedBeans.add(cacheKey);
                    //创建代理
                    Object proxy = createProxy(bean.getClass(), beanName, specificInterceptors,
new SingletonTargetSource(bean));
                    this.proxyTypes.put(cacheKey, proxy.getClass());
                    return proxy;
                }

                this.nonAdvisedBeans.add(cacheKey);
                return bean;
            }
```

函数中我们已经看到了代理创建的雏形。当然，真正开始之前还需要经过一些判断，比如是否已经处理过或者是否是需要跳过的 bean，而真正创建代理的代码是从 getAdvicesAndAdvisorsForBean 开始的。

创建代理主要包含了两个步骤。

1. 获取增强方法或者增强器。
2. 根据获取的增强进行代理。

核心逻辑的时序图如图 7-2 所示。

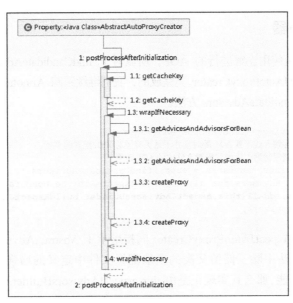

图 7-2　AbstractAutoProxyCreator 的 postProcessAfterInitialization 函数执行时序图

虽然看似简单，但是每个步骤中都经历了大量复杂的逻辑。首先来看看获取增强方法的实现逻辑。

```
    @Override
    protected Object[] getAdvicesAndAdvisorsForBean(Class beanClass, String beanName,
TargetSource targetSource) {
        List advisors = findEligibleAdvisors(beanClass, beanName);
        if (advisors.isEmpty()) {
            return DO_NOT_PROXY;
        }
        return advisors.toArray();
    }
    protected List<Advisor> findEligibleAdvisors(Class beanClass, String beanName) {
        List<Advisor> candidateAdvisors = findCandidateAdvisors();
        List<Advisor> eligibleAdvisors = findAdvisorsThatCanApply(candidateAdvisors,
beanClass, beanName);
        extendAdvisors(eligibleAdvisors);
        if (!eligibleAdvisors.isEmpty()) {
            eligibleAdvisors = sortAdvisors(eligibleAdvisors);
        }
        return eligibleAdvisors;
    }
```

对于指定 bean 的增强方法的获取一定是包含两个步骤的，获取所有的增强以及寻找所有增强中适用于 bean 的增强并应用，那么 findCandidateAdvisors 与 findAdvisorsThatCanApply 便是做了这两件事情。当然，如果无法找到对应的增强器便返回 DO_NOT_PROXY，其中 DO_NOT_PROXY=null。

7.3.1 获取增强器

由于我们分析的是使用注解进行的 AOP，所以对于 findCandidateAdvisors 的实现其实是由 AnnotationAwareAspectJAutoProxyCreator 类完成的，我们继续跟踪 AnnotationAwareAspectJAuto-ProxyCreator 的 findCandidateAdvisors 方法。

```
    @Override
    protected List<Advisor> findCandidateAdvisors() {
        //当使用注解方式配置AOP的时候并不是丢弃了对XML配置的支持，
        //在这里调用父类方法加载配置文件中的AOP声明
        List<Advisor> advisors = super.findCandidateAdvisors();
        // Build Advisors for all AspectJ aspects in the bean factory.
        advisors.addAll(this.aspectJAdvisorsBuilder.buildAspectJAdvisors());
        return advisors;
    }
```

AnnotationAwareAspectJAutoProxyCreator 间接继承了 AbstractAdvisorAutoProxyCreator，在实现获取增强的方法中除了保留父类的获取配置文件中定义的增强外，同时添加了获取 Bean 的注解增强的功能，那么其实现正是由 this.aspectJAdvisorsBuilder.buildAspectJAdvisors() 来实现的。

在真正研究代码之前读者可以尝试着自己去想象一下解析思路，看看自己的实现与 Spring 是否有差别呢？或者我们一改以往的方式，先来了解函数提供的大概功能框架，读者可以在头脑中尝试实现这些功能点，看看是否有思路。

7.3 创建 AOP 代理

1. 获取所有 beanName，这一步骤中所有在 beanFacotry 中注册的 bean 都会被提取出来。
2. 遍历所有 beanName，并找出声明 AspectJ 注解的类，进行进一步的处理。
3. 对标记为 AspectJ 注解的类进行增强器的提取。
4. 将提取结果加入缓存。

现在我们来看看函数实现，对 Spring 中所有的类进行分析，提取 Advisor。

```
public List<Advisor> buildAspectJAdvisors() {
    List<String> aspectNames = null;

    synchronized (this) {
        aspectNames = this.aspectBeanNames;
        if (aspectNames == null) {
            List<Advisor> advisors = new LinkedList<Advisor>();
            aspectNames = new LinkedList<String>();
            //获取所有的 beanName
            String[] beanNames =BeanFactoryUtils.beanNamesForTypeIncludingAncestors
(this.beanFactory, Object.class, true, false);
            //循环所有的 beanName 找出对应的增强方法
            for (String beanName : beanNames) {
                //不合法的 bean 则略过，由子类定义规则，默认返回 true
                if (!isEligibleBean(beanName)) {
                    continue;
                }
                //获取对应的 bean 的类型
                Class beanType = this.beanFactory.getType(beanName);
                if (beanType == null) {
                    continue;
                }
                //如果存在 Aspect 注解
                if (this.advisorFactory.isAspect(beanType)) {
                    aspectNames.add(beanName);
                    AspectMetadata amd = new AspectMetadata(beanType, beanName);
                    if (amd.getAjType().getPerClause().getKind() == PerClauseKind.
SINGLETON) {

                        MetadataAwareAspectInstanceFactory factory =
                            new BeanFactoryAspectInstanceFactory(this.bean-
Factory, beanName);
                        //解析标记 AspectJ 注解中的增强方法
                        List<Advisor> classAdvisors = this.advisorFactory.
getAdvisors(factory);

                        if (this.beanFactory.isSingleton(beanName)) {
                            this.advisorsCache.put(beanName, classAdvisors);
                        }
                        else {
                            this.aspectFactoryCache.put(beanName, factory);
                        }
                        advisors.addAll(classAdvisors);
                    }else {
                        // Per target or per this.
                        if (this.beanFactory.isSingleton(beanName)) {
                            throw new IllegalArgumentException("Bean with name '"
+ beanName +
```

第 7 章 AOP

```
                                        "' is a singleton, but aspect instantiation model is not singleton");
                                    }
                                    MetadataAwareAspectInstanceFactory factory =
                                        new PrototypeAspectInstanceFactory(this.beanFactory, beanName);
                                    this.aspectFactoryCache.put(beanName, factory);
                                    advisors.addAll(this.advisorFactory.getAdvisors(factory));
                                }
                            }
                        }
                        this.aspectBeanNames = aspectNames;
                        return advisors;
                    }
                }

                if (aspectNames.isEmpty()) {
                    return Collections.EMPTY_LIST;
                }

                //记录在缓存中
                List<Advisor> advisors = new LinkedList<Advisor>();
                for (String aspectName : aspectNames) {
                    List<Advisor> cachedAdvisors = this.advisorsCache.get(aspectName);
                    if (cachedAdvisors != null) {
                        advisors.addAll(cachedAdvisors);
                    }
                    else {
                        MetadataAwareAspectInstanceFactory factory = this.aspectFactoryCache.get(aspectName);
                        advisors.addAll(this.advisorFactory.getAdvisors(factory));
                    }
                }
                return advisors;
            }
```

至此，我们已经完成了 Advisor 的提取，在上面的步骤中最为重要也最为繁杂的就是增强器的获取。而这一功能委托给了 getAdvisors 方法去实现（this.advisorFactory.getAdvisors(factory)）。

```
            public List<Advisor> getAdvisors(MetadataAwareAspectInstanceFactory maaif) {
                //获取标记为 AspectJ 的类
                final Class<?> aspectClass = maaif.getAspectMetadata().getAspectClass();
                //获取标记为 AspectJ 的 name
                final String aspectName = maaif.getAspectMetadata().getAspectName();
                //验证
                validate(aspectClass);

                final MetadataAwareAspectInstanceFactory lazySingletonAspectInstanceFactory =
                    new LazySingletonAspectInstanceFactoryDecorator(maaif);

                final List<Advisor> advisors = new LinkedList<Advisor>();
                ReflectionUtils.doWithMethods(aspectClass, new ReflectionUtils.MethodCallback() {
                    public void doWith(Method method) throws IllegalArgumentException {
                        //声明为 Pointcut 的方法不处理
                        if (AnnotationUtils.getAnnotation(method, Pointcut.class) == null) {
```

```
                    Advisor advisor = getAdvisor(method, lazySingletonAspectInstance
Factory, advisors.size(), aspectName);
                    if (advisor != null) {
                        advisors.add(advisor);
                    }
                }
            }
        });

        if (!advisors.isEmpty() && lazySingletonAspectInstanceFactory.getAspect Metadata
().isLazilyInstantiated()) {
            //如果寻找的增强器不为空而且又配置了增强延迟初始化,那么需要在首位加入同步实例化增强器
            Advisor instantiationAdvisor = new SyntheticInstantiationAdvisor (lazySingleton-
AspectInstanceFactory);
            advisors.add(0, instantiationAdvisor);
        }

        //获取 DeclareParents 注解
        for (Field field : aspectClass.getDeclaredFields()) {
            Advisor advisor = getDeclareParentsAdvisor(field);
            if (advisor != null) {
                advisors.add(advisor);
            }
        }

        return advisors;
    }
```

函数中首先完成了对增强器的获取,包括获取注解以及根据注解生成增强的步骤,然后考虑到在配置中可能会将增强配置成延迟初始化,那么需要在首位加入同步实例化增强器以保证增强使用之前的实例化,最后是对 DeclareParents 注解的获取,下面将详细介绍一下每个步骤。

1. 普通增强器的获取

普通增强器的获取逻辑通过 getAdvisor 方法实现,实现步骤包括对切点的注解的获取以及根据注解信息生成增强。

```
public Advisor getAdvisor(Method candidateAdviceMethod, MetadataAwareAspectInstanceFactory aif,
        int declarationOrderInAspect, String aspectName) {

    validate(aif.getAspectMetadata().getAspectClass());
    //切点信息的获取
    AspectJExpressionPointcut ajexp =
            getPointcut(candidateAdviceMethod, aif.getAspectMetadata().getAspectClass());
    if (ajexp == null) {
        return null;
    }
    //根据切点信息生成增强器
    return new InstantiationModelAwarePointcutAdvisorImpl(
            this, ajexp, aif, candidateAdviceMethod, declarationOrderInAspect,
aspectName);
}
```

1. 切点信息的获取。所谓获取切点信息就是指定注解的表达式信息的获取，如 @Before("test()")。

```
private AspectJExpressionPointcut getPointcut(Method candidateAdviceMethod, Class<?> candidateAspectClass) {
    //获取方法上的注解
    AspectJAnnotation<?> aspectJAnnotation =
            AbstractAspectJAdvisorFactory.findAspectJAnnotationOnMethod (candidateAdviceMethod);
    if (aspectJAnnotation == null) {
        return null;
    }
    //使用 AspectJExpressionPointcut 实例封装获取的信息
    AspectJExpressionPointcut ajexp =
            new AspectJExpressionPointcut(candidateAspectClass, new String[0], new Class[0]);
    //提取得到的注解中的表达式，如：
    //@Pointcut("execution(* *.*test*(..))")中的 execution(* *.*test*(..))
    ajexp.setExpression(aspectJAnnotation.getPointcutExpression());
    return ajexp;
}

protected static AspectJAnnotation findAspectJAnnotationOnMethod(Method method) {
    //设置敏感的注解类
    Class<? extends Annotation>[] classesToLookFor = new Class[] {
            Before.class, Around.class, After.class, AfterReturning.class,
            AfterThrowing.class, Pointcut.class};
    for (Class<? extends Annotation> c : classesToLookFor) {
        AspectJAnnotation foundAnnotation = findAnnotation(method, c);
        if (foundAnnotation != null) {
            return foundAnnotation;
        }
    }
    return null;
}
//获取指定方法上的注解并使用 AspectJAnnotation 封装
private static <A extends Annotation> AspectJAnnotation<A> findAnnotation(Method method, Class<A> toLookFor) {
    A result = AnnotationUtils.findAnnotation(method, toLookFor);
    if (result != null) {
        return new AspectJAnnotation<A>(result);
    }
    else {
        return null;
    }
}
```

2. 根据切点信息生成增强。所有的增强都由 Advisor 的实现类 InstantiationModelAware-PointcutAdvisorImpl 统一封装的。

```
public InstantiationModelAwarePointcutAdvisorImpl(AspectJAdvisorFactory af, AspectJExpressionPointcut ajexp,
        MetadataAwareAspectInstanceFactory aif, Method method, int declarationOrderInAspect, String aspectName) {
```

```
            //test()
            this.declaredPointcut = ajexp;
            //public void test.AspectJTest.beforeTest()
            this.method = method;

            this.atAspectJAdvisorFactory = af;

            this.aspectInstanceFactory = aif;
            //0
            this.declarationOrder = declarationOrderInAspect;
            //test.AspectJTest
            this.aspectName = aspectName;

            if (aif.getAspectMetadata().isLazilyInstantiated()) {
                // Static part of the pointcut is a lazy type.
                Pointcut preInstantiationPointcut =
                        Pointcuts.union(aif.getAspectMetadata().getPerClausePointcut(),
this.declaredPointcut);

                this.pointcut = new PerTargetInstantiationModelPointcut(this.declaredPointcut,
preInstantiationPointcut, aif);
                this.lazy = true;
            }else {
                // A singleton aspect.
                this.instantiatedAdvice = **instantiateAdvice**(this.declaredPointcut);
                this.pointcut = declaredPointcut;
                this.lazy = false;
            }
        }
```

在封装过程中只是简单地将信息封装在类的实例中，所有的信息单纯地赋值，在实例初始化的过程中还完成了对于增强器的初始化。因为不同的增强所体现的逻辑是不同的，比如@Before（"test()"）与@After（"test()"）标签的不同就是增强器增强的位置不同，所以就需要不同的增强器来完成不同的逻辑，而根据注解中的信息初始化对应的增强器就是在 instantiateAdvice 函数中实现的。

```
        private Advice instantiateAdvice(AspectJExpressionPointcut pcut) {
            return this.atAspectJAdvisorFactory.**getAdvice**(
                    this.method, pcut, this.aspectInstanceFactory, this.declarationOrder,
this.aspectName);
        }
        public Advice getAdvice(Method candidateAdviceMethod, AspectJExpressionPointcut ajexp,
                MetadataAwareAspectInstanceFactory aif, int declarationOrderInAspect,
String aspectName) {

            Class<?> candidateAspectClass = aif.getAspectMetadata().getAspectClass();
            validate(candidateAspectClass);

            AspectJAnnotation<?> aspectJAnnotation =
                    AbstractAspectJAdvisorFactory.findAspectJAnnotationOnMethod (candidate
AdviceMethod);
            if (aspectJAnnotation == null) {
```

```
            return null;
        }

        // If we get here, we know we have an AspectJ method.
        // Check that it's an AspectJ-annotated class
        if (!isAspect(candidateAspectClass)) {
            throw new AopConfigException("Advice must be declared inside an aspect type: " +
                    "Offending method '" + candidateAdviceMethod + "' in class [" +
                    candidateAspectClass.getName() + "]");
        }

        if (logger.isDebugEnabled()) {
            logger.debug("Found AspectJ method: " + candidateAdviceMethod);
        }

        AbstractAspectJAdvice SpringAdvice;
        //根据不同的注解类型封装不同的增强器
        switch (aspectJAnnotation.getAnnotationType()) {
            case AtBefore:
                SpringAdvice = new AspectJMethodBeforeAdvice(candidateAdviceMethod, ajexp, aif);
                break;
            case AtAfter:
                SpringAdvice = new AspectJAfterAdvice(candidateAdviceMethod, ajexp, aif);
                break;
            case AtAfterReturning:
                SpringAdvice = new AspectJAfterReturningAdvice(candidateAdviceMethod, ajexp, aif);
                AfterReturning afterReturningAnnotation = (AfterReturning) aspectJAnnotation.getAnnotation();
                if (StringUtils.hasText(afterReturningAnnotation.returning())) {
                    SpringAdvice.setReturningName(afterReturningAnnotation.returning());
                }
                break;
            case AtAfterThrowing:
                SpringAdvice = new AspectJAfterThrowingAdvice(candidateAdviceMethod, ajexp, aif);
                AfterThrowing afterThrowingAnnotation = (AfterThrowing) aspectJAnnotation.getAnnotation();
                if (StringUtils.hasText(afterThrowingAnnotation.throwing())) {
                    SpringAdvice.setThrowingName(afterThrowingAnnotation.throwing());
                }
                break;
            case AtAround:
                SpringAdvice = new AspectJAroundAdvice(candidateAdviceMethod, ajexp, aif);
                break;
            case AtPointcut:
                if (logger.isDebugEnabled()) {
                    logger.debug("Processing pointcut '" + candidateAdviceMethod.getName() + "'");
                }
                return null;
            default:
                throw new UnsupportedOperationException(
                        "Unsupported advice type on method " + candidateAdviceMethod);
```

```
            }
            // Now to configure the advice...
            SpringAdvice.setAspectName(aspectName);
            SpringAdvice.setDeclarationOrder(declarationOrderInAspect);
            String[] argNames = this.parameterNameDiscoverer.getParameterNames
(candidateAdvice Method);
            if (argNames != null) {
                SpringAdvice.setArgumentNamesFromStringArray(argNames);
            }
            SpringAdvice.calculateArgumentBindings();
            return SpringAdvice;
    }
```

从函数中可以看到，Spring 会根据不同的注解生成不同的增强器，例如 AtBefore 会对应 AspectJMethodBeforeAdvice，而在 AspectJMethodBeforeAdvice 中完成了增强方法的逻辑。我们尝试分析几个常用的增强器实现。

- MethodBeforeAdviceInterceptor。

我们首先查看 MethodBeforeAdviceInterceptor 类的内部实现。

```
public class MethodBeforeAdviceInterceptor implements MethodInterceptor, Serializable {

    private MethodBeforeAdvice advice;


    /**
     * Create a new MethodBeforeAdviceInterceptor for the given advice.
     * @param advice the MethodBeforeAdvice to wrap
     */
    public MethodBeforeAdviceInterceptor(MethodBeforeAdvice advice) {
        Assert.notNull(advice, "Advice must not be null");
        this.advice = advice;
    }

    public Object invoke(MethodInvocation mi) throws Throwable {
        this.advice.before(mi.getMethod(), mi.getArguments(), mi.getThis() );
        return mi.proceed();
    }

}
```

其中的属性 MethodBeforeAdvice 代表着前置增强的 AspectJMethodBeforeAdvice，跟踪 before 方法：

```
    public void before(Method method, Object[] args, Object target) throws Throwable {
        invokeAdviceMethod(getJoinPointMatch(), null, null);
    }

    protected Object invokeAdviceMethod(JoinPointMatch jpMatch, Object returnValue,
Throwable ex) throws Throwable {
        return invokeAdviceMethodWithGivenArgs(argBinding(getJoinPoint(), jpMatch,
returnValue, ex));
    }
```

```
    protected Object invokeAdviceMethodWithGivenArgs(Object[] args) throws Throwable {
        Object[] actualArgs = args;
        if (this.aspectJAdviceMethod.getParameterTypes().length == 0) {
            actualArgs = null;
        }
        try {
            ReflectionUtils.makeAccessible(this.aspectJAdviceMethod);
            //激活增强方法
            return this.aspectJAdviceMethod.invoke(this.aspectInstanceFactory.getAspectInstance(), actualArgs);
        }
        catch (IllegalArgumentException ex) {
            throw new AopInvocationException("Mismatch on arguments to advice method [" +
                    this.aspectJAdviceMethod + "]; pointcut expression [" +
                    this.pointcut.getPointcutExpression() + "]", ex);
        }
        catch (InvocationTargetException ex) {
            throw ex.getTargetException();
        }
    }
```

invokeAdviceMethodWithGivenArgs 方法中的 aspectJAdviceMethod 正是对于前置增强的方法，在这里实现了调用。

- **AspectJAfterAdvice**。

后置增强与前置增强有稍许不一致的地方。回顾之前讲过的前置增强，大致的结构是在拦截器链中放置 MethodBeforeAdviceInterceptor，而在 MethodBeforeAdviceInterceptor 中又放置了 AspectJMethodBeforeAdvice，并在调用 invoke 时首先串联调用。但是在后置增强的时候却不一样，没有提供中间的类，而是直接在拦截器链中使用了中间的 AspectJAfterAdvice。

```
public class AspectJAfterAdvice extends AbstractAspectJAdvice implements MethodInterceptor,
AfterAdvice {

    public AspectJAfterAdvice(
            Method aspectJBeforeAdviceMethod, AspectJExpressionPointcut pointcut,
AspectInstanceFactory aif) {

        super(aspectJBeforeAdviceMethod, pointcut, aif);
    }

    public Object invoke(MethodInvocation mi) throws Throwable {
        try {
            return mi.proceed();
        }
        finally {
            //激活增强方法
            invokeAdviceMethod(getJoinPointMatch(), null, null);
        }
    }

    public boolean isBeforeAdvice() {
        return false;
```

```java
    }
    public boolean isAfterAdvice() {
        return true;
    }
}
```

2. 增加同步实例化增强器

如果寻找的增强器不为空而且又配置了增强延迟初始化,那么就需要在首位加入同步实例化增强器。同步实例化增强器 SyntheticInstantiationAdvisor 如下:

```java
protected static class SyntheticInstantiationAdvisor extends DefaultPointcutAdvisor {
            public SyntheticInstantiationAdvisor(final MetadataAwareAspectInstanceFactory aif) {
                super(aif.getAspectMetadata().getPerClausePointcut(), new MethodBeforeAdvice() {
                //目标方法前调用,类似@Before
                    public void before(Method method, Object[] args, Object target) {
                        //简单初始化 aspect
                        aif.getAspectInstance();
                    }
                });
            }
}
```

3. 获取 DeclareParents 注解

DeclareParents 主要用于引介增强的注解形式的实现,而其实现方式与普通增强很类似,只不过使用 DeclareParentsAdvisor 对功能进行封装。

```java
private Advisor getDeclareParentsAdvisor(Field introductionField) {
            DeclareParents declareParents = (DeclareParents) introductionField.getAnnotation
(DeclareParents.class);
            if (declareParents == null) {
                // Not an introduction field
                return null;
            }

            if (DeclareParents.class.equals(declareParents.defaultImpl())) {
                // This is what comes back if it wasn't set. This seems bizarre...
                // TODO this restriction possibly should be relaxed
                throw new IllegalStateException("defaultImpl must be set on Declare Parents");
            }

            return new DeclareParentsAdvisor(
                    introductionField.getType(), declareParents.value(), declareParents.defaultImpl());
    }
```

7.3.2 寻找匹配的增强器

前面的函数中已经完成了所有增强器的解析,但是对于所有增强器来讲,并不一定都适用于当前的 Bean，还要挑取出适合的增强器，也就是满足我们配置的通配符的增强器。具体实现在 findAdvisorsThatCanApply 中。

```java
protected List<Advisor> findAdvisorsThatCanApply(
        List<Advisor> candidateAdvisors, Class beanClass, String beanName) {

    ProxyCreationContext.setCurrentProxiedBeanName(beanName);
    try {
        //过滤已经得到的 advisors
        return AopUtils.findAdvisorsThatCanApply(candidateAdvisors, beanClass);
    }
    finally {
        ProxyCreationContext.setCurrentProxiedBeanName(null);
    }
}
```

继续看 findAdvisorsThatCanApply：

```java
public static List<Advisor> findAdvisorsThatCanApply(List<Advisor> candidateAdvisors,
        Class<?> clazz) {
    if (candidateAdvisors.isEmpty()) {
        return candidateAdvisors;
    }
    List<Advisor> eligibleAdvisors = new LinkedList<Advisor>();
    //首先处理引介增强
    for (Advisor candidate : candidateAdvisors) {
        if (candidate instanceof IntroductionAdvisor && canApply(candidate, clazz)) {
            eligibleAdvisors.add(candidate);
        }
    }
    boolean hasIntroductions = !eligibleAdvisors.isEmpty();
    for (Advisor candidate : candidateAdvisors) {
        //引介增强已经处理
        if (candidate instanceof IntroductionAdvisor) {
            continue;
        }
        //对于普通 bean 的处理
        if (canApply(candidate, clazz, hasIntroductions)) {
            eligibleAdvisors.add(candidate);
        }
    }
    return eligibleAdvisors;
}
```

findAdvisorsThatCanApply 函数的主要功能是寻找所有增强器中适用于当前 class 的增强器。引介增强与普通的增强处理是不一样的，所以分开处理。而对于真正的匹配在 canApply 中实现。

```java
public static boolean canApply(Advisor advisor, Class<?> targetClass, boolean hasIntroductions) {
    if (advisor instanceof IntroductionAdvisor) {
        return ((IntroductionAdvisor) advisor).getClassFilter().matches(targetClass);
    }else if (advisor instanceof PointcutAdvisor) {
        PointcutAdvisor pca = (PointcutAdvisor) advisor;
        return canApply(pca.getPointcut(), targetClass, hasIntroductions);
```

```java
        }else {
            // It doesn't have a pointcut so we assume it applies.
            return true;
        }
    }

    public static boolean canApply(Pointcut pc, Class<?> targetClass, boolean hasIntroductions) {
        Assert.notNull(pc, "Pointcut must not be null");
        if (!pc.getClassFilter().matches(targetClass)) {
            return false;
        }

        MethodMatcher methodMatcher = pc.getMethodMatcher();
        IntroductionAwareMethodMatcher introductionAwareMethodMatcher = null;
        if (methodMatcher instanceof IntroductionAwareMethodMatcher) {
            introductionAwareMethodMatcher = (IntroductionAwareMethodMatcher) methodMatcher;
        }

        Set<Class> classes = new HashSet<Class>(ClassUtils.getAllInterfacesForClassAsSet (targetClass));
        classes.add(targetClass);
        //classes:[interface test.IITestBean, class test.TestBean]
        for (Class<?> clazz : classes) {
            Method[] methods = clazz.getMethods();
            for (Method method : methods) {
                if ((introductionAwareMethodMatcher != null &&
                        introductionAwareMethodMatcher.matches(method, targetClass, hasIntroductions)) ||
                        methodMatcher.matches(method, targetClass)) {
                    return true;
                }
            }
        }

        return false;
    }
```

7.3.3 创建代理

在获取了所有对应 bean 的增强器后，便可以进行代理的创建了。

```java
    protected Object createProxy(
            Class<?> beanClass, String beanName, Object[] specificInterceptors,
TargetSource targetSource) {

        ProxyFactory proxyFactory = new ProxyFactory();
        //获取当前类中相关属性
        proxyFactory.copyFrom(this);
        //决定对于给定的 bean 是否应该使用 targetClass 而不是它的接口代理，
        //检查 proxyTargeClass 设置以及 preserveTargetClass 属性
        if (!shouldProxyTargetClass(beanClass, beanName)) {
            // Must allow for introductions; can't just set interfaces to
            // the target's interfaces only.
```

```
                Class<?>[] targetInterfaces = ClassUtils.getAllInterfacesForClass(beanClass,
this.proxyClassLoader);
                for (Class<?> targetInterface : targetInterfaces) {
                    //添加代理接口
                    proxyFactory.addInterface(targetInterface);
                }
            }

            Advisor[] advisors = buildAdvisors(beanName, specificInterceptors);
            for (Advisor advisor : advisors) {
                //加入增强器
                proxyFactory.addAdvisor(advisor);
            }
            //设置要代理的类
            proxyFactory.setTargetSource(targetSource);
            //定制代理
            customizeProxyFactory(proxyFactory);
            //用来控制代理工厂被配置之后，是否还允许修改通知
            //缺省值为 false（即在代理被配置之后，不允许修改代理的配置）
            proxyFactory.setFrozen(this.freezeProxy);
            if (advisorsPreFiltered()) {
                proxyFactory.setPreFiltered(true);
            }

            return proxyFactory.getProxy(this.proxyClassLoader);
}
```

对于代理类的创建及处理，Spring 委托给了 ProxyFactory 去处理，而在此函数中主要是对 ProxyFactory 的初始化操作，进而对真正的创建代理做准备，这些初始化操作包括如下内容。

1. 获取当前类中的属性。
2. 添加代理接口。
3. 封装 Advisor 并加入到 ProxyFactory 中。
4. 设置要代理的类。
5. 当然在 Spring 中还为子类提供了定制的函数 customizeProxyFactory，子类可以在此函数中进行对 ProxyFactory 的进一步封装。
6. 进行获取代理操作。

其中，封装 Advisor 并加入到 ProxyFactory 中以及创建代理是两个相对繁琐的过程，可以通过 ProxyFactory 提供的 addAdvisor 方法直接将增强器置入代理创建工厂中，但是将拦截器封装为增强器还是需要一定的逻辑的。

```
protected Advisor[] buildAdvisors(String beanName, Object[] specificInterceptors) {
    //解析注册的所有 interceptorName
    Advisor[] commonInterceptors = resolveInterceptorNames();

    List<Object> allInterceptors = new ArrayList<Object>();
    if (specificInterceptors != null) {
        //加入拦截器
        allInterceptors.addAll(Arrays.asList(specificInterceptors));
        if (commonInterceptors != null) {
```

```java
                    if (this.applyCommonInterceptorsFirst) {
                        allInterceptors.addAll(0, Arrays.asList(commonInterceptors));
                    }
                    else {
                        allInterceptors.addAll(Arrays.asList(commonInterceptors));
                    }
                }
            }
            if (logger.isDebugEnabled()) {
                int nrOfCommonInterceptors = (commonInterceptors != null ? commonInterceptors.length : 0);
                int nrOfSpecificInterceptors = (specificInterceptors != null ? specificInterceptors.length : 0);
                logger.debug("Creating implicit proxy for bean '" + beanName + "' with " + nrOfCommonInterceptors +
                        " common interceptors and " + nrOfSpecificInterceptors + " specific interceptors");
            }

            Advisor[] advisors = new Advisor[allInterceptors.size()];
            for (int i = 0; i < allInterceptors.size(); i++) {
                //拦截器进行封装转化为Advisor
                advisors[i] = this.advisorAdapterRegistry.wrap(allInterceptors.get(i));
            }
            return advisors;
        }

        public Advisor wrap(Object adviceObject) throws UnknownAdviceTypeException {
            //如果要封装的对象本身就是Advisor类型的，那么无须再做过多处理
            if (adviceObject instanceof Advisor) {
                return (Advisor) adviceObject;
            }
            //因为此封装方法只对Advisor与Advice两种类型的数据有效，如果不是将不能封装
            if (!(adviceObject instanceof Advice)) {
                throw new UnknownAdviceTypeException(adviceObject);
            }
            Advice advice = (Advice) adviceObject;
            if (advice instanceof MethodInterceptor) {
                //如果是MethodInterceptor类型则使用DefaultPointcutAdvisor封装
                return new DefaultPointcutAdvisor(advice);
            }
            //如果存在Advisor的适配器那么也同样需要进行封装
            for (AdvisorAdapter adapter : this.adapters) {
                // Check that it is supported.
                if (adapter.supportsAdvice(advice)) {
                    return new DefaultPointcutAdvisor(advice);
                }
            }
            throw new UnknownAdviceTypeException(advice);
        }
```

由于Spring中涉及过多的拦截器、增强器、增强方法等方式来对逻辑进行增强，所以非常有必要统一封装成Advisor来进行代理的创建，完成了增强的封装过程，那么解析最重要的一步就是代理的创建与获取了。

```java
public Object getProxy(ClassLoader classLoader) {
    return createAopProxy().getProxy(classLoader);
}
```

1. 创建代理

```java
protected final synchronized AopProxy createAopProxy() {
    if (!this.active) {
        activate();
    }
    //创建代理
    return getAopProxyFactory().createAopProxy(this);
}
public AopProxy createAopProxy(AdvisedSupport config) throws AopConfigException {
    if (config.isOptimize() || config.isProxyTargetClass() || hasNoUserSuppliedProxyInterfaces(config)) {
        Class targetClass = config.getTargetClass();
        if (targetClass == null) {
            throw new AopConfigException("TargetSource cannot determine target class: " +
                    "Either an interface or a target is required for proxy creation.");
        }
        if (targetClass.isInterface()) {
            return new JdkDynamicAopProxy(config);
        }
        if (!cglibAvailable) {
            throw new AopConfigException(
                    "Cannot proxy target class because CGLIB2 is not available. " +
                    "Add CGLIB to the class path or specify proxy interfaces.");
        }
        return CglibProxyFactory.createCglibProxy(config);
    }
    else {
        return new JdkDynamicAopProxy(config);
    }
}
```

到此已经完成了代理的创建，不管我们之前是否阅读过 Spring 的源代码，但是都或多或少地听过对于 Spring 的代理中 JDKProxy 的实现和 CglibProxy 的实现。Spring 是如何选取的呢？网上的介绍有很多，现在我们就从源代码的角度分析，看看到底 Spring 是如何选择代理方式的。

从 if 中的判断条件可以看到 3 个方面影响着 Spring 的判断。

- optimize：用来控制通过 CGLIB 创建的代理是否使用激进的优化策略。除非完全了解 AOP 代理如何处理优化，否则不推荐用户使用这个设置。目前这个属性仅用于 CGLIB 代理，对于 JDK 动态代理（默认代理）无效。
- proxyTargetClass：这个属性为 true 时，目标类本身被代理而不是目标类的接口。如果这个属性值被设为 true，CGLIB 代理将被创建，设置方式为<aop:aspectj-autoproxy-proxy-target-class="true"/>。

- hasNoUserSuppliedProxyInterfaces：是否存在代理接口。

下面是对 JDK 与 Cglib 方式的总结。
- 如果目标对象实现了接口，默认情况下会采用 JDK 的动态代理实现 AOP。
- 如果目标对象实现了接口，可以强制使用 CGLIB 实现 AOP。
- 如果目标对象没有实现接口，必须采用 CGLIB 库，Spring 会自动在 JDK 动态代理和 CGLIB 之间转换。

如何强制使用 CGLIB 实现 AOP？
- 添加 CGLIB 库，Spring_HOME/cglib/*.jar。
- 在 Spring 配置文件中加入<aop:aspectj-autoproxy proxy-target-class="true"/>。

JDK 动态代理和 CGLIB 字节码生成的区别？
- JDK 动态代理只能对实现了接口的类生成代理，而不能针对类。
- CGLIB 是针对类实现代理，主要是对指定的类生成一个子类，覆盖其中的方法，因为是继承，所以该类或方法最好不要声明成 final。

2. 获取代理

确定了使用哪种代理方式后便可以进行代理的创建了，但是创建之前有必要回顾一下两种方式的使用方法。

1. JDK 代理使用示例。

创建业务接口，业务对外提供的接口，包含着业务可以对外提供的功能。

```java
public interface UserService {

    /**
     * 目标方法
     */
    public abstract void add();

}
```

创建业务接口实现类。

```java
public class UserServiceImpl implements UserService {

    /* (non-Javadoc)
     * @see dynamic.proxy.UserService#add()
     */
    public void add() {
        System.out.println("--------------------add---------------");
    }
}
```

创建自定义的 InvocationHandler，用于对接口提供的方法进行增强。

```java
public class MyInvocationHandler implements InvocationHandler {

    // 目标对象
    private Object target;
```

```java
/**
 * 构造方法
 * @param target 目标对象
 */
public MyInvocationHandler(Object target) {
    super();
    this.target = target;
}


/**
 * 执行目标对象的方法
 */
public Object invoke(Object proxy, Method method, Object[] args) throws Throwable {

    // 在目标对象的方法执行之前简单打印一下
    System.out.println("------------------before-----------------");

    // 执行目标对象的方法
    Object result = method.invoke(target, args);

    // 在目标对象的方法执行之后简单打印一下
    System.out.println("------------------after-----------------");

    return result;
}

/**
 * 获取目标对象的代理对象
 * @return 代理对象
 */
public Object getProxy() {
    return Proxy.newProxyInstance(Thread.currentThread().getContextClassLoader(),
            target.getClass().getInterfaces(), this);
}
}
```

最后进行测试，验证对于接口的增强是否起到作用。

```java
public class ProxyTest {

    @Test
    public void testProxy() throws Throwable {
        // 实例化目标对象
        UserService userService = new UserServiceImpl();

        // 实例化InvocationHandler
        MyInvocationHandler invocationHandler = new MyInvocationHandler(userService);

        // 根据目标对象生成代理对象
        UserService proxy = (UserService) invocationHandler.getProxy();

        // 调用代理对象的方法
        proxy.add();
```

        }
    }
执行结果如下:
```
-------------------before--------------
-------------------add----------------
-------------------after---------------
```
用起来很简单,其实这基本上就是 AOP 的一个简单实现了,在目标对象的方法执行之前和执行之后进行了增强。Spring 的 AOP 实现其实也是用了 Proxy 和 InvocationHandler 这两个东西的。

我们再次来回顾一下使用 JDK 代理的方式,在整个创建过程中,对于 InvocationHandler 的创建是最为核心的,在自定义的 InvocationHandler 中需要重写 3 个函数。

- 构造函数,将代理的对象传入。
- invoke 方法,此方法中实现了 AOP 增强的所有逻辑。
- getProxy 方法,此方法千篇一律,但是必不可少。

那么,我们看看 Spring 中的 JDK 代理实现是不是也是这么做的呢?继续之前的跟踪,到达 JdkDynamicAopProxy 的 getProxy。

```java
public Object getProxy(ClassLoader classLoader) {
    if (logger.isDebugEnabled()) {
        logger.debug("Creating JDK dynamic proxy: target source is " + this.advised.getTargetSource());
    }
    Class[] proxiedInterfaces = AopProxyUtils.completeProxiedInterfaces(this.advised);
    findDefinedEqualsAndHashCodeMethods(proxiedInterfaces);
    return Proxy.newProxyInstance(classLoader, proxiedInterfaces, this);
}
```

通过之前的示例我们知道,JDKProxy 的使用关键是创建自定义的 InvocationHandler,而 InvocationHandler 中包含了需要覆盖的函数 getProxy,而当前的方法正是完成了这个操作。再次确认一下 JdkDynamicAopProxy 也确实实现了 InvocationHandler 接口,那么我们就可以推断出,在 JdkDynamicAopProxy 中一定会有个 invoke 函数,并且 JdkDynamicAopProxy 会把 AOP 的核心逻辑写在其中。查看代码,果然有这样一个函数:

```java
public Object invoke(Object proxy, Method method, Object[] args) throws Throwable {
    MethodInvocation invocation;
    Object oldProxy = null;
    boolean setProxyContext = false;

    TargetSource targetSource = this.advised.targetSource;
    Class targetClass = null;
    Object target = null;

    try {
        //equals方法的处理
        if (!this.equalsDefined && AopUtils.isEqualsMethod(method)) {
            return equals(args[0]);
        }
```

```
//hash 方法的处理
if (!this.hashCodeDefined && AopUtils.isHashCodeMethod(method)) {
    return hashCode();
}
/*
 * Class 类的 isAssignableFrom(Class cls)方法：
 * 如果调用这个方法的class 或接口与参数 cls 表示的类或接口相同，
 * 或者是参数 cls 表示的类或接口的父类，则返回 true。
 * 形象地：自身类.class.isAssignableFrom(自身类或子类.class)   返回 true
 *     例：
 *     System.out.println(ArrayList.class.isAssignableFrom(Object.class));
 *         //false
 *     System.out.println(Object.class.isAssignableFrom(ArrayList.class));
 *         //true
 */
if (!this.advised.opaque && method.getDeclaringClass().isInterface() &&
        method.getDeclaringClass().isAssignableFrom(Advised.class)) {
    return AopUtils.invokeJoinpointUsingReflection(this.advised, method, args);
}

Object retVal;
//有时候目标对象内部的自我调用将无法实施切面中的增强则需要通过此属性暴露代理
if (this.advised.exposeProxy) {
    oldProxy = AopContext.setCurrentProxy(proxy);
    setProxyContext = true;
}

target = targetSource.getTarget();
if (target != null) {
    targetClass = target.getClass();
}

//获取当前方法的拦截器链
List<Object> chain = this.advised.getInterceptorsAndDynamicInterceptionAdvice
(method, targetClass);

if (chain.isEmpty()) {
    //如果没有发现任何拦截器那么直接调用切点方法
    retVal = AopUtils.invokeJoinpointUsingReflection(target, method, args);
}else {
    //将拦截器封装在 ReflectiveMethodInvocation，
    //以便于使用其 proceed 进行链接表用拦截器
    invocation = new ReflectiveMethodInvocation(proxy, target, method,
args, targetClass, chain);
    //执行拦截器链
    retVal = invocation.proceed();
}

//返回结果
if (retVal != null && retVal == target && method.getReturnType().IsInstance
(proxy) &&
        !RawTargetAccess.class.isAssignableFrom(method.getDeclaringClass())) {
    retVal = proxy;
}
```

```
            return retVal;
        }
        finally {
            if (target != null && !targetSource.isStatic()) {
                // Must have come from TargetSource.
                targetSource.releaseTarget(target);
            }
            if (setProxyContext) {
                // Restore old proxy.
                AopContext.setCurrentProxy(oldProxy);
            }
        }
    }
```

上面的函数中最主要的工作就是创建了一个拦截器链，并使用 ReflectiveMethodInvocation 类进行了链的封装，而在 ReflectiveMethodInvocation 类的 proceed 方法中实现了拦截器的逐一调用，那么我们继续来探究，在 proceed 方法中是怎么实现前置增强在目标方法前调用后置增强在目标方法后调用的逻辑呢？

```
    public Object proceed() throws Throwable {
        //    执行完所有增强后执行切点方法
        if (this.currentInterceptorIndex == this.interceptorsAndDynamicMethodMatchers.size() - 1) {
            return invokeJoinpoint();
        }

        //获取下一个要执行的拦截器
        Object interceptorOrInterceptionAdvice =
            this.interceptorsAndDynamicMethodMatchers.get(++this.currentInterceptorIndex);

        if (interceptorOrInterceptionAdvice instanceof InterceptorAndDynamicMethodMatcher) {
            //动态匹配
            InterceptorAndDynamicMethodMatcher dm =
                (InterceptorAndDynamicMethodMatcher) interceptorOrInterceptionAdvice;
            if (dm.methodMatcher.matches(this.method, this.targetClass, this.arguments)) {
                return dm.interceptor.invoke(this);
            }else {
                //不匹配则不执行拦截器
                return proceed();
            }
        }else {
            /*普通拦截器，直接调用拦截器,比如：
             * ExposeInvocationInterceptor、
             * DelegatePerTargetObjectIntroductionInterceptor、
             * MethodBeforeAdviceInterceptor
             * AspectJAroundAdvice、
             * AspectJAfterAdvice
             */
            //将 this 作为参数传递以保证当前实例中调用链的执行
            return ((MethodInterceptor) interceptorOrInterceptionAdvice).invoke(this);
        }
    }
```

在 proceed 方法中，或许代码逻辑并没有我们想象得那么复杂，ReflectiveMethodInvocation

中的主要职责是维护了链接调用的计数器，记录着当前调用链接的位置，以便链可以有序地进行下去，那么在这个方法中并没有我们之前设想的维护各种增强的顺序，而是将此工作委托给了各个增强器，使各个增强器在内部进行逻辑实现。

2. CGLIB 使用示例。

CGLIB 是一个强大的高性能的代码生成包。它广泛地被许多 AOP 的框架使用，例如 Spring AOP 和 dynaop，为它们提供方法的 Interception（拦截）。最流行的 OR Mapping 工具 Hibernate 也使用 CGLIB 来代理单端 single-ended（多对一和一对一）关联（对集合的延迟抓取是采用其他机制实现的）。EasyMock 和 jMock 是通过使用模仿（moke）对象来测试 Java 代码的包。它们都通过使用 CGLIB 来为那些没有接口的类创建模仿（moke）对象。

CGLIB 包的底层通过使用一个小而快的字节码处理框架 ASM，来转换字节码并生成新的类。除了 CGLIB 包，脚本语言例如 Groovy 和 BeanShell，也是使用 ASM 来生成 Java 的字节码。当然不鼓励直接使用 ASM，因为它要求你必须对 JVM 内部结构（包括 class 文件的格式和指令集）都很熟悉。

我们先快速地了解 CGLIB 的使用示例。

```java
import java.lang.reflect.Method;
import net.sf.cglib.proxy.Enhancer;
import net.sf.cglib.proxy.MethodInterceptor;
import net.sf.cglib.proxy.MethodProxy;

public class EnhancerDemo {
    public static void main(String[] args) {
        Enhancer enhancer = new Enhancer();
        enhancer.setSuperclass(EnhancerDemo.class);
        enhancer.setCallback(new MethodInterceptorImpl());

        EnhancerDemo demo = (EnhancerDemo) enhancer.create();
        demo.test();
        System.out.println(demo);
    }

    public void test() {
        System.out.println("EnhancerDemo test()");
    }

    private static class MethodInterceptorImpl implements MethodInterceptor {
        @Override
        public Object intercept(Object obj, Method method, Object[] args,
            MethodProxy proxy) throws Throwable {
            System.err.println("Before invoke " + method);
            Object result = proxy.invokeSuper(obj, args);
            System.err.println("After invoke " + method);
            return result;
        }
    }
}
```

7.3 创建 AOP 代理

运行结果如下：
```
Before invoke public void EnhancerDemo.test()
EnhancerDemo test()
After invokepublic void EnhancerDemo.test()
Before invoke public java.lang.String java.lang.Object.toString()
Before invoke public native int java.lang.Object.hashCode()
After invokepublic native int java.lang.Object.hashCode()
After invokepublic java.lang.String java.lang.Object.toString()
EnhancerDemo$$EnhancerByCGLIB$$bc9b2066@1621e42
```

可以看到 System.out.println(demo)，demo 首先调用了 toString()方法，然后又调用了 hashCode，生成的对象为 EnhancerDemo$$EnhancerByCGLIB$$bc9b2066 的实例，这个类是运行时由 CGLIB 产生的。

完成 CGLIB 代理的类是委托给 Cglib2AopProxy 类去实现的，我们进入这个类一探究竟。

按照前面提供的示例，我们容易判断出来，Cglib2AopProxy 的入口应该是在 getProxy，也就是说在 Cglib2AopProxy 类的 getProxy 方法中实现了 Enhancer 的创建及接口封装。

```
public Object getProxy(ClassLoader classLoader) {
    if (logger.isDebugEnabled()) {
        logger.debug("Creating CGLIB2 proxy: target source is " + this.advised.getTargetSource());
    }

    try {
        Class rootClass = this.advised.getTargetClass();
        Assert.state(rootClass != null, "Target class must be available for creating a CGLIB proxy");

        Class proxySuperClass = rootClass;
        if (ClassUtils.isCglibProxyClass(rootClass)) {
            proxySuperClass = rootClass.getSuperclass();
            Class[] additionalInterfaces = rootClass.getInterfaces();
            for (Class additionalInterface : additionalInterfaces) {
                this.advised.addInterface(additionalInterface);
            }
        }

        //验证 Class
        validateClassIfNecessary(proxySuperClass);

        //创建及配置 Enhancer
        Enhancer enhancer = createEnhancer();
        if (classLoader != null) {
            enhancer.setClassLoader(classLoader);
            if (classLoader instanceof SmartClassLoader &&
                    ((SmartClassLoader) classLoader).isClassReloadable(proxySuperClass)) {
                enhancer.setUseCache(false);
            }
        }
        enhancer.setSuperclass(proxySuperClass);
        enhancer.setStrategy(new UndeclaredThrowableStrategy(UndeclaredThrowableException.class));
```

```java
            enhancer.setInterfaces(AopProxyUtils.completeProxiedInterfaces(this.advised));
            enhancer.setInterceptDuringConstruction(false);

            //设置拦截器
            Callback[] callbacks = getCallbacks(rootClass);
            enhancer.setCallbacks(callbacks);
            enhancer.setCallbackFilter(new ProxyCallbackFilter(
                    this.advised.getConfigurationOnlyCopy(), this.fixedInterceptorMap, this.fixedInterceptorOffset));

            Class[] types = new Class[callbacks.length];
            for (int x = 0; x < types.length; x++) {
                types[x] = callbacks[x].getClass();
            }
            enhancer.setCallbackTypes(types);

            //生成代理类以及创建代理
            Object proxy;
            if (this.constructorArgs != null) {
                proxy = enhancer.create(this.constructorArgTypes, this.constructorArgs);
            }
            else {
                proxy = enhancer.create();
            }

            return proxy;
        }
        catch (CodeGenerationException ex) {
            throw new AopConfigException("Could not generate CGLIB subclass of class [" +
                    this.advised.getTargetClass() + "]: " +
                    "Common causes of this problem include using a final class or a non-visible class",
                    ex);
        }
        catch (IllegalArgumentException ex) {
            throw new AopConfigException("Could not generate CGLIB subclass of class [" +
                    this.advised.getTargetClass() + "]: " +
                    "Common causes of this problem include using a final class or a non-visible class",
                    ex);
        }
        catch (Exception ex) {
            // TargetSource.getTarget() failed
            throw new AopConfigException("Unexpected AOP exception", ex);
        }
    }
```

以上函数完整地阐述了一个创建 Spring 中的 Enhancer 的过程，读者可以参考 Enhancer 的文档查看每个步骤的含义，这里最重要的是通过 getCallbacks 方法设置拦截器链。

```java
    private Callback[] getCallbacks(Class rootClass) throws Exception {
        //对于 expose-proxy 属性的处理
        boolean exposeProxy = this.advised.isExposeProxy();
        boolean isFrozen = this.advised.isFrozen();
        boolean isStatic = this.advised.getTargetSource().isStatic();
```

7.3 创建 AOP 代理

```java
        //将拦截器封装在 DynamicAdvisedInterceptor 中
        Callback aopInterceptor = new DynamicAdvisedInterceptor(this.advised);

        // Choose a "straight to target" interceptor. (used for calls that are
        // unadvised but can return this). May be required to expose the proxy.
        Callback targetInterceptor;
        if (exposeProxy) {
            targetInterceptor = isStatic ?
                    new StaticUnadvisedExposedInterceptor (this.advised.getTargetSource().getTarget()) :
                    new DynamicUnadvisedExposedInterceptor(this.advised.getTargetSource());
        }else {
            targetInterceptor = isStatic ?
                    new StaticUnadvisedInterceptor(this.advised.getTargetSource().getTarget()) :
                    new DynamicUnadvisedInterceptor(this.advised.getTargetSource());
        }

        // Choose a "direct to target" dispatcher (used for
        // unadvised calls to static targets that cannot return this).
        Callback targetDispatcher = isStatic ?
                new StaticDispatcher(this.advised.getTargetSource().getTarget()) : new SerializableNoOp();

        Callback[] mainCallbacks = new Callback[]{
                //将拦截器链加入 Callback 中
                aopInterceptor,
                targetInterceptor, // invoke target without considering advice, if optimized
                new SerializableNoOp(), // no override for methods mapped to this
                targetDispatcher, this.advisedDispatcher,
                new EqualsInterceptor(this.advised),
                new HashCodeInterceptor(this.advised)
        };

        Callback[] callbacks;

        // If the target is a static one and the advice chain is frozen,
        // then we can make some optimisations by sending the AOP calls
        // direct to the target using the fixed chain for that method.
        if (isStatic && isFrozen) {
            Method[] methods = rootClass.getMethods();
            Callback[] fixedCallbacks = new Callback[methods.length];
            this.fixedInterceptorMap = new HashMap<String, Integer>(methods.length);

            // TODO: small memory optimisation here (can skip creation for
            // methods with no advice)
            for (int x = 0; x < methods.length; x++) {
                List<Object> chain = this.advised.getInterceptorsAndDynamicInterceptionAdvice(methods[x], rootClass);
                fixedCallbacks[x] = new FixedChainStaticTargetInterceptor(
                        chain, this.advised.getTargetSource().getTarget(), this.advised.getTargetClass());
                this.fixedInterceptorMap.put(methods[x].toString(), x);
```

```
            }
                // Now copy both the callbacks from mainCallbacks
                // and fixedCallbacks into the callbacks array.
                callbacks = new Callback[mainCallbacks.length + fixedCallbacks.length];
                System.arraycopy(mainCallbacks, 0, callbacks, 0, mainCallbacks.length);
                System.arraycopy(fixedCallbacks, 0, callbacks, mainCallbacks.length,
fixedCallbacks.length);
                this.fixedInterceptorOffset = mainCallbacks.length;
            }
        else {
            callbacks = mainCallbacks;
        }
        return callbacks;
}
```

在 getCallback 中 Spring 考虑了很多情况,但是对于我们来说,只需要理解最常用的就可以了,比如将 advised 属性封装在 DynamicAdvisedInterceptor 并加入在 callbacks 中,这么做的目的是什么呢,如何调用呢?在前面的示例中,我们了解到 CGLIB 中对于方法的拦截是通过将自定义的拦截器(实现 MethodInterceptor 接口)加入 Callback 中并在调用代理时直接激活拦截器中的 intercept 方法来实现的,那么在 getCallback 中正是实现了这样一个目的,DynamicAdvisedInterceptor 继承自 MethodInterceptor,加入 Callback 中后,在再次调用代理时会直接调用 DynamicAdvisedInterceptor 中的 intercept 方法,由此推断,对于 CGLIB 方式实现的代理,其核心逻辑必然在 DynamicAdvisedInterceptor 中的 intercept 中。

```
        public Object intercept(Object proxy, Method method, Object[] args, MethodProxy method
Proxy) throws Throwable {
                Object oldProxy = null;
                boolean setProxyContext = false;
                Class targetClass = null;
                Object target = null;
                try {
                    if (this.advised.exposeProxy) {
                        // Make invocation available if necessary.
                        oldProxy = AopContext.setCurrentProxy(proxy);
                        setProxyContext = true;
                    }
                    target = getTarget();
                    if (target != null) {
                        targetClass = target.getClass();
                    }
                    //获取拦截器链
                    List<Object> chain = this.advised.getInterceptorsAndDynamicInterceptionAdvice
(method, targetClass);
                    Object retVal;
                    if (chain.isEmpty() && Modifier.isPublic(method.getModifiers())) {
                        //如果拦截器链为空则直接激活原方法
                        retVal = methodProxy.invoke(target, args);
                    }else {
                        //进入链
                        retVal = new CglibMethodInvocation(proxy, target, method, args,
targetClass, chain, methodProxy).proceed();
```

```
        }
        retVal = massageReturnTypeIfNecessary(proxy, target, method, retVal);
        return retVal;
    }
    finally {
        if (target != null) {
            releaseTarget(target);
        }
        if (setProxyContext) {
            // Restore old proxy.
            AopContext.setCurrentProxy(oldProxy);
        }
    }
}
```

上述的实现与 JDK 方式实现代理中的 invoke 方法大同小异，都是首先构造链，然后封装此链进行串联调用，稍有些区别就是在 JDK 中直接构造 ReflectiveMethodInvocation，而在 cglib 中使用 CglibMethodInvocation。CglibMethodInvocation 继承自 ReflectiveMethodInvocation，但是 proceed 方法并没有重写。

7.4 静态 AOP 使用示例

加载时织入（Load-Time Weaving，LTW）指的是在虚拟机载入字节码文件时动态织入 AspectJ 切面。Spring 框架的值添加为 AspectJ LTW 在动态织入过程中提供了更细粒度的控制。使用 Java（5+）的代理能使用一个叫"Vanilla"的 AspectJ LTW，这需要在启动 JVM 的时候将某个 JVM 参数设置为开。这种 JVM 范围的设置在一些情况下或许不错，但通常情况下显得有些粗颗粒。而用 Spring 的 LTW 能让你在 per-ClassLoader 的基础上打开 LTW，这显然更加细粒度并且对"单 JVM 多应用"的环境更具意义（例如在一个典型应用服务器环境中）。另外，在某些环境下，这能让你使用 LTW 而不对应用服务器的启动脚本做任何改动，不然则需要添加 -javaagent:path/to/aspectjweaver.jar 或者（以下将会提及）-javaagent:path/to/Spring-agent.jar。开发人员只需简单修改应用上下文的一个或几个文件就能使用 LTW,而不需依靠那些管理者部署配置，比如启动脚本的系统管理员。

我们还是以之前的 AOP 示例为基础，如果想从动态代理的方式改成静态代理的方式需要做如下改动。

1. Spring 全局配置文件的修改，加入 LWT 开关。

```
<?xml version="1.0" encoding="UTF-8"?>
<beans xmlns="http://www.Springframework.org/schema/beans"
       xmlns:xsi="http://www.w3.org/2001/XMLSchema-instance"
       xmlns:aop="http://www.Springframework.org/schema/aop"
       xmlns:context="http://www.Springframework.org/schema/context"
       xsi:schemaLocation="http://www.Springframework.org/schema/beans
                           http://www.Springframework.org/schema/beans/Spring- beans
-3.0.xsd
```

```
                        http://www.Springframework.org/schema/aop
                        http://www.Springframework.org/schema/aop/Spring-aop -3.0.xsd
                        http://www.Springframework.org/schema/context http: //www.
Springframework.org/schema/context/Spring-context-3.0.xsd
                        ">
        <aop:aspectj-autoproxy />

        <bean id="test" class="test.TestBean"/>
        <bean class="test.AspectJTest"/>
        <context:load-time-weaver />
</beans>
```

2. 加入 aop.xml。在 class 目录下的 META-INF（没有则自己建立）文件夹下建立 aop.xml，内容如下：

```
<!DOCTYPE aspectj PUBLIC  "-//AspectJ//DTD//EN" "http://www.eclipse.org/aspectj/dtd/aspectj.dtd">
<aspectj>
    <weaver>
        <!-- only weave classes in our application-specific packages -->
        <include within="test.*" />
    </weaver>

    <aspects>
        <!-- weave in just this aspect -->
        <aspect name="test.AspectJTest" />
    </aspects>
</aspectj>
```

主要是告诉 AspectJ 需要对哪个包进行织入，并使用哪些增强器。

3. 加入启动参数。如果是在 Eclipse 中启动的话需要加上启动参数，如图 7-3 所示。

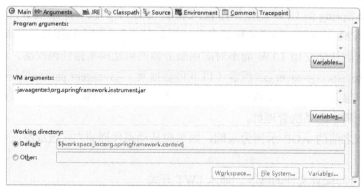

图 7-3　Eclipse 使用 AspectJ 的配置

4. 测试。
```
public static void main(String[] args) {
        ApplicationContext bf = new ClassPathXmlApplicationContext("aspectTest.xml");
        IITestBean bean=(IITestBean) bf.getBean("test");
        bean.testBeanM();
}
```

测试结果与动态 AOP 并无差别，打印出结果：
```
beforeTest
test
afterTest
```

7.5 创建 AOP 静态代理

AOP 的静态代理主要是在虚拟机启动时通过改变目标对象字节码的方式来完成对目标对象的增强，它与动态代理相比具有更高的效率，因为在动态代理调用的过程中，还需要一个动态创建代理类并代理目标对象的步骤，而静态代理则是在启动时便完成了字节码增强，当系统再次调用目标类时与调用正常的类并无差别，所以在效率上会相对高些。

7.5.1 Instrumentation 使用

Java 在 1.5 引入 java.lang.instrument，你可以由此实现一个 Java agent，通过此 agent 来修改类的字节码即改变一个类。本节会通过 Java Instrument 实现一个简单的 profiler。当然 instrument 并不限于 profiler，instrument 可以做很多事情，它类似一种更低级、更松耦合的 AOP，可以从底层来改变一个类的行为。你可以由此产生无限的遐想。接下来要做的事情，就是计算一个方法所花的时间，通常我们会在代码中按以下方式编写。

在方法开头加入 long stime = System.nanoTime();，在方法结尾通过 System.nanoTime()-stime 得出方法所花时间。你不得不在想监控的每个方法中写入重复的代码，好一点的情况，你可以用 AOP 来做这事，但总是感觉有点别扭，这种 profiler 的代码还是要打包在你的项目中，Java Instrument 使得这一切更干净。

1. 写 ClassFileTransformer 类

```java
package org.toy;
import java.lang.instrument.ClassFileTransformer;
import java.lang.instrument.IllegalClassFormatException;
import java.security.ProtectionDomain;
import javassist.CannotCompileException;
import javassist.ClassPool;
import javassist.CtBehavior;
import javassist.CtClass;
import javassist.NotFoundException;
import javassist.expr.ExprEditor;
import javassist.expr.MethodCall;
public class PerfMonXformer implements ClassFileTransformer {
    public byte[] transform(ClassLoader loader, String className,
            Class<?> classBeingRedefined, ProtectionDomain protectionDomain,
            byte[] classfileBuffer) throws IllegalClassFormatException {
        byte[] transformed = null;
        System.out.println("Transforming " + className);
        ClassPool pool = ClassPool.getDefault();
        CtClass cl = null;
```

```
            try {
                cl = pool.makeClass(new java.io.ByteArrayInputStream(
                        classfileBuffer));
                if (cl.isInterface() == false) {
                    CtBehavior[] methods = cl.getDeclaredBehaviors();
                    for (int i = 0; i < methods.length; i++) {
                        if (methods[i].isEmpty() == false) {
                         //修改 method 字节码
                            doMethod(methods[i]);
                        }
                    }
                    transformed = cl.toBytecode();
                }
            } catch (Exception e) {
                System.err.println("Could not instrument  " + className
                        + ",  exception : " + e.getMessage());
            } finally {
                if (cl != null) {
                    cl.detach();
                }
            }
            return transformed;
        }

        private void doMethod(CtBehavior method) throws NotFoundException,
                CannotCompileException {
            method.insertBefore("long stime = System.nanoTime();");
            method.insertAfter("System.out.println(/"leave "+method.getName()+" and time: /"+(System.nanoTime()-stime));");

        }
    }
```

2. 编写 agent 类

```
    package org.toy;
    import java.lang.instrument.Instrumentation;
    import java.lang.instrument.ClassFileTransformer;
    public class PerfMonAgent {
        static private Instrumentation inst = null;
        /**
         * This method is called before the application's main-method is called,
         * when this agent is specified to the Java VM.
         **/
        public static void premain(String agentArgs, Instrumentation _inst) {
            System.out.println("PerfMonAgent.premain() was called.");
            // Initialize the static variables we use to track information.
            inst = _inst;
            // Set up the class-file transformer.
            ClassFileTransformer trans = new PerfMonXformer();
            System.out.println("Adding a PerfMonXformer instance to the JVM.");
            inst.addTransformer(trans);
        }
    }
```

上面两个类就是 agent 的核心了，JVM 启动时在应用加载前会调用 PerfMonAgent.premain，然后 PerfMonAgent.premain 中实例化了一个定制的 ClassFileTransforme，即 PerfMonXformer 并通过 inst.addTransformer(trans)把 PerfMonXformer 的实例加入 Instrumentation 实例（由 JVM 传入），这就使得应用中的类加载时，PerfMonXformer.transform 都会被调用，你在此方法中可以改变加载的类。真的有点神奇，为了改变类的字节码，我使用了 JBoss 的 Javassist，虽然你不一定要这么用，但 JBoss 的 Javassist 真的很强大，能让你很容易地改变类的字节码。在上面的方法中我通过改变类的字节码，在每个类的方法入口中加入了 long stime = System.nanoTime()，在方法的出口加入了：

```
System.out.println("methodClassName.methodName:"+(System.nanoTime()-stime));
```

3. 打包 agent

对于 agent 的打包，有点讲究。

- JAR 的 META-INF/MANIFEST.MF 加入 Premain-Class: xx，xx 在此语境中就是我们的 agent 类，即 org.toy.PerfMonAgent。
- 如果你的 agent 类引入别的包，需使用 Boot-Class-Path: xx，xx 在此语境中就是上面提到的 JBoss javassit，即/home/pwlazy/.m2/repository/javassist/javassist/3.8.0 .GA/ javassist-3.8.0.GA.jar。

下面附上 Maven 的 POM。

```xhtml
[xhtml]view plaincopyprint?
<project xmlns="http://maven.apache.org/POM/4.0.0" xmlns:xsi="http://www.w3.org/2001/XMLSchema-instance"
    xsi:schemaLocation="http://maven.apache.org/POM/4.0.0 http://maven.apache.org/maven-v4_0_0.xsd">
    <modelVersion>4.0.0</modelVersion>
    <groupId>org.toy</groupId>
    <artifactId>toy-inst</artifactId>
    <packaging>jar</packaging>
    <version>1.0-SNAPSHOT</version>
    <name>toy-inst</name>
    <url>http://maven.apache.org</url>
    <dependencies>
      <dependency>
        <groupId>javassist</groupId>
        <artifactId>javassist</artifactId>
        <version>3.8.0.GA</version>
      </dependency>
      <dependency>
        <groupId>junit</groupId>
        <artifactId>junit</artifactId>
        <version>3.8.1</version>
        <scope>test</scope>
      </dependency>
    </dependencies>

    <build>
     <plugins>
       <plugin>
         <groupId>org.apache.maven.plugins</groupId>
```

```xml
            <artifactId>maven-jar-plugin</artifactId>
            <version>2.2</version>
            <configuration>
              <archive>
                <manifestEntries>
                  <Premain-Class>org.toy.PerfMonAgent</Premain-Class>
                  <Boot-Class-Path>/home/pwlazy/.m2/repository/javassist/javassist/3.8.0.GA/javassist-3.8.0.GA.jar</Boot-Class-Path>
                </manifestEntries>
              </archive>
            </configuration>
          </plugin>

          <plugin>
            <artifactId>maven-compiler-plugin </artifactId >
                <configuration>
                    <source> 1.6 </source >
                    <target> 1.6 </target>
                </configuration>
          </plugin>
       </plugins>


  </build>

</project>
```

4. 打包应用

```java
package org.toy;
public class App {
    public static void main(String[] args) {
        new App().test();
    }
    public void test() {
        System.out.println("Hello World!!");
    }
}
```

Java 选项中有-javaagent:xx，其中 xx 就是你的 agent JAR，Java 通过此选项加载 agent，由 agent 来监控 classpath 下的应用。

最后的执行结果：

```
PerfMonAgent.premain() was called.
Adding a PerfMonXformer instance to the JVM.
Transforming org/toy/App
Hello World!!
java.io.PrintStream.println:314216
org.toy.App.test:540082
Transforming java/lang/Shutdown
Transforming java/lang/Shutdown$Lock
java.lang.Shutdown.runHooks:29124
java.lang.Shutdown.sequence:132768
```

由执行结果可以看出，执行顺序以及通过改变 org.toy.App 的字节码加入监控代码确实生效了。你也可以发现，通过 Instrment 实现 agent 使得监控代码和应用代码完全隔离了。

通过之前的两个小示例我们似乎已经有所体会，在 Spring 中的静态 AOP 直接使用了 AspectJ 提供的方法，而 AspectJ 又是在 Instrument 基础上进行的封装。就以上面的例子来看，至少在 AspectJ 中会有如下功能。

- 读取 META-INF/aop.xml。
- 将 aop.xml 中定义的增强器通过自定义的 ClassFileTransformer 织入对应的类中。

当然这都是 AspectJ 所做的事情，并不在我们讨论的范畴，Spring 是直接使用 AspectJ，也就是将动态代理的任务直接委托给了 AspectJ，那么，Spring 怎么嵌入 AspectJ 的呢?同样我们还是从配置文件入手。

7.5.2 自定义标签

在 Spring 中如果需要使用 AspectJ 的功能,首先要做的第一步就是在配置文件中加入配置: `<context:load-time-weaver/>`。我们根据之前介绍的自定义命名空间的知识便可以推断，引用 AspectJ 的入口便是这里，可以通过查找 load-time-weaver 来找到对应的自定义命名处理类。

通过 Eclipse 提供的字符串搜索功能，我们找到了 ContextNamespaceHandler，在其中有这样一段函数。

```
public void init() {
        registerBeanDefinitionParser("property-placeholder", new PropertyPlaceholderBeanDefinitionParser());
        registerBeanDefinitionParser("property-override", new PropertyOverrideBeanDefinitionParser());
        registerBeanDefinitionParser("annotation-config", new AnnotationConfigBeanDefinitionParser());
        registerBeanDefinitionParser("component-scan", new ComponentScanBeanDefinitionParser());
        registerBeanDefinitionParser("load-time-weaver", new LoadTimeWeaverBeanDefinitionParser());
        registerBeanDefinitionParser("Spring-configured", new SpringConfiguredBeanDefinitionParser());
        registerBeanDefinitionParser("mbean-export", new MBeanExportBeanDefinitionParser());
        registerBeanDefinitionParser("mbean-server", new MBeanServerBeanDefinitionParser());
}
```

继续跟进 LoadTimeWeaverBeanDefinitionParser，作为 BeanDefinitionParser 接口的实现类，它的核心逻辑是从 parse 函数开始的，而经过父类的封装，LoadTimeWeaverBeanDefinitionParser 类的核心实现被转移到了 doParse 函数中，如下：

```
protected void doParse(Element element, ParserContext parserContext, BeanDefinitionBuilder builder) {
        builder.setRole(BeanDefinition.ROLE_INFRASTRUCTURE);

        if (isAspectJWeavingEnabled(element.getAttribute(ASPECTJ_WEAVING_ATTRIBUTE), parserContext)) {
            RootBeanDefinition weavingEnablerDef = new RootBeanDefinition();
            // ASPECTJ_WEAVING_ENABLER_CLASS_NAME =
```

```
                    // "org.Springframework.context.weaving.AspectJWeavingEnabler";
            weavingEnablerDef.setBeanClassName(ASPECTJ_WEAVING_ENABLER_CLASS_NAME);
            parserContext.getReaderContext().registerWithGeneratedName(weavingEnablerDef);

            if (isBeanConfigurerAspectEnabled(parserContext.getReaderContext().getBeanClassLoader())) {
                new SpringConfiguredBeanDefinitionParser().parse(element, parserContext);
            }
        }
    }
```

其实之前在分析动态 AOP 也就是在分析配置<aop:aspectj-autoproxy />中已经提到了自定义配置的解析流程，对于<aop:aspectj-autoproxy/>的解析无非是以标签作为标志，进而进行相关处理类的注册，那么对于自定义标签<context:load-time-weaver />其实是起到了同样的作用。

上面函数的核心作用其实就是注册一个对于 ApectJ 处理的类 org.Springframework.context.weaving.AspectJWeavingEnabler，它的注册流程总结起来如下。

1. 是否开启 AspectJ。

之前虽然反复提到了在配置文件中加入了<context:load-time-weaver/>便相当于加入了 AspectJ 开关。但是，并不是配置了这个标签就意味着开启 AspectJ 功能，这个标签中还有一个属性 aspectj-weaving，这个属性有 3 个备选值，on、off 和 autodetect，默认为 autodetect，也就是说，如果我们只是使用了<context:load-time-weaver/>，那么 Spring 会帮助我们检测是否可以使用 AspectJ 功能，而检测的依据便是文件 META-INF/aop.xml 是否存在，看看在 Spring 中的实现方式。

```
protected boolean isAspectJWeavingEnabled(String value, ParserContext parserContext) {
    if ("on".equals(value)) {
        return true;
    }
    else if ("off".equals(value)) {
        return false;
    }
    else {
        //自动检测
        ClassLoader cl = parserContext.getReaderContext().getResourceLoader().getClassLoader();
        return (cl.getResource(AspectJWeavingEnabler.ASPECTJ_AOP_XML_RESOURCE) != null);
    }
}
```

2. 将 org.Springframework.context.weaving.AspectJWeavingEnabler 封装在 BeanDefinition 中注册。

当通过 AspectJ 功能验证后便可以进行 AspectJWeavingEnabler 的注册了，注册的方式很简单，无非是将类路径注册在新初始化的 RootBeanDefinition 中，在 RootBeanDefinition 的获取时会转换成对应的 class。

尽管在 init 方法中注册了 AspectJWeavingEnabler，但是对于标签本身 Spring 也会以 bean 的形式保存，也就是当 Spring 解析到<context:load-time-weaver/>标签的时候也会产生一个 bean，而这个 bean 中的信息是什么呢？

在 LoadTimeWeaverBeanDefinitionParser 类中有这样的函数：

```
@Override
protected String getBeanClassName(Element element) {
        if (element.hasAttribute(WEAVER_CLASS_ATTRIBUTE)) {
            return element.getAttribute(WEAVER_CLASS_ATTRIBUTE);
        }
        return DEFAULT_LOAD_TIME_WEAVER_CLASS_NAME;
}

    @Override
protected String resolveId(Element element, AbstractBeanDefinition definition,
ParserContext parserContext) {
        return ConfigurableApplicationContext.LOAD_TIME_WEAVER_BEAN_NAME;
}
```

其中，可以看到：

```
WEAVER_CLASS_ATTRIBUTE="weaver-class"
DEFAULT_LOAD_TIME_WEAVER_CLASS_NAME =
        "org.Springframework.context.weaving.DefaultContextLoadTimeWeaver";
ConfigurableApplicationContext.LOAD_TIME_WEAVER_BEAN_NAME="loadTimeWeaver"
```

单凭以上的信息我们至少可以推断，当 Spring 在读取到自定义标签<context:load-time-weaver/>后会产生一个 bean，而这个 bean 的 id 为 loadTimeWeaver，class 为 org.Springframework.context.weaving.DefaultContextLoadTimeWeaver，也就是完成了 DefaultContextLoadTimeWeaver 类的注册。

完成了以上的注册功能后，并不意味这在 Spring 中就可以使用 AspectJ 了，因为我们还有一个很重要的步骤忽略了，就是 LoadTimeWeaverAwareProcessor 的注册。在 AbstractApplicationContext 中的 prepareBeanFactory 函数中有这样一段代码：

```
//增加对 AspectJ 的支持
        if (beanFactory.containsBean(LOAD_TIME_WEAVER_BEAN_NAME)) {
            beanFactory.addBeanPostProcessor(new LoadTimeWeaverAwareProcessor
(beanFactory));
            // Set a temporary ClassLoader for type matching.
            beanFactory.setTempClassLoader(new ContextTypeMatchClassLoader (beanFactory.
getBeanClassLoader()));
        }
```

在 AbstractApplicationContext 中的 prepareBeanFactory 函数是在容器初始化时候调用的，也就是说只有注册了 LoadTimeWeaverAwareProcessor 才会激活整个 AspectJ 的功能。

7.5.3 织入

当我们完成了所有的 AspectJ 的准备工作后便可以进行织入分析了，首先还是从 LoadTimeWeaverAwareProcessor 开始。

LoadTimeWeaverAwareProcessor 实现 BeanPostProcessor 方法，那么对于 BeanPostProcessor 接口来讲，postProcessBeforeInitialization 与 postProcessAfterInitialization 有着其特殊意义，也就是说在所有 bean 的初始化之前与之后都会分别调用对应的方法，那么在 LoadTimeWeaverAwareProcessor

中的 postProcessBeforeInitialization 函数中完成了什么样的逻辑呢？

```
    public Object postProcessBeforeInitialization(Object bean, String beanName) throws
BeansException {
            if (bean instanceof LoadTimeWeaverAware) {
                LoadTimeWeaver ltw = this.loadTimeWeaver;
                if (ltw == null) {
                    Assert.state(this.beanFactory != null,
                            "BeanFactory required if no LoadTimeWeaver explicitly specified");
                    ltw = this.beanFactory.getBean(
                            ConfigurableApplicationContext.LOAD_TIME_WEAVER_BEAN_NAME,
LoadTimeWeaver.class);
                }
                ((LoadTimeWeaverAware) bean).setLoadTimeWeaver(ltw);
            }
            return bean;
    }
```

我们综合之前讲解的所有信息，将所有相关信息串联起来一起分析这个函数。

在 LoadTimeWeaverAwareProcessor 中的 postProcessBeforeInitialization 函数中，因为最开始的 if 判断注定这个后处理器只对 LoadTimeWeaverAware 类型的 bean 起作用，而纵观所有的 bean，实现 LoadTimeWeaver 接口的类只有 AspectJWeavingEnabler。

当在 Spring 中调用 AspectJWeavingEnabler 时，this.loadTimeWeaver 尚未被初始化，那么，会直接调用 beanFactory.getBean 方法获取对应的 DefaultContextLoadTimeWeaver 类型的 bean，并将其设置为 AspectJWeavingEnabler 类型 bean 的 loadTimeWeaver 属性中。当然 AspectJWeavingEnabler 同样实现了 BeanClassLoaderAware 以及 Ordered 接口，实现 BeanClassLoaderAware 接口保证了在 bean 初始化的时候调用 AbstractAutowireCapableBeanFactory 的 invokeAwareMethods 的时候将 beanClassLoader 赋值给当前类。而实现 Ordered 接口则保证在实例化 bean 时当前 bean 会被最先初始化。

而 DefaultContextLoadTimeWeaver 类又同时实现了 LoadTimeWeaver、BeanClassLoaderAware 以及 DisposableBean。其中 DisposableBean 接口保证在 bean 销毁时会调用 destroy 方法进行 bean 的清理，而 BeanClassLoaderAware 接口则保证在 bean 的初始化调用 AbstractAutowireCapable-BeanFactory 的 invokeAwareMethods 时调用 setBeanClassLoader 方法。

```
    public void setBeanClassLoader(ClassLoader classLoader) {
            LoadTimeWeaver serverSpecificLoadTimeWeaver = createServerSpecificLoadTimeWeaver
(classLoader);
            if (serverSpecificLoadTimeWeaver != null) {
                if (logger.isInfoEnabled()) {
                    logger.info("Determined server-specific load-time weaver: " +
                            serverSpecificLoadTimeWeaver.getClass().getName());
                }
                this.loadTimeWeaver = serverSpecificLoadTimeWeaver;
            }else if (InstrumentationLoadTimeWeaver.isInstrumentationAvailable()) {
                //检查当前虚拟机中的 Instrumentation 实例是否可用
                logger.info("Found Spring's JVM agent for instrumentation");
                this.loadTimeWeaver = new InstrumentationLoadTimeWeaver(classLoader);
            }else {
```

```
        try {
            this.loadTimeWeaver = new ReflectiveLoadTimeWeaver(classLoader);
            logger.info("Using a reflective load-time weaver for class loader: " +
                    this.loadTimeWeaver.getInstrumentableClassLoader(). getClass().getName());
        }
        catch (IllegalStateException ex) {
            throw new IllegalStateException(ex.getMessage() + " Specify a custom LoadTimeWeaver or start your " +
                    "Java virtual machine with Spring's agent: -javaagent:org.Springframework.instrument.jar");
        }
    }
}
```

上面的函数中有一句很容易被忽略但是很关键的代码：

`this.loadTimeWeaver = new InstrumentationLoadTimeWeaver(classLoader);`

这句代码不仅仅是实例化了一个 InstrumentationLoadTimeWeaver 类型的实例，而且在实例化过程中还做了一些额外的操作。

在实例化的过程中会对当前的 this.instrumentation 属性进行初始化，而初始化的代码如下：this.instrumentation = getInstrumentation()，也就是说在 InstrumentationLoadTimeWeaver 实例化后其属性 Instrumentation 已经被初始化为代表着当前虚拟机的实例了。综合我们讲过的例子，对于注册转换器，如 addTransformer 函数等，便可以直接使用此属性进行操作了。

也就是经过以上程序的处理后，在 Spring 中的 bean 之间的关系如下。

- AspectJWeavingEnabler 类型的 bean 中的 loadTimeWeaver 属性被初始化为 DefaultContextLoadTimeWeaver 类型的 bean。
- DefaultContextLoadTimeWeaver 类型的 bean 中的 loadTimeWeaver 属性被初始化为 InstrumentationLoadTimeWeaver。

因为 AspectJWeavingEnabler 类同样实现了 BeanFactoryPostProcessor，所以当所有 bean 解析结束后会调用其 postProcessBeanFactory 方法。

```
public void postProcessBeanFactory(ConfigurableListableBeanFactory beanFactory) throws BeansException {
        enableAspectJWeaving(this.loadTimeWeaver, this.beanClassLoader);
    }

    public static void enableAspectJWeaving(LoadTimeWeaver weaverToUse, ClassLoader beanClassLoader) {
        if (weaverToUse == null) {
//此时已经被初始化为 DefaultContextLoadTimeWeaver
            if (InstrumentationLoadTimeWeaver.isInstrumentationAvailable()) {
                weaverToUse = new InstrumentationLoadTimeWeaver(beanClassLoader);
            }
            else {
                throw new IllegalStateException("No LoadTimeWeaver available");
            }
        }
```

```java
//使用DefaultContextLoadTimeWeaver类型的bean中的loadTimeWeaver属性注册转换器
        weaverToUse.addTransformer(new AspectJClassBypassingClassFileTransformer(
                new ClassPreProcessorAgentAdapter()));
    }

    private static class AspectJClassBypassingClassFileTransformer implements ClassFileTransformer {

        private final ClassFileTransformer delegate;

        public AspectJClassBypassingClassFileTransformer(ClassFileTransformer delegate) {
            this.delegate = delegate;
        }

        public byte[] transform(ClassLoader loader, String className, Class<?> classBeingRedefined, ProtectionDomain protectionDomain, byte[] classfileBuffer) throws IllegalClassFormatException {
            if (className.startsWith("org.aspectj") || className.startsWith("org/aspectj")) {
                return classfileBuffer;
            }
            //委托给AspectJ代理继续处理
            return this.delegate.transform(loader, className, classBeingRedefined, protectionDomain, classfileBuffer);
        }
    }
```

AspectJClassBypassingClassFileTransformer 的作用仅仅是告诉 AspectJ 以 org.aspectj 开头的或者 org/aspectj 开头的类不进行处理。

第 2 部分 企业应用

第 8 章　数据库连接 JDBC
第 9 章　整合 MyBatis
第 10 章　事务
第 11 章　SpringMVC
第 12 章　远程服务
第 13 章　Spring 消息

第 8 章　数据库连接 JDBC

JDBC（Java Data Base Connectivity，Java 数据库连接）是一种用于执行 SQL 语句的 Java API，可以为多种关系数据库提供统一访问，它由一组用 Java 语言编写的类和接口组成。JDBC 为数据库开发人员提供了一个标准的 API，据此可以构建更高级的工具和接口，使数据库开发人员能够用纯 Java API 编写数据库应用程序，并且可跨平台运行，并且不受数据库供应商的限制。

JDBC 连接数据库的流程及其原理如下。

1. 在开发环境中加载指定数据库的驱动程序。接下来的实验中，使用的数据库是 MySQL，所以需要去下载 MySQL 支持 JDBC 的驱动程序（最新的版本是 mysql-connector-java-5.1.18-bin.jar），将下载得到的驱动程序加载进开发环境中（开发环境是 MyEclipse，具体示例时会讲解如何加载）。

2. 在 Java 程序中加载驱动程序。在 Java 程序中，可以通过"Class.forName("指定数据库的驱动程序")"的方式来加载添加到开发环境中的驱动程序，例如加载 MySQL 的数据驱动程序的代码为 Class.forName("com.mysql.jdbc.Driver")。

3. 创建数据连接对象。通过 DriverManager 类创建数据库连接对象 Connection。DriverManager 类作用于 Java 程序和 JDBC 驱动程序之间，用于检查所加载的驱动程序是否可以建立连接，然后通过它的 getConnection 方法根据数据库的 URL、用户名和密码，创建一个 JDBC Connection 对象，例如：Connection connection = DriverManager.geiConnection("连接数据库的 URL","用户名","密码")。其中，URL=协议名+IP 地址（域名）+端口+数据库名称；用户名和密码是指登录数据库时所使用的用户名和密码。具体示例创建 MySQL 的数据库连接代码如下：

```
Connection connectMySQL = DriverManager.geiConnection("jdbc:mysql://localhost:3306/myuser","root" ,"root" );
```

4. 创建 Statement 对象。Statement 类的主要是用于执行静态 SQL 语句并返回它所生成结果的对象。通过 Connection 对象的 createStatement()方法可以创建一个 Statement 对象。例如：Statement statament = connection.createStatement()。具体示例创建 Statement 对象代码如下：

```
Statement statamentMySQL =connectMySQL.createStatement();
```

5. 调用 Statement 对象的相关方法执行相对应的 SQL 语句。通过 execuUpdate()方法来对数据更新，包括插入和删除等操作，例如向 staff 表中插入一条数据的代码：

```
statement.excuteUpdate( "INSERT INTO staff(name, age, sex,address, depart, worklen,wage)" + " VALUES ('Tom1', 321, 'M', 'china','Personnel','3','3000' ) ") ;
```

通过调用 Statement 对象的 executeQuery() 方法进行数据的查询，而查询结果会得到 ResulSet 对象，ResulSet 表示执行查询数据库后返回的数据的集合，ResultSet 对象具有可以指向当前数据行的指针。通过该对象的 next() 方法，使得指针指向下一行，然后将数据以列号或者字段名取出。如果当 next() 方法返回 null，则表示下一行中没有数据存在。使用示例代码如下：

```
ResultSet resultSel = statement.executeQuery( "select * from staff" );
```

6. 关闭数据库连接。使用完数据库或者不需要访问数据库时，通过 Connection 的 close() 方法及时关闭数据连接。

8.1 Spring 连接数据库程序实现（JDBC）

Spring 中的 JDBC 连接与直接使用 JDBC 去连接还是有所差别的，Spring 对 JDBC 做了大量封装，消除了冗余代码，使得开发量大大减小。下面通过一个小例子让大家简单认识 Spring 中的 JDBC 操作。

1. 创建数据表结构

```
CREATE TABLE 'user' (
  'id' int(11) NOT NULL auto_increment,
  'name' varchar(255) default NULL,
  'age' int(11) default NULL,
  'sex' varchar(255) default NULL,
  PRIMARY KEY  ('id')
) ENGINE=InnoDB DEFAULT CHARSET=utf8;
```

2. 创建对应数据表的 PO

```
public class User {

    private int id;
    private String name;
    private int age;
private String sex;

        //省略 set/get 方法
}
```

3. 创建表与实体间的映射

```
public class UserRowMapper implements RowMapper {

    @Override
    public Object mapRow(ResultSet set, int index) throws SQLException {
        User person = new User(set.getInt("id"), set.getString("name"), set
                .getInt("age"), set.getString("sex"));
        return person;
```

4. 创建数据操作接口

```
public interface UserService {

    public  void save(User user);

    public  List<User> getUsers();

}
```

5. 创建数据操作接口实现类

```
public class UserServiceImpl implements UserService {

    private JdbcTemplate jdbcTemplate;

    // 设置数据源
    public void setDataSource(DataSource dataSource) {
        this.jdbcTemplate = new JdbcTemplate(dataSource);
    }

    public void save(User user) {
        jdbcTemplate.update("insert into user(name,age,sex)values(?,?,?)",
                new Object[] { user.getName(), user.getAge(),
                        user.getSex() }, new int[] { java.sql.Types.VARCHAR,
                        java.sql.Types.INTEGER, java.sql.Types.VARCHAR });
    }

    @SuppressWarnings("unchecked")
    public List<User> getUsers() {
        List<User> list = jdbcTemplate.query("select * from user", new UserRowMapper());
        return list;
    }

}
```

6. 创建 Spring 配置文件

```xml
<?xml version="1.0" encoding="UTF-8"?>
<beans xmlns="http://www.Springframework.org/schema/beans"
    xmlns:xsi="http://www.w3.org/2001/XMLSchema-instance"
    xsi:schemaLocation="http://www.Springframework.org/schema/beans
            http://www.Springframework.org/schema/beans/Spring-beans-2.5.xsd
            ">
    <!--配置数据源 -->
    <bean id="dataSource" class="org.apache.commons.dbcp.BasicDataSource"
        destroy-method="close">
        <property name="driverClassName" value="com.mysql.jdbc.Driver" />
        <property name="url"
            value="jdbc:mysql://localhost:3306/lexueba" />
        <property name="username" value="root" />
        <property name="password" value="haojia0421xixi" />
        <!-- 连接池启动时的初始值 -->
```

```xml
            <property name="initialSize" value="1" />
            <!-- 连接池的最大值 -->
            <property name="maxActive" value="300" />
            <!-- 最大空闲值.当经过一个高峰时间后,连接池可以慢慢将已经用不到的连接慢慢释放一部分,一直减
少到maxIdle为止 -->
            <property name="maxIdle" value="2" />
            <!-- 最小空闲值.当空闲的连接数少于阀值时,连接池就会预申请去一些连接,以免洪峰来时来不及申请 -->
            <property name="minIdle" value="1" />
    </bean>

    <!-- 配置业务bean：PersonServiceBean -->
    <bean id="userService" class="service.UserServiceImpl">
        <!-- 向属性dataSource注入数据源 -->
        <property name="dataSource" ref="dataSource"></property>
    </bean>
</beans>
```

7．测试

```java
public class SpringJDBCTest {

    public static void main(String[] args) {
        ApplicationContext act = new ClassPathXmlApplicationContext("bean.xml");

        UserService userService = (UserService) act.getBean("userService");
        User user = new User();
        user.setName("张三");
        user.setAge(20);
        user.setSex("男");
        // 保存一条记录
        userService.save(user);

        List<User> person1 = userService.getUsers();
        System.out.println("++++++++得到所有User");
        for (User person2 : person1) {
            System.out.println(person2.getId() + "   " + person2.getName()
                    + "   " + person2.getAge() + "   " + person2.getSex());
        }

    }
}
```

8.2 save/update 功能的实现

我们以上面的例子为基础开始分析Spring中对JDBC的支持,首先寻找整个功能的切入点,在示例中我们可以看到所有的数据库操作都封装在了 UserServiceImpl 中,而 UserServiceImpl 中的所有数据库操作又以其内部属性 jdbcTemplate 为基础。这个 jdbcTemplate 可以作为源码分析的切入点,我们一起看看它是如何实现又是如何被初始化的。

在 UserServiceImpl 中 jdbcTemplate 的初始化是从 setDataSource 函数开始的,DataSource 实例通过参数注入,DataSource 的创建过程是引入第三方的连接池,这里不做过多介绍。DataSource 是整个数据库操作的基础,里面封装了整个数据库的连接信息。我们首先以保存实

体类为例进行代码跟踪。
```java
public void save(User user) {
    jdbcTemplate.update("insert into user(name,age,sex)values(?,?,?)",
            new Object[] { user.getName(), user.getAge(),
                user.getSex() }, new int[] { java.sql.Types.VARCHAR,
                java.sql.Types.INTEGER, java.sql.Types.VARCHAR });
}
```

对于保存一个实体类来讲，在操作中我们只需要提供 SQL 语句以及语句中对应的参数和参数类型，其他操作便可以交由 Spring 来完成了，这些工作到底包括什么呢？进入 jdbcTemplate 中的 update 方法。

```java
public int update(String sql, Object[] args, int[] argTypes) throws DataAccessException {
    return update(sql, newArgTypePreparedStatementSetter(args, argTypes));
}

public int update(String sql, PreparedStatementSetter pss) throws DataAccessException {
    return update(new SimplePreparedStatementCreator(sql), pss);
}
```

进入 update 方法后，Spring 并不是急于进入核心处理操作，而是先做足准备工作，使用 ArgTypePreparedStatementSetter 对参数与参数类型进行封装，同时又使用 Simple PreparedStatement Creator 对 SQL 语句进行封装。至于为什么这么封装，暂且留下悬念。

经过了数据封装后便可以进入了核心的数据处理代码了。

```java
protected int update(final PreparedStatementCreator psc, final PreparedStatementSetter pss)
        throws DataAccessException {

    logger.debug("Executing prepared SQL update");
    return execute(psc, new PreparedStatementCallback<Integer>() {
        public Integer doInPreparedStatement(PreparedStatement ps) throws SQLException {
            try {
                if (pss != null) {
                    //设置 PreparedStatement 所需的全部参数
                    pss.setValues(ps);
                }
                int rows = ps.executeUpdate();
                if (logger.isDebugEnabled()) {
                    logger.debug("SQL update affected " + rows + " rows");
                }
                return rows;
            }
            finally {
                if (pss instanceof ParameterDisposer) {
                    ((ParameterDisposer) pss).cleanupParameters();
                }
            }
        }
    });
}
```

如果读者了解过其他操作方法，可以知道 execute 方法是最基础的操作，而其他操作比如 update、query 等方法则是传入不同的 PreparedStatementCallback 参数来执行不同的逻辑。

8.2.1 基础方法 execute

execute 作为数据库操作的核心入口，将大多数数据库操作相同的步骤统一封装，而将个性化的操作使用参数 PreparedStatementCallback 进行回调。

```java
public <T> T execute(PreparedStatementCreator psc, PreparedStatementCallback<T> action)
        throws DataAccessException {
    Assert.notNull(psc, "PreparedStatementCreator must not be null");
    Assert.notNull(action, "Callback object must not be null");
    if (logger.isDebugEnabled()) {
        String sql = getSql(psc);
        logger.debug("Executing prepared SQL statement" + (sql != null ? " [" + sql + "]" : ""));
    }
    //获取数据库连接
    Connection con = DataSourceUtils.getConnection(getDataSource());
    PreparedStatement ps = null;
    try {
        Connection conToUse = con;
        if (this.nativeJdbcExtractor != null &&
                this.nativeJdbcExtractor.isNativeConnectionNecessaryForNativePreparedStatements()) {
            conToUse = this.nativeJdbcExtractor.getNativeConnection(con);
        }
        ps = psc.createPreparedStatement(conToUse);
        //应用用户设定的输入参数
        applyStatementSettings(ps);
        PreparedStatement psToUse = ps;
        if (this.nativeJdbcExtractor != null) {
            psToUse = this.nativeJdbcExtractor.getNativePreparedStatement(ps);
        }
        //调用回调函数
        T result = action.doInPreparedStatement(psToUse);
        handleWarnings(ps);
        return result;
    }
    catch (SQLException ex) {
        //释放数据库连接避免当 异常转换器没有被初始化的时候出现潜在的连接池死锁
        if (psc instanceof ParameterDisposer) {
            ((ParameterDisposer) psc).cleanupParameters();
        }
        String sql = getSql(psc);
        psc = null;
        JdbcUtils.closeStatement(ps);
        ps = null;
        DataSourceUtils.releaseConnection(con, getDataSource());
        con = null;
        throw getExceptionTranslator().translate("PreparedStatementCallback", sql, ex);
    }
    finally {
        if (psc instanceof ParameterDisposer) {
            ((ParameterDisposer) psc).cleanupParameters();
        }
        JdbcUtils.closeStatement(ps);
        DataSourceUtils.releaseConnection(con, getDataSource());
    }
}
```

以上方法对常用操作进行了封装，包括如下几项内容。

1. 获取数据库连接

获取数据库连接也并非直接使用 dataSource.getConnection()方法那么简单，同样也考虑了诸多情况。

```java
public static Connection doGetConnection(DataSource dataSource) throws SQLException {
        Assert.notNull(dataSource, "No DataSource specified");

        ConnectionHolder conHolder = (ConnectionHolder) TransactionSynchronizationManager.getResource(dataSource);
        if (conHolder != null && (conHolder.hasConnection() || conHolder.isSynchronizedWithTransaction())) {
            conHolder.requested();
            if (!conHolder.hasConnection()) {
                logger.debug("Fetching resumed JDBC Connection from DataSource");
                conHolder.setConnection(dataSource.getConnection());
            }
            return conHolder.getConnection();
        }

        logger.debug("Fetching JDBC Connection from DataSource");
        Connection con = dataSource.getConnection();

        //当前线程支持同步
        if (TransactionSynchronizationManager.isSynchronizationActive()) {
            logger.debug("Registering transaction synchronization for JDBC Connection");
            //在事务中使用同一数据库连接
            ConnectionHolder holderToUse = conHolder;
            if (holderToUse == null) {
                holderToUse = new ConnectionHolder(con);
            }
            else {
                holderToUse.setConnection(con);
            }
            //记录数据库连接
            holderToUse.requested();
            TransactionSynchronizationManager.registerSynchronization(
                    new ConnectionSynchronization(holderToUse, dataSource));
            holderToUse.setSynchronizedWithTransaction(true);
            if (holderToUse != conHolder) {
                TransactionSynchronizationManager.bindResource(dataSource, holderToUse);
            }
        }

        return con;
}
```

在数据库连接方面，Spring 主要考虑的是关于事务方面的处理。基于事务处理的特殊性，Spring 需要保证线程中的数据库操作都是使用同一个事务连接。

2. 应用用户设定的输入参数

```
protected void applyStatementSettings(Statement stmt) throws SQLException {
        int fetchSize = getFetchSize();
        if (fetchSize > 0) {
            stmt.setFetchSize(fetchSize);
        }
        int maxRows = getMaxRows();
        if (maxRows > 0) {
            stmt.setMaxRows(maxRows);
        }
        DataSourceUtils.applyTimeout(stmt, getDataSource(), getQueryTimeout());
}
```

setFetchSize 最主要是为了减少网络交互次数设计的。访问 ResultSet 时,如果它每次只从服务器上读取一行数据,则会产生大量的开销。setFetchSize 的意思是当调用 rs.next 时,ResultSet 会一次性从服务器上取得多少行数据回来,这样在下次 rs.next 时,它可以直接从内存中获取数据而不需要网络交互,提高了效率。 这个设置可能会被某些 JDBC 驱动忽略,而且设置过大也会造成内存的上升。

setMaxRows 将此 Statement 对象生成的所有 ResultSet 对象可以包含的最大行数限制设置为给定数。

3. 调用回调函数

处理一些通用方法外的个性化处理,也就是 PreparedStatementCallback 类型的参数的 doInPreparedStatement 方法的回调。

4. 警告处理

```
protected void handleWarnings(Statement stmt) throws SQLException {
        //当设置为忽略警告时只尝试打印日志
        if (isIgnoreWarnings()) {
            if (logger.isDebugEnabled()) {
                //如果日志开启的情况下打印日志
                SQLWarning warningToLog = stmt.getWarnings();
                while (warningToLog != null) {
                    logger.debug("SQLWarning ignored: SQL state '" + warningToLog.getSQLState() + "', error code '" +
                            warningToLog.getErrorCode() + "', message [" + warningToLog.getMessage() + "]");
                    warningToLog = warningToLog.getNextWarning();
                }
            }
        }
        else {
            handleWarnings(stmt.getWarnings());
        }
}
```

这里用到了一个类 SQLWarning，SQLWarning 提供关于数据库访问警告信息的异常。这些警告直接链接到导致报告警告的方法所在的对象。警告可以从 Connection、Statement 和 ResultSet 对象中获得。试图在已经关闭的连接上获取警告将导致抛出异常。类似地，试图在已经关闭的语句上或已经关闭的结果集上获取警告也将导致抛出异常。注意，关闭语句时还会关闭它可能生成的结果集。

很多人不是很理解什么情况下会产生警告而不是异常，在这里给读者提示个最常见的警告 DataTruncation：DataTruncation 直接继承 SQLWarning，由于某种原因意外地截断数据值时会以 DataTruncation 警告形式报告异常。

对于警告的处理方式并不是直接抛出异常，出现警告很可能会出现数据错误，但是，并不一定会影响程序执行，所以用户可以自己设置处理警告的方式，如默认的是忽略警告，当出现警告时只打印警告日志，而另一种方式只直接抛出异常。

5. 资源释放

数据库的连接释放并不是直接调用了 Connection 的 API 中的 close 方法。考虑到存在事务的情况，如果当前线程存在事务，那么说明在当前线程中存在共用数据库连接，这种情况下直接使用 ConnectionHolder 中的 released 方法进行连接数减一，而不是真正的释放连接。

```
public static void releaseConnection(Connection con, DataSource dataSource) {
    try {
        doReleaseConnection(con, dataSource);
    }
    catch (SQLException ex) {
        logger.debug("Could not close JDBC Connection", ex);
    }
    catch (Throwable ex) {
        logger.debug("Unexpected exception on closing JDBC Connection", ex);
    }
}
    public static void doReleaseConnection(Connection con, DataSource dataSource) throws SQLException {
        if (con == null) {
            return;
        }

        if (dataSource != null) {
            //当前线程存在事务的情况下说明存在共用数据库连接直接使用ConnectionHolder中的
// released方法进行连接数减一而不是真正的释放连接
            ConnectionHolder conHolder = (ConnectionHolder) TransactionSynchronizationManager.getResource(dataSource);
            if (conHolder != null && connectionEquals(conHolder, con)) {
                // It's the transactional Connection: Don't close it.
                conHolder.released();
                return;
            }
        }
```

```
            if (!(dataSource instanceof SmartDataSource) || ((SmartDataSource) dataSource).
shouldClose(con)) {
                logger.debug("Returning JDBC Connection to DataSource");
                con.close();
            }
        }
```

8.2.2 Update 中的回调函数

PreparedStatementCallback 作为一个接口，其中只有一个函数 doInPreparedStatement，这个函数是用于调用通用方法 execute 的时候无法处理的一些个性化处理方法，在 update 中的函数实现：

```
public Integer doInPreparedStatement(PreparedStatement ps) throws SQLException {
    try {
        if (pss != null) {
            //设置 PreparedStatement 所需的全部参数
            pss.setValues(ps);
        }
        int rows = ps.executeUpdate();
        if (logger.isDebugEnabled()) {
            logger.debug("SQL update affected " + rows + " rows");
        }
        return rows;
    }
    finally {
        if (pss instanceof ParameterDisposer) {
            ((ParameterDisposer) pss).cleanupParameters();
        }
    }
}
```

其中用于真正执行 SQL 的 ps.executeUpdate 没有太多需要讲解的，因为我们平时在直接使用 JDBC 方式进行调用的时候会经常使用此方法。但是，对于设置输入参数的函数 pss.setValues(ps)，我们有必要去深入研究一下。在没有分析源码之前，我们至少可以知道其功能，不妨再回顾下 Spring 中使用 SQL 的执行过程，直接使用：

```
jdbcTemplate.update("insert into user(name,age,sex)values(?,?,?)",
        new Object[] { user.getName(), user.getAge(),
            user.getSex() }, new int[] { java.sql.Types.VARCHAR,
                java.sql.Types.INTEGER, java.sql.Types.VARCHAR });
```

SQL 语句对应的参数，对应参数的类型清晰明了，这都归功于 Spring 为我们做了封装，而真正的 JDBC 调用其实非常繁琐，你需要这么做：

```
        PreparedStatement updateSales = con.prepareStatement("insert into user(name,age,
sex)values(?,?,?)");
    updateSales.setString(1, user.getName());
    updateSales.setInt(2, user.getAge());
    updateSales.setString(3, user.getSex());
```

那么看看 Spring 是如何做到封装上面的操作呢？

首先，所有的操作都是以 pss.setValues(ps)为入口的。还记得我们之前的分析路程吗？这个

pss 所代表的当前类正是 ArgPreparedStatementSetter。其中的 setValues 如下：

```java
public void setValues(PreparedStatement ps) throws SQLException {
    int parameterPosition = 1;
    if (this.args != null) {
        //遍历每个参数以作类型匹配及转换
        for (int i = 0; i < this.args.length; i++) {
            Object arg = this.args[i];
            //如果是集合类则需要进入集合类内部递归解析集合内部属性
            if (arg instanceof Collection && this.argTypes[i] != Types.ARRAY) {
                Collection entries = (Collection) arg;
                for (Iterator it = entries.iterator(); it.hasNext();) {
                    Object entry = it.next();
                    if (entry instanceof Object[]) {
                        Object[] valueArray = ((Object[])entry);
                        for (int k = 0; k < valueArray.length; k++) {
                            Object argValue = valueArray[k];
                            doSetValue(ps, parameterPosition, this.argTypes[i], argValue);
                            parameterPosition++;
                        }
                    }else {
                        doSetValue(ps, parameterPosition, this.argTypes[i], entry);
                        parameterPosition++;
                    }
                }
            }else {
                //解析当前属性
                doSetValue(ps, parameterPosition, this.argTypes[i], arg);
                parameterPosition++;
            }
        }
    }
}
```

对单个参数及类型的匹配处理：

```java
protected void doSetValue(PreparedStatement ps, int parameterPosition, int argType, Object argValue)
        throws SQLException {
    StatementCreatorUtils.setParameterValue(ps, parameterPosition, argType, argValue);
}
public static void setParameterValue(
        PreparedStatement ps, int paramIndex, int sqlType, Object inValue)
        throws SQLException {

    setParameterValueInternal(ps, paramIndex, sqlType, null, null, inValue);
}

private static void setParameterValueInternal(
        PreparedStatement ps, int paramIndex, int sqlType, String typeName, Integer scale, Object inValue)
        throws SQLException {

    String typeNameToUse = typeName;
```

```java
                int sqlTypeToUse = sqlType;
                Object inValueToUse = inValue;


                if (inValue instanceof SqlParameterValue) {
                    SqlParameterValue parameterValue = (SqlParameterValue) inValue;
                    if (logger.isDebugEnabled()) {
                        logger.debug("Overriding type info with runtime info from SqlParameterValue: column index " + paramIndex +
                                ", SQL type " + parameterValue.getSqlType() +
                                ", Type name " + parameterValue.getTypeName());
                    }
                    if (parameterValue.getSqlType() != SqlTypeValue.TYPE_UNKNOWN) {
                        sqlTypeToUse = parameterValue.getSqlType();
                    }
                    if (parameterValue.getTypeName() != null) {
                        typeNameToUse = parameterValue.getTypeName();
                    }
                    inValueToUse = parameterValue.getValue();
                }

                if (logger.isTraceEnabled()) {
                    logger.trace("Setting SQL statement parameter value: column index " + paramIndex +
                            ", parameter value [" + inValueToUse +
                            "], value class [" + (inValueToUse != null ? inValueToUse.getClass().getName() : "null") +
                            "], SQL type " + (sqlTypeToUse == SqlTypeValue.TYPE_UNKNOWN ? "unknown" : Integer.toString(sqlTypeToUse)));
                }

                if (inValueToUse == null) {
                    setNull(ps, paramIndex, sqlTypeToUse, typeNameToUse);
                }
                else {
                    setValue(ps, paramIndex, sqlTypeToUse, typeNameToUse, scale, inValueToUse);
                }
            }
```

8.3 query 功能的实现

在之前的章节中我们介绍了 update 方法的功能实现，那么在数据库操作中查找操作也是使用率非常高的函数，同样我们也需要了解它的实现过程。使用方法如下：

```java
List<User> list = jdbcTemplate.query("select * from user where age=?",new Object[]{20},new int[]{java.sql.Types.INTEGER} ,new UserRowMapper());
```

跟踪 jdbcTemplate 中的 query 方法。

```java
public <T> List<T> query(String sql, Object[] args, int[] argTypes, RowMapper<T> rowMapper) throws DataAccessException {
    return query(sql, args, argTypes, new RowMapperResultSetExtractor<T> (rowMapper));
}
```

```java
    public <T> T query(String sql, Object[] args, int[] argTypes, ResultSetExtractor<T> rse)
throws DataAccessException {
        return query(sql, newArgTypePreparedStatementSetter(args, argTypes), rse);
}
```

上面函数中与 update 方法中都同样使用了 **newArgTypePreparedStatementSetter**

```java
    public <T> T query(String sql, PreparedStatementSetter pss, ResultSetExtractor<T> rse)
throws DataAccessException {
        return query(new SimplePreparedStatementCreator(sql), pss, rse);
}
    public <T> T query(
            PreparedStatementCreator psc, final PreparedStatementSetter pss, final
ResultSetExtractor<T> rse)
            throws DataAccessException {

        Assert.notNull(rse, "ResultSetExtractor must not be null");
        logger.debug("Executing prepared SQL query");

        return execute(psc, new PreparedStatementCallback<T>() {
            public T doInPreparedStatement(PreparedStatement ps) throws SQLException {
                ResultSet rs = null;
                try {
                    if (pss != null) {
                        pss.setValues(ps);
                    }
                    rs = ps.executeQuery();
                    ResultSet rsToUse = rs;
                    if (nativeJdbcExtractor != null) {
                        rsToUse = nativeJdbcExtractor.getNativeResultSet(rs);
                    }
                    return rse.extractData(rsToUse);
                }
                finally {
                    JdbcUtils.closeResultSet(rs);
                    if (pss instanceof ParameterDisposer) {
                        ((ParameterDisposer) pss).cleanupParameters();
                    }
                }
            }
        });
}
```

可以看到整体套路与 update 差不多的，只不过在回调类 PreparedStatementCallback 的实现中使用的是 ps.executeQuery() 执行查询操作，而且在返回方法上也做了一些额外的处理。

rse.extractData(rsToUse) 方法负责将结果进行封装并转换至 POJO，rse 当前代表的类为 RowMapperResultSetExtractor，而在构造 RowMapperResultSetExtractor 的时候我们又将自定义的 rowMapper 设置了进去。调用代码如下：

```java
    public List<T> extractData(ResultSet rs) throws SQLException {
        List<T> results = (this.rowsExpected > 0 ? new ArrayList<T>(this.rowsExpected
) : new ArrayList<T>());
            int rowNum = 0;
            while (rs.next()) {
```

8.3 query 功能的实现

```
            results.add(this.rowMapper.mapRow(rs, rowNum++));
        }
        return results;
    }
```

上面的代码中并没有什么复杂的逻辑，只是对返回结果遍历并以此使用 rowMapper 进行转换。

之前讲了 update 方法以及 query 方法，使用这两个函数示例的 SQL 都是带有参数的，也就是带有 "?" 的，那么还有另一种情况是不带有 "?" 的，Spring 中使用的是另一种处理方式。例如：

```
            List<User> list = jdbcTemplate.query("select * from user", new UserRowMapper());
```

跟踪进入：

```
public <T> List<T> query(String sql, RowMapper<T> rowMapper) throws DataAccessException {
    return query(sql, new RowMapperResultSetExtractor<T>(rowMapper));
}

public <T> T query(final String sql, final ResultSetExtractor<T> rse) throws DataAccessException {
    Assert.notNull(sql, "SQL must not be null");
    Assert.notNull(rse, "ResultSetExtractor must not be null");
    if (logger.isDebugEnabled()) {
        logger.debug("Executing SQL query [" + sql + "]");
    }
    class QueryStatementCallback implements StatementCallback<T>, SqlProvider {
        public T doInStatement(Statement stmt) throws SQLException {
            ResultSet rs = null;
            try {
                rs = stmt.executeQuery(sql);
                ResultSet rsToUse = rs;
                if (nativeJdbcExtractor != null) {
                    rsToUse = nativeJdbcExtractor.getNativeResultSet(rs);
                }
                return rse.extractData(rsToUse);
            }
            finally {
                JdbcUtils.closeResultSet(rs);
            }
        }
        public String getSql() {
            return sql;
        }
    }
    return execute(new QueryStatementCallback());
}
```

与之前的 query 方法最大的不同是少了参数及参数类型的传递，自然也少了 PreparedStatementSetter 类型的封装。既然少了 PreparedStatementSetter 类型的传入，调用的 execute 方法自然也会有所改变了。

```
public <T> T execute(StatementCallback<T> action) throws DataAccessException {
    Assert.notNull(action, "Callback object must not be null");

    Connection con = DataSourceUtils.getConnection(getDataSource());
```

```
            Statement stmt = null;
            try {
                Connection conToUse = con;
                if (this.nativeJdbcExtractor != null &&
                        this.nativeJdbcExtractor.isNativeConnectionNecessary ForNative
Statements()) {
                    conToUse = this.nativeJdbcExtractor.getNativeConnection(con);
                }
                stmt = conToUse.createStatement();
                applyStatementSettings(stmt);
                Statement stmtToUse = stmt;
                if (this.nativeJdbcExtractor != null) {
                    stmtToUse = this.nativeJdbcExtractor.getNativeStatement(stmt);
                }
                T result = action.doInStatement(stmtToUse);
                handleWarnings(stmt);
                return result;
            }
            catch (SQLException ex) {
                // Release Connection early, to avoid potential connection pool deadlock
                // in the case when the exception translator hasn't been initialized yet.
                JdbcUtils.closeStatement(stmt);
                stmt = null;
                DataSourceUtils.releaseConnection(con, getDataSource());
                con = null;
                throw getExceptionTranslator().translate("StatementCallback", getSql(action), ex);
            }
            finally {
                JdbcUtils.closeStatement(stmt);
                DataSourceUtils.releaseConnection(con, getDataSource());
            }
        }
```

这个 exexute 与之前的 execute 并无太大差别，都是做一些常规的处理，诸如获取连接、释放连接等，但是，有一个地方是不一样的，就是 statement 的创建。这里直接使用 connection 创建，而带有参数的 SQL 使用的是 PreparedStatementCreator 类来创建的。一个是普通的 Statement，另一个是 PreparedStatement，两者究竟是何区别呢？

PreparedStatement 接口继承 Statement，并与之在两方面有所不同。

- PreparedStatement 实例包含已编译的 SQL 语句。这就是使语句"准备好"。包含 PreparedStatement 对象中的 SQL 语句可具有一个或多个 IN 参数。IN 参数的值在 SQL 语句创建时未被指定。相反的，该语句为每个 IN 参数保留一个问号（"?"）作为占位符。每个问号的值必须在该语句执行之前，通过适当的 setXXX 方法来提供。

- 由于 PreparedStatement 对象已预编译过，所以其执行速度要快于 Statement 对象。因此，多次执行的 SQL 语句经常创建为 PreparedStatement 对象，以提高效率。

作为 Statement 的子类，PreparedStatement 继承了 Statement 的所有功能。另外，它还添加了一整套方法，用于设置发送给数据库以取代 IN 参数占位符的值。同时，三种方法 execute、executeQuery 和 executeUpdate 已被更改以使之不再需要参数。这些方法的 Statement 形式（接受 SQL 语句参数的形式）不应该用于 PreparedStatement 对象。

8.4 queryForObject

Spring 中不仅仅为我们提供了 query 方法，还在此基础上做了封装，提供了不同类型的 query 方法，如图 8-1 所示。

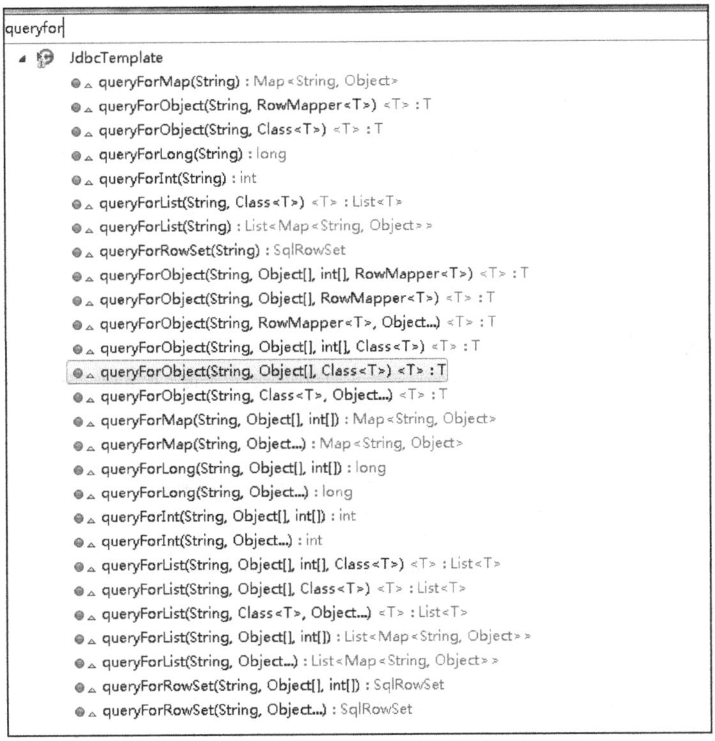

图 8-1 Spring 中的 query 相关方法

我们以 queryForObject 为例，来讨论一下 Spring 是如何在返回结果的基础上进行封装的。
```
public <T> T queryForObject(String sql, Class<T> requiredType) throws DataAccessException {
        return queryForObject(sql, getSingleColumnRowMapper(requiredType));
}

public <T> T queryForObject(String sql, RowMapper<T> rowMapper) throws DataAccessException {
        List<T> results = query(sql, rowMapper);
        return DataAccessUtils.requiredSingleResult(results);
}
```
其实最大的不同还是对于 RowMapper 的使用。SingleColumnRowMapper 类中的 mapRow：
```
public T mapRow(ResultSet rs, int rowNum) throws SQLException {
        //验证返回结果数
        ResultSetMetaData rsmd = rs.getMetaData();
```

```java
            int nrOfColumns = rsmd.getColumnCount();
            if (nrOfColumns != 1) {
                throw new IncorrectResultSetColumnCountException(1, nrOfColumns);
            }

            //抽取第一个结果进行处理
            Object result = getColumnValue(rs, 1, this.requiredType);
            if (result != null && this.requiredType != null && !this.requiredType.isInstance(result)) {
                //转换到对应的类型
                try {
                    return (T) convertValueToRequiredType(result, this.requiredType);
                }
                catch (IllegalArgumentException ex) {
                    throw new TypeMismatchDataAccessException(
                            "Type mismatch affecting row number " + rowNum + " and column type '" +
                            rsmd.getColumnTypeName(1) + "': " + ex.getMessage());
                }
            }
            return (T) result;
    }
```

对应的类型转换函数：

```java
    protected Object convertValueToRequiredType(Object value, Class requiredType) {
            if (String.class.equals(requiredType)) {
                return value.toString();
            }
            else if (Number.class.isAssignableFrom(requiredType)) {
                if (value instanceof Number) {
                    // Convert original Number to target Number class.
                    //转换原始 Number 类型的实体到 Number 类
                    return NumberUtils.convertNumberToTargetClass(((Number) value), requiredType);
                }else {
                    //转换 string 类型的值到 Number 类
                    return NumberUtils.parseNumber(value.toString(), requiredType);
                }
            }
            else {
                throw new IllegalArgumentException(
                        "Value [" + value + "] is of type [" + value.getClass().getName() +
                        "] and cannot be converted to required type [" + requiredType.getName() + "]");
            }
    }
```

第 9 章　整合 MyBatis

　　MyBatis 本是 Apache 的一个开源项目 iBatis，2010 年这个项目由 Apache Software Foundation 迁移到了 Google Code，并且改名为 MyBatis。

　　MyBatis 是支持普通 SQL 查询、存储过程和高级映射的优秀持久层框架。MyBatis 消除了几乎所有的 JDBC 代码和参数的手工设置以及结果集的检索。MyBatis 使用简单的 XML 或注解用于配置和原始映射，将接口和 Java 的 POJOs（Plain Old Java Objects，普通的 Java 对象）映射成数据库中的记录。

9.1　MyBatis 独立使用

　　尽管我们接触更多的是 MyBatis 与 Spring 的整合使用，但是 MyBatis 有它自己的独立使用方法，了解其独立使用的方法套路对分析 Spring 整合 MyBatis 非常有帮助，因为 Spring 无非就是将这些功能进行封装以简化我们的开发流程。MyBatis 独立使用包括以下几步。

1．建立 PO

用于对数据库中数据的映射，使程序员更关注于对 Java 类的使用而不是数据库的操作。

```
public class User {
private Integer id;
    private String name;
    private Integer age;
     //省略 set/get 方法

    public User(String name, Integer age) {
super();
        this.name = name;
        this.age = age;
    }

    public User() {
        super();
```

```
    } //必须要有这个无参构造方法,不然根据 UserMapper.xml 中的配置,在查询数据库时,将不能反射构造
      //出 User 实例
}
```

2. 建立 Mapper

数据库操作的映射文件,也就是我们常常说的 DAO,用于映射数据库的操作,可以通过配置文件指定方法对应的 SQL 语句或者直接使用 Java 提供的注解方式进行 SQL 指定。

```
public interface UserMapper {
    public void insertUser(User user);
    public User getUser(Integer id);
}
```

3. 建立配置文件

配置文件主要用于配置程序中可变性高的设置,一个偏大的程序一定会存在一些经常会变化的变量,如果每次变化都需要改变源码那会是非常糟糕的设计,所以,我们看到各种各样的框架或者应用的时候都免不了要配置配置文件,MyBatis 中的配置文件主要封装在 configuration 中,配置文件的基本结构如图 9-1 所示。

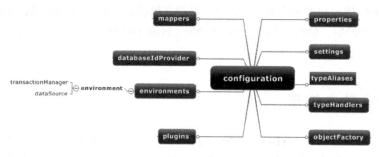

图 9-1　配置文件结构

- configuration:根元素。
- properties:定义配置外在化。
- settings:一些全局性的配置。
- typeAliases:为一些类定义别名。
- typeHandlers:定义类型处理,也就是定义 Java 类型与数据库中的数据类型之间的转换关系。
- objectFactory:用于指定结果集对象的实例是如何创建的。
- plugins:MyBatis 的插件,插件可以修改 MyBatis 内部的运行规则。
- environments:环境。
- environment:配置 MyBatis 的环境。
- transactionManager:事务管理器。
- dataSource:数据源。

- mappers：指定映射文件或映射类。

读者如果对上面的各个配置具体使用方法感兴趣，可以进一步查阅相关资料，这里只举出最简单的实例以方便读者快速回顾 MyBatis。

```xml
<?xml version="1.0" encoding="UTF-8"?>
<!DOCTYPE configuration
    PUBLIC "-//mybatis.org//DTD Config 3.0//EN"
    "http://mybatis.org/dtd/mybatis-3-config.dtd">

<configuration>
    <settings>
        <!-- changes from the defaults for testing -->
        <setting name="cacheEnabled" value="false" />
        <setting name="useGeneratedKeys" value="true" />
        <setting name="defaultExecutorType" value="REUSE" />
    </settings>
    <typeAliases>
        <typeAlias alias="User" type="bean.User"/>
    </typeAliases>
    <environments default="development">
        <environment id="development">
            <transactionManager type="jdbc"/>
            <dataSource type="POOLED">
                <property name="driver" value="com.mysql.jdbc.Driver"/>
                <property name="url" value="jdbc:mysql://localhost/lexueba"/>
                <property name="username" value="root"/>
                <property name="password" value="haojia0421xixi"/>
            </dataSource>
        </environment>
    </environments>
    <mappers>
        <mapper resource="resource/UserMapper.xml" />
    </mappers>
</configuration>
```

4. 建立映射文件

对应于 MyBaits 全局配置中的 mappers 配置属性，主要用于建立对应数据库操作接口的 SQL 映射。MyBatis 会将这里设定的 SQL 与对应的 Java 接口相关联，以保证在 MyBatis 中调用接口的时候会到数据库中执行相应的 SQL 来简化开发。

```xml
<?xml version="1.0" encoding="UTF-8" ?>
<!DOCTYPE mapper
    PUBLIC "-//mybatis.org//DTD Mapper 3.0//EN"
    "http://mybatis.org/dtd/mybatis-3-mapper.dtd">
<mapper namespace="Mapper.UserMapper">
<!-- 这里 namespace 必须是 UserMapper 接口的路径，不然要运行的时候要报错 "is not known to the MapperRegistry" -->
    <insert id="insertUser" parameterType="User" >
        insert into user(name,age) values(#{name},#{age})
        <!-- 这里 sql 结尾不能加分号，否则报 "ORA-00911" 的错误 -->
    </insert>

    <!-- 这里的 id 必须和 UserMapper 接口中的接口方法名相同，不然运行的时候也要报错 -->
```

```xml
<select id="getUser" resultType="User" parameterType="java.lang.Integer"    >
    select * from user where id=#{id}
</select>
</mapper>
```

5. 建立测试类

至此我们已经完成了 MyBatis 的建立过程，接下来的工作就是对之前的所有工作进行测试，以便直接查看 MyBatis 为我们提供的效果。

```java
public class MyBatisUtil {
    private  final static SqlSessionFactory sqlSessionFactory;
    static {
        String resource = "resource/mybatis-config.xml";
        Reader reader = null;
        try {
            reader = Resources.getResourceAsReader(resource);
        } catch (IOException e) {
            System.out.println(e.getMessage());

        }
        sqlSessionFactory = new SqlSessionFactoryBuilder().build(reader);
    }

    public static SqlSessionFactory getSqlSessionFactory() {
        return sqlSessionFactory;
    }
}
public class TestMapper {
    static SqlSessionFactory sqlSessionFactory = null;
    static {
        sqlSessionFactory = MyBatisUtil.getSqlSessionFactory();
    }

    @Test
    public void testAdd() {
        SqlSession sqlSession = sqlSessionFactory.openSession();
        try {
            UserMapper userMapper = sqlSession.getMapper(UserMapper.class);
            User user = new User("tom",new Integer(5));
            userMapper.insertUser(user);
            sqlSession.commit();//这里一定要提交,不然数据进不去数据库中
        } finally {
            sqlSession.close();
        }
    }

    @Test
    public void getUser() {
        SqlSession sqlSession = sqlSessionFactory.openSession();
        try {
            UserMapper userMapper = sqlSession.getMapper(UserMapper.class);
            User user = userMapper.getUser(1);
            System.out.println("name: "+user.getName()+"|age: "+user.getAge());
        } finally {
```

```
            sqlSession.close();
        }
    }
}
```

注意，这里在数据库设定了 id 自增策略，所以插入的数据会直接在数据库中赋值，当执行测试后如果数据表为空，那么在表中会出现一条我们插入的数据，并会在查询时将此数据查出。

9.2 Spring 整合 MyBatis

了解了 MyBatis 的独立使用过程后，我们再看看它与 Spring 整合的使用方式，比对之前的示例来找出 Spring 究竟为我们做了哪些操作来简化程序员的业务开发。由于在上面示例基础上作更改，所以，User 与 UserMapper 保持不变。

1. Spring 配置文件

配置文件是 Spring 的核心，Spring 的所有操作也都是由配置文件开始的，所以，我们的示例也首先从配置文件开始。

```xml
<?xml version="1.0" encoding="UTF-8"?>
<beans xmlns="http://www.Springframework.org/schema/beans"
    xmlns:xsi="http://www.w3.org/2001/XMLSchema-instance"
    xsi:schemaLocation="http://www.Springframework.org/schema/beans http://www.Springframework.
org/ schema/beans/Spring-beans-3.0.xsd">

    <bean id="dataSource" class="org.apache.commons.dbcp.BasicDataSource">
        <property name="driverClassName" value="com.mysql.jdbc.Driver"></property>
        <property name="url" value="jdbc:mysql://localhost:3306/lexueba?useUnicode=true&amp;characterEncoding=UTF-8&amp;zeroDateTimeBehavior=convertToNull"></property>
        <property name="username" value="root"></property>
        <property name="password" value="haojia0421xixi"></property>
        <property name="maxActive" value="100"></property>
        <property name="maxIdle" value="30"></property>
        <property name="maxWait" value="500"></property>
        <property name="defaultAutoCommit" value="true"></property>
    </bean>

    <bean id="sqlSessionFactory" class="org.mybatis.Spring.SqlSessionFactoryBean">
        <property name="configLocation" value="classpath:test/mybatis/MyBatis-Configuration.xml"></property>
        <property name="dataSource" ref="dataSource" />
    </bean>


    <bean id="userMapper" class="org.mybatis.Spring.mapper.MapperFactoryBean">
        <property name="mapperInterface" value="test.mybatis.dao.UserMapper"></property>

        <property name="sqlSessionFactory" ref="sqlSessionFactory"></property>
    </bean>

</beans>
```

对比之前独立使用 MyBatis 的配置文件，我们发现，之前在 environments 中设置的 dataSource 被转移到了 Spring 的核心配置文件中管理。而且，针对于 MyBatis，注册了 org.mybatis.Spring.SqlSessionFactoryBean 类型 bean，以及用于映射接口的 org.mybatis.Spring.mapper.MapperFactoryBean，这两个 bean 的作用我们会在稍后分析。

之前我们了解到，MyBatis 提供的配置文件包含了诸多属性，虽然大多数情况我们都会保持 MyBatis 原有的风格，将 MyBatis 的配置文件独立出来，并在 Spring 中的 org.mybatis.Spring.SqlSessionFactoryBean 类型的 bean 中通过 configLocation 属性引入，但是，这并不代表 Spring 不支持直接配置。以上面示例为例，你完全可以省去 MyBatis-Configuration.xml，而将其中的配置以属性的方式注入到 SqlSessionFactoryBean 中，至于每个属性名称以及用法，我们会在后面的章节中进行详细的分析。

2. MyBatis 配置文件

对比独立使用 MyBatis 时的配置文件，当前的配置文件除了移除 environments 配置外并没有太多的变化。

```xml
<?xml version="1.0" encoding="UTF-8" ?>
<!DOCTYPE configuration PUBLIC "-//mybatis.org//DTD Config 3.0//EN"
"http://mybatis.org/dtd/mybatis-3-config.dtd">
<configuration>
 <typeAliases>
        <typeAlias alias="User" type="test.mybatis.bean.User"/>
    </typeAliases>
    <mappers>
        <mapper resource="test/mybatis/UserMapper.xml"/>
    </mappers>
</configuration>
```

3. 映射文件（保持不变）

```xml
<?xml version="1.0" encoding="UTF-8" ?>
<!DOCTYPE mapper PUBLIC "-//mybatis.org//DTD Mapper 3.0//EN" "http://mybatis.org/dtd/mybatis-3-mapper.dtd">
<mapper namespace="test.mybatis.dao.UserMapper">
    <insert id="insertUser" parameterType="User" >
        insert into user(name,age) values(#{name},#{age})
    </insert>

    <select id="getUser" resultType="User" parameterType="java.lang.String"    >
        select * from user where name=#{name}
    </select>
</mapper>
```

4. 测试

至此，我们已经完成了 Spring 与 MyBatis 的整合，我们发现，对于 MyBatis 方面的配置文件，除了将 dataSource 配置移到 Spring 配置文件中管理外，并没有太多变化，而在 Spring 的配置文件中又增加了用于处理 MyBatis 的两个 bean。

Spring 整合 MyBatis 的优势主要在于使用上,我们来看看 Spring 中使用 MyBatis 的用法。

```
public class UserServiceTest {
    public static void main(String[] args) {
        ApplicationContext context = new ClassPathXmlApplicationContext("test/ mybatis/ applicationContext.xml");
        UserMapper userDao = (UserMapper)context.getBean("userMapper");
        System.out.println(userDao.getUser("1"));
    }
}
```

测试中我们看到,在 Spring 中使用 MyBatis 非常方便,用户甚至无法察觉自己正在使用 MyBatis,而这一切相对于独立使用 MyBatis 时必须要做的各种冗余操作来说无非是大大简化了我们的工作量。

9.3 源码分析

通过 Spring 整合 MyBatis 的示例,我们感受到了 Spring 为用户更加快捷地进行开发所做的努力,开发人员的工作效率由此得到了显著的提升。但是,相对于使用来说,我们更想知道其背后所隐藏的秘密,Spring 整合 MyBatis 是何如实现的呢?通过分析整合示例中的配置文件,我们可以知道配置的 bean 其实是成树状结构的,而在树的最顶层是类型为 org.mybatis.Spring.SqlSessionFactoryBean 的 bean,它将其他相关 bean 组装在了一起,那么,我们的分析就从此类开始。

9.3.1 sqlSessionFactory 创建

通过配置文件我们分析,对于配置文件的读取解析,Spring 应该通过 org.mybatis.Spring.SqlSessionFactoryBean 封装了 MyBatis 中的实现。我们进入这个类,首先查看这个类的层次结构,如图 9-2 所示。

图 9-2 SqlSessionFactoryBean 类的层次结构图

根据这个类的层次结构找出我们感兴趣的两个接口,FactoryBean 和 InitializingBean。

- InitializingBean:实现此接口的 bean 会在初始化时调用其 afterPropertiesSet 方法来进行 bean 的逻辑初始化。
- FactoryBean:一旦某个 bean 实现次接口,那么通过 getBean 方法获取 bean 时其实是获取此类的 getObject()返回的实例。

我们首先以 InitializingBean 接口的 afterPropertiesSet()方法作为突破点。

1. SqlSessionFactoryBean 的初始化

查看 org.mybatis.Spring.SqlSessionFactoryBean 类型的 bean 在初始化时做了哪些逻辑实现。

```
public void afterPropertiesSet() throws Exception {
    notNull(dataSource, "Property 'dataSource' is required");
    notNull(sqlSessionFactoryBuilder, "Property 'sqlSessionFactoryBuilder' is required");

    this.sqlSessionFactory = buildSqlSessionFactory();
}
```

很显然，此函数主要目的就是对于 sqlSessionFactory 的初始化，通过之前展示的独立使用 MyBatis 的示例，我们了解到 SqlSessionFactory 是所有 MyBatis 功能的基础。

```
protected SqlSessionFactory buildSqlSessionFactory() throws IOException {

    Configuration configuration;

    XMLConfigBuilder xmlConfigBuilder = null;
    if (this.configLocation != null) {
        xmlConfigBuilder = new XMLConfigBuilder(this.configLocation.getInputStream(), null, this.configurationProperties);
        configuration = xmlConfigBuilder.getConfiguration();
    } else {
        if (this.logger.isDebugEnabled()) {
            this.logger.debug("Property 'configLocation' not specified, using default MyBatis Configuration");
        }
        configuration = new Configuration();
        configuration.setVariables(this.configurationProperties);
    }

    if (this.objectFactory != null) {
        configuration.setObjectFactory(this.objectFactory);
    }

    if (this.objectWrapperFactory != null) {
        configuration.setObjectWrapperFactory(this.objectWrapperFactory);
    }

    if (hasLength(this.typeAliasesPackage)) {
        String[] typeAliasPackageArray = tokenizeToStringArray(this.typeAliasesPackage,
            ConfigurableApplicationContext.CONFIG_LOCATION_DELIMITERS);
        for (String packageToScan : typeAliasPackageArray) {
            configuration.getTypeAliasRegistry().registerAliases(packageToScan,
                typeAliasesSuperType == null ? Object.class : typeAliasesSuperType);
            if (this.logger.isDebugEnabled()) {
                this.logger.debug("Scanned package: '" + packageToScan + "' for aliases");
            }
        }
    }

    if (!isEmpty(this.typeAliases)) {
        for (Class<?> typeAlias : this.typeAliases) {
```

```java
      configuration.getTypeAliasRegistry().registerAlias(typeAlias);
      if (this.logger.isDebugEnabled()) {
        this.logger.debug("Registered type alias: '" + typeAlias + "'");
      }
    }
  }

  if (!isEmpty(this.plugins)) {
    for (Interceptor plugin : this.plugins) {
      configuration.addInterceptor(plugin);
      if (this.logger.isDebugEnabled()) {
        this.logger.debug("Registered plugin: '" + plugin + "'");
      }
    }
  }

  if (hasLength(this.typeHandlersPackage)) {
    String[] typeHandlersPackageArray = tokenizeToStringArray(this.typeHandlersPackage,
        ConfigurableApplicationContext.CONFIG_LOCATION_DELIMITERS);
    for (String packageToScan : typeHandlersPackageArray) {
      configuration.getTypeHandlerRegistry().register(packageToScan);
      if (this.logger.isDebugEnabled()) {
        this.logger.debug("Scanned package: '" + packageToScan + "' for type handlers");
      }
    }
  }

  if (!isEmpty(this.typeHandlers)) {
    for (TypeHandler<?> typeHandler : this.typeHandlers) {
      configuration.getTypeHandlerRegistry().register(typeHandler);
      if (this.logger.isDebugEnabled()) {
        this.logger.debug("Registered type handler: '" + typeHandler + "'");
      }
    }
  }

  if (xmlConfigBuilder != null) {
    try {
      xmlConfigBuilder.parse();

      if (this.logger.isDebugEnabled()) {
        this.logger.debug("Parsed configuration file: '" + this.configLocation + "'");
      }
    } catch (Exception ex) {
      throw new NestedIOException("Failed to parse config resource: " + this. config
Location, ex);
    } finally {
      ErrorContext.instance().reset();
    }
  }

  if (this.transactionFactory == null) {
    this.transactionFactory = new SpringManagedTransactionFactory();
  }
```

```
        Environment environment = new Environment(this.environment, this.transactionFactory,
this.dataSource);
        configuration.setEnvironment(environment);

    if (this.databaseIdProvider != null) {
      try {
        configuration.setDatabaseId(this.databaseIdProvider.getDatabaseId (this.dataSource));
      } catch (SQLException e) {
        throw new NestedIOException("Failed getting a databaseId", e);
      }
    }

    if (!isEmpty(this.mapperLocations)) {
      for (Resource mapperLocation : this.mapperLocations) {
        if (mapperLocation == null) {
          continue;
        }

        try {
          XMLMapperBuilder xmlMapperBuilder = new XMLMapperBuilder(mapperLocation.getInputStream(),
              configuration, mapperLocation.toString(), configuration.getSqlFragments());
          xmlMapperBuilder.parse();
        } catch (Exception e) {
          throw new NestedIOException("Failed to parse mapping resource: '" + mapperLocation + "'", e);
        } finally {
          ErrorContext.instance().reset();
        }

        if (this.logger.isDebugEnabled()) {
          this.logger.debug("Parsed mapper file: '" + mapperLocation + "'");
        }
      }
    } else {
      if (this.logger.isDebugEnabled()) {
        this.logger.debug("Property 'mapperLocations' was not specified or no matching resources found");
      }
    }

    return this.sqlSessionFactoryBuilder.build(configuration);
}
```

从函数中可以看到，尽管我们还是习惯于将 MyBatis 的配置与 Spring 的配置独立出来，但是，这并不代表 Spring 中的配置不支持直接配置。也就是说，在上面提供的示例中，你完全可以取消配置中的 configLocation 属性，而把其中的属性直接写在 SqlSessionFactoryBean 中。

```
<bean id="sqlSessionFactory" class="org.mybatis.Spring.SqlSessionFactoryBean">
    <property name="configLocation" value="classpath:test/mybatis/MyBatis- Configuration.xml"></property>
    <property name="dataSource" ref="dataSource" />
    <property name="typeAliasesPackage" value="aaaaa"/>
```

```
... ...
</bean>
```

从这个函数中可以得知，配置文件还可以支持其他多种属性的配置，如 configLocation、objectFactory、objectWrapperFactory、typeAliasesPackage、typeAliases、typeHandlersPackage、plugins、typeHandlers、transactionFactory、databaseIdProvider、mapperLocations。

其实，如果只按照常用的配置，那么我们只需要在函数最开始按照如下方式处理 configuration：

```
xmlConfigBuilder = new XMLConfigBuilder(this.configLocation.getInputStream(), null, this.configurationProperties);
        configuration = xmlConfigBuilder.getConfiguration();
```

根据 configLocation 构造 XMLConfigBuilder 并进行解析，但是，为了体现 Spring 更强大的兼容性，Spring 还整合了 MyBatis 中其他属性的注入，并通过实例 configuration 来承载每一步所获取的信息并最终使用 sqlSessionFactoryBuilder 实例根据解析到的 configuration 创建 SqlSessionFactory 实例。

2．获取 SqlSessionFactoryBean 实例

由于 SqlSessionFactoryBean 实现了 FactoryBean 接口，所以当通过 getBean 方法获取对应实例时，其实是获取该类的 getObject()函数返回的实例，也就是获取初始化后的 sqlSessionFactory 属性。

```
public SqlSessionFactory getObject() throws Exception {
  if (this.sqlSessionFactory == null) {
    afterPropertiesSet();
  }

  return this.sqlSessionFactory;
}
```

9.3.2 MapperFactoryBean 的创建

为了使用 MyBatis 功能，示例中的 Spring 配置文件提供了两个 bean，除了之前分析的 SqlSssionFactoryBean 类型的 bean 以外，还有一个是 MapperFactoryBean 类型的 bean。

结合两个测试用例综合分析，对于单独使用 MyBatis 的时候调用数据库接口的方式是：
```
UserMapper userMapper = sqlSession.getMapper(UserMapper.class);
```
而在这一过程中，其实是 MyBatis 在获取映射的过程中根据配置信息为 UserMapper 类型动态创建了代理类。而对于 Spring 的创建方式：
```
UserMapper userMapper = (UserMapper)context.getBean("userMapper");
```
Spring 中获取的名为 userMapper 的 bean，其实是与单独使用 MyBatis 完成了一样的功能，那么我们可以推断，在 bean 的创建过程中一定是使用了 MyBatis 中的原生方法 sqlSession.getMapper(UserMapper.class)进行了再一次封装。结合配置文件，我们把分析目标转向 org.mybatis.Spring.mapper. MapperFactoryBean，初步推测其中的逻辑应该在此类中实现。同样，还是

首先查看的类层次结构图 MapperFactoryBean，如图 9-3 所示。

图 9-3　MapperFactoryBean 类的层次结构图

同样，在实现的接口中发现了我们感兴趣的两个接口 InitializingBean 与 FactoryBean。我们的分析还是从 bean 的初始化开始。

1. MapperFactoryBean 的初始化

因为实现了 InitializingBean 接口，Spring 会保证在 bean 初始化时首先调用 afterPropertiesSet 方法来完成其初始化逻辑。追踪父类，发现 afterPropertiesSet 方法是在 DaoSupport 类中实现，代码如下：

```
public final void afterPropertiesSet() throws IllegalArgumentException, BeanInitialization
Exception {
        // Let abstract subclasses check their configuration.
        checkDaoConfig();

        // Let concrete implementations initialize themselves.
        try {
            initDao();
        }
        catch (Exception ex) {
            throw new BeanInitializationException("Initialization of DAO failed", ex);
        }
}
```

但从函数名称来看我们大体推测，MapperFactoryBean 的初始化包括对 DAO 配置的验证以及对 DAO 的初始工作，其中 initDao()方法是模板方法，设计为留给子类做进一步逻辑处理。而 checkDaoConfig()才是我们分析的重点。

```
 @Override
 protected void checkDaoConfig() {
    super.checkDaoConfig();

    notNull(this.mapperInterface, "Property 'mapperInterface' is required");

    Configuration configuration = getSqlSession().getConfiguration();
    if (this.addToConfig && !configuration.hasMapper(this.mapperInterface)) {
      try {
        configuration.addMapper(this.mapperInterface);
```

```
            } catch (Throwable t) {
                logger.error("Error while adding the mapper '" + this.mapperInterface + "' to configuration.", t);
                throw new IllegalArgumentException(t);
            } finally {
                ErrorContext.instance().reset();
            }
        }
    }
```

super.checkDaoConfig()在 SqlSessionDaoSupport 类中实现，代码如下：

```
protected void checkDaoConfig() {
    notNull(this.sqlSession, "Property 'sqlSessionFactory' or 'sqlSessionTemplate' are required");
}
```

结合代码我们了解到对于 DAO 配置的验证，Spring 做了以下几个方面的工作。

- 父类中对于 sqlSession 不为空的验证。

sqlSession 作为根据接口创建映射器代理的接触类一定不可以为空，而 sqlSession 的初始化工作是在设定其 sqlSessionFactory 属性时完成的。

```
public void setSqlSessionFactory(SqlSessionFactory sqlSessionFactory) {
    if (!this.externalSqlSession) {
        this.sqlSession = new SqlSessionTemplate(sqlSessionFactory);
    }
}
```

也就是说，对于下面的配置如果忽略了对于 sqlSessionFactory 属性的设置，那么在此时就会被检测出来。

```
<bean id="userMapper" class="org.mybatis.Spring.mapper.MapperFactoryBean">
    <property name="mapperInterface" value="test.mybatis.dao.UserMapper"></property>
    <property name="sqlSessionFactory" ref="sqlSessionFactory"></property>
</bean>
```

- 映射接口的验证。

接口是映射器的基础，sqlSession 会根据接口动态创建相应的代理类，所以接口必不可少。

- 映射文件存在性验证。

对于函数前半部分的验证我们都很容易理解，无非是对配置文件中的属性是否存在做验证，但是后面部分是完成了什么方面的验证呢？如果读者读过 MyBatis 源码，你就会知道，在 MyBatis 实现过程中并没有手动调用 configuration.addMapper 方法，而是在映射文件读取过程中一旦解析到如<mapper namespace="Mapper.UserMapper">，便会自动进行类型映射的注册。那么，Spring 中为什么会把这个功能单独拿出来放在验证里呢？这是不是多此一举呢？

在上面的函数中，configuration.addMapper(this.mapperInterface)其实就是将 UserMapper 注册到映射类型中，如果你可以保证这个接口一定存在对应的映射文件，那么其实这个验证并没有必要。但是，由于这个是我们自行决定的配置，无法保证这里配置的接口一定存在对应的映射文件，所以这里非常有必要进行验证。在执行此代码的时候，MyBatis 会检查嵌入的映射接口是否存在对应的映射文件，如果没有回抛出异常，Spring 正是在用这种方式来完成接口对应的映射文件存在性验证。

2. 获取 MapperFactoryBean 的实例

由于 MapperFactoryBean 实现了 FactoryBean 接口，所以当通过 getBean 方法获取对应实例的时候其实是获取该类的 getObject()函数返回的实例。

```
public T getObject() throws Exception {
    return getSqlSession().getMapper(this.mapperInterface);
}
```

这段代码正是我们在提供 MyBatis 独立使用的时候的一个代码调用。Spring 通过 FactoryBean 进行了封装。

9.3.3　MapperScannerConfigurer

我们在 applicationContext.xml 中配置了 userMapper 供需要时使用。但如果需要用到的映射器较多的话，采用这种配置方式就显得很低效。为了解决这个问题，我们可以使用 MapperScannerConfigurer，让它扫描特定的包，自动帮我们成批地创建映射器。这样一来，就能大大减少配置的工作量，比如我们将 applicationContext.xml 文件中的配置改成如下：

```xml
<?xml version="1.0" encoding="UTF-8"?>
<beans xmlns="http://www.Springframework.org/schema/beans"
    xmlns:xsi="http://www.w3.org/2001/XMLSchema-instance"
    xsi:schemaLocation="http://www.Springframework.org/schema/beans http://www.Springframework.org/schema/beans/Spring-beans-3.0.xsd">

    <bean id="dataSource" class="org.apache.commons.dbcp.BasicDataSource">
        <property name="driverClassName" value="com.mysql.jdbc.Driver"></property>
        <property name="url" value="jdbc:mysql://localhost:3306/lexueba?useUnicode=true&amp;characterEncoding=UTF-8&amp;zeroDateTimeBehavior=convertToNull"></property>
        <property name="username" value="root"></property>
        <property name="password" value="haojia0421xixi"></property>
        <property name="maxActive" value="100"></property>
        <property name="maxIdle" value="30"></property>
        <property name="maxWait" value="500"></property>
        <property name="defaultAutoCommit" value="true"></property>
    </bean>

    <bean id="sqlSessionFactory" class="org.mybatis.Spring.SqlSessionFactoryBean">
        <property name="configLocation" value="classpath:test/mybatis/MyBatis-Configuration.xml"></property>
        <property name="dataSource" ref="dataSource" />
        <property name="typeAliasesPackage" value="aaaaa"/>
    </bean>

    <!-- 注释掉原有代码
    <bean id="userMapper" class="org.mybatis.Spring.mapper.MapperFactoryBean">
        <property name="mapperInterface" value="test.mybatis.dao.UserMapper"></property>
```

```
            <property name="sqlSessionFactory" ref="sqlSessionFactory"></property>
        </bean>
    -->

    <bean class="org.mybatis.Spring.mapper.MapperScannerConfigurer">
        <property name="basePackage" value="test.mybatis.dao" />
    </bean>

</beans>
```

在上面的配置中，我们屏蔽掉了原始的代码（userMapper 的创建）而增加了 MapperScannerConfigurer 的配置，basePackage 属性是让你为映射器接口文件设置基本的包路径。你可以使用分号或逗号作为分隔符设置多于一个的包路径。每个映射器将会在指定的包路径中递归地被搜索到。被发现的映射器将会使用 Spring 对自动侦测组件默认的命名策略来命名。也就是说，如果没有发现注解，它就会使用映射器的非大写的非完全限定类名。但是如果发现了 @Component 或 JSR-330@Named 注解，它会获取名称。

通过上面的配置，Spring 就会帮助我们对 test. mybatis.dao 下面的所有接口进行自动的注入，而不需要为每个接口重复在 Spring 配置文件中进行声明了。那么，这个功能又是如何做到的呢？MapperScanner Configurer 中又有哪些核心操作呢？同样，首先查看类的层次结构图，如图 9-4 所示。

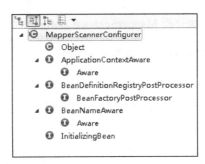

图 9-4　MapperScannerConfigurer 类的层次结构图

我们又看到了令人感兴趣的接口 InitializingBean，马上查找类的 afterPropertiesSet 方法来看看类的初始化逻辑。

```
public void afterPropertiesSet() throws Exception {
    notNull(this.basePackage, "Property 'basePackage' is required");
}
```

很遗憾，分析并没有想我们之前那样顺利，afterPropertiesSet()方法除了一句对 basePackage 属性的验证代码外并没有太多的逻辑实现。好吧，让我们回过头再次查看 MapperScanner Configurer 类层次结构图中感兴趣的接口。于是，我们发现了 BeanDefinitionRegistryPostProcessor 与 BeanFactoryPostProcessor，Spring 在初始化的过程中同样会保证这两个接口的调用。

首先查看 MapperScannerConfigurer 类中对于 BeanFactoryPostProcessor 接口的实现：

```java
public void postProcessBeanFactory(ConfigurableListableBeanFactory beanFactory) {
    // left intentionally blank
}
```

没有任何逻辑实现，只能说明我们找错地方了，继续找，查看 MapperScannerConfigurer 类中对于 BeanDefinitionRegistryPostProcessor 接口的实现。

```java
public void postProcessBeanDefinitionRegistry(BeanDefinitionRegistry registry) throws BeansException {
    if (this.processPropertyPlaceHolders) {
        processPropertyPlaceHolders();
    }

    ClassPathMapperScanner scanner = new ClassPathMapperScanner(registry);
    scanner.setAddToConfig(this.addToConfig);
    scanner.setAnnotationClass(this.annotationClass);
    scanner.setMarkerInterface(this.markerInterface);
    scanner.setSqlSessionFactory(this.sqlSessionFactory);
    scanner.setSqlSessionTemplate(this.sqlSessionTemplate);
    scanner.setSqlSessionFactoryBeanName(this.sqlSessionFactoryBeanName);
    scanner.setSqlSessionTemplateBeanName(this.sqlSessionTemplateBeanName);
    scanner.setResourceLoader(this.applicationContext);
    scanner.setBeanNameGenerator(this.nameGenerator);
    scanner.registerFilters();
    scanner.scan(StringUtils.tokenizeToStringArray(this.basePackage, ConfigurableApplicationContext.CONFIG_LOCATION_DELIMITERS));
}
```

Bingo! 这次找对地方了。大致看一下代码实现，正是完成了对指定路径扫描的逻辑。那么，我们就以此为入口，详细地分析 MapperScannerConfigurer 所提供的逻辑实现。

1. processPropertyPlaceHolders 属性的处理

首先，难题就是 processPropertyPlaceHolders 属性的处理。或许读者并未过多接触此属性，我们只能查看 processPropertyPlaceHolders() 函数来反推此属性所代表的功能。

```java
/*
 * BeanDefinitionRegistries are called early in application startup, before
 * BeanFactoryPostProcessors. This means that PropertyResourceConfigurers will not have been
 * loaded and any property substitution of this class' properties will fail. To avoid this, find
 * any PropertyResourceConfigurers defined in the context and run them on this class' bean
 * definition. Then update the values.
 */
private void processPropertyPlaceHolders() {
    Map<String, PropertyResourceConfigurer> prcs = applicationContext.getBeansOfType(PropertyResourceConfigurer.class);

    if (!prcs.isEmpty() && applicationContext instanceof GenericApplicationContext) {
        BeanDefinition mapperScannerBean = ((GenericApplicationContext) applicationContext)
            .getBeanFactory().getBeanDefinition(beanName);

        // PropertyResourceConfigurer does not expose any methods to explicitly perform
        // property placeholder substitution. Instead, create a BeanFactory that just
        // contains this mapper scanner and post process the factory.
```

9.3 源码分析

```
DefaultListableBeanFactory factory = new DefaultListableBeanFactory();
factory.registerBeanDefinition(beanName, mapperScannerBean);

for (PropertyResourceConfigurer prc : prcs.values()) {
  prc.postProcessBeanFactory(factory);
}

PropertyValues values = mapperScannerBean.getPropertyValues();

this.basePackage = updatePropertyValue("basePackage", values);
this.sqlSessionFactoryBeanName = updatePropertyValue("sqlSessionFactoryBeanName", values);
this.sqlSessionTemplateBeanName = updatePropertyValue("sqlSessionTemplateBeanName", values);
    }
  }
```

不知读者是否悟出了此函数的作用呢？或许此函数的说明会给我们一些提示：BeanDefinitionRegistries 会在应用启动的时候调用，并且会早于 BeanFactoryPostProcessors 的调用，这就意味着 PropertyResourceConfigurers 还没有被加载所有对于属性文件的引用将会失效。为避免此种情况发生，此方法手动地找出定义的 PropertyResourceConfigurers 并进行提前调用以保证对于属性的引用可以正常工作。

我想读者已经有所感悟，结合之前讲过的 PropertyResourceConfigurer 的用法，举例说明一下，如要创建配置文件如 test.properties，并添加属性对：

```
basePackage=test.mybatis.dao
```

然后在 Spring 配置文件中加入属性文件解析器：

```xml
<bean id="mesHandler" class="org.Springframework.beans.factory.config.Property PlaceholderConfigurer">
    <property name="locations">
        <list>
            <value>config/test.properties</value>
        </list>
    </property>
</bean>
```

修改 MapperScannerConfigurer 类型的 bean 的定义：

```xml
<bean class="org.mybatis.Spring.mapper.MapperScannerConfigurer">
  <property name="basePackage" value="${basePackage}" />
</bean>
```

此时你会发现，这个配置并没有达到预期的效果，因为在解析${basePackage}的时候 PropertyPlaceholderConfigurer 还没有被调用，也就是属性文件中的属性还没有加载至内存中，Spring 还不能直接使用它。为了解决这个问题，Spring 提供了 processPropertyPlaceHolders 属性，你需要这样配置 MapperScannerConfigurer 类型的 bean。

```xml
<bean class="org.mybatis.Spring.mapper.MapperScannerConfigurer">
  <property name="basePackage" value="test.mybatis.dao" />
  <property name="processPropertyPlaceHolders" value="true" />
</bean>
```

通过 processPropertyPlaceHolders 属性的配置，将程序引入我们正在分析的 processProperty

PlaceHolders 函数中来完成属性文件的加载。至此，我们终于理清了这个属性的作用，再次回顾这个函数所做的事情。

1. 找到所有已经注册的 PropertyResourceConfigurer 类型的 bean。
2. 模拟 Spring 中的环境来用处理器。这里通过使用 new DefaultListableBeanFactory()来模拟 Spring 中的环境（完成处理器的调用后便失效），将映射的 bean，也就是 MapperScannerConfigurer 类型 bean 注册到环境中来进行后理器的调用，处理器 PropertyPlaceholderConfigurer 调用完成的功能，即找出所有 bean 中应用属性文件的变量并替换。也就是说，在处理器调用后，模拟环境中模拟的 MapperScannerConfigurer 类型的 bean 如果有引入属性文件中的属性那么已经被替换了，这时，再将模拟 bean 中相关的属性提取出来应用在真实的 bean 中。

2. 根据配置属性生成过滤器

在 postProcessBeanDefinitionRegistry 方法中可以看到，配置中支持很多属性的设定，但是我们感兴趣的或者说影响扫描结果的并不多，属性设置后通过在 scanner.registerFilters()代码中生成对应的过滤器来控制扫描结果。

```
    public void registerFilters() {
        boolean acceptAllInterfaces = true;

//对于 annotationClass 属性的处理
        if (this.annotationClass != null) {
            addIncludeFilter(new AnnotationTypeFilter(this.annotationClass));
            acceptAllInterfaces = false;
        }

//对于 markerInterface 属性的处理
        if (this.markerInterface != null) {
            addIncludeFilter(new AssignableTypeFilter(this.markerInterface) {
              @Override
              protected boolean matchClassName(String className) {
                return false;
              }
            });
            acceptAllInterfaces = false;
        }

        if (acceptAllInterfaces) {
          // default include filter that accepts all classes
            addIncludeFilter(new TypeFilter() {
              public boolean match(MetadataReader metadataReader, MetadataReaderFactory metadataReaderFactory) throws IOException {
                return true;
              }
            });
        }

//不扫描 package-info.java 文件
        addExcludeFilter(new TypeFilter() {
```

```
            public boolean match(MetadataReader metadataReader, MetadataReaderFactory metada
ReaderFactory) throws IOException {
                String className = metadataReader.getClassMetadata().getClassName();
                return className.endsWith("package-info");
            }
        });
    }
```

代码中得知,根据之前属性的配置生成了对应的过滤器。

1. annotationClass 属性处理。

如果 annotationClass 不为空,表示用户设置了此属性,那么就要根据此属性生成过滤器以保证达到用户想要的效果,而封装此属性的过滤器就是 AnnotationTypeFilter。AnnotationTypeFilter 保证在扫描对应 Java 文件时只接受标记有注解为 annotationClass 的接口。

2. markerInterface 属性处理。

如果 markerInterface 不为空,表示用户设置了此属性,那么就要根据此属性生成过滤器以保证达到用户想要的效果,而封装此属性的过滤器就是实现 AssignableTypeFilter 接口的局部类。表示扫描过程中只有实现 markerInterface 接口的接口才会被接受。

3. 全局默认处理。

在上面两个属性中如果存在其中任何属性,acceptAllInterfaces 的值将会变改变,但是如果用户没有设定以上两个属性,那么,Spring 会为我们增加一个默认的过滤器实现 TypeFilter 接口的局部类,旨在接受所有接口文件。

4. package-info.java 处理。

对于命名为 package-info 的 Java 文件,默认不作为逻辑实现接口,将其排除掉,使用 TypeFilter 接口的局部类实现 match 方法。

从上面的函数我们看出,控制扫描文件 Spring 通过不同的过滤器完成,这些定义的过滤器记录在了 includeFilters 和 excludeFilters 属性中。

```
    public void addIncludeFilter(TypeFilter includeFilter) {
        this.includeFilters.add(includeFilter);
    }
    public void addExcludeFilter(TypeFilter excludeFilter) {
        this.excludeFilters.add(0, excludeFilter);
    }
```

至于过滤器为什么会在扫描过程中起作用,我们在讲解扫描实现时候再继续深入研究。

3. 扫描 Java 文件

设置了相关属性以及生成了对应的过滤器后便可以进行文件的扫描了,扫描工作是由 ClassPathMapperScanner 类型的实例 scanner 中的 scan 方法完成的。

```
    public int scan(String... basePackages) {
        int beanCountAtScanStart = this.registry.getBeanDefinitionCount();
        
        doScan(basePackages);
        
        //如果配置了 includeAnnotationConfig,则注册对应注解的处理器以保证注解功能的正常使用。
```

```
                if (this.includeAnnotationConfig) {
                    AnnotationConfigUtils.registerAnnotationConfigProcessors(this.registry);
                }

                return this.registry.getBeanDefinitionCount() - beanCountAtScanStart;
            }
```

scan 是个全局方法，扫描工作通过 doScan(basePackages) 委托给了 doScan 方法，同时，还包括了 includeAnnotationConfig 属性的处理，AnnotationConfigUtils.registerAnnotation ConfigProcessors (this.registry) 代码主要是完成对于注解处理器的简单注册，比如 AutowiredAnnotationBeanPostProcessor、RequiredAnnotationBeanPostProcessor 等，这里不再赘述，我们重点研究文件扫描功能的实现。

ClassPathMapperScanner.java

```
        public Set<BeanDefinitionHolder> doScan(String... basePackages) {
            Set<BeanDefinitionHolder> beanDefinitions = super.doScan(basePackages);

            if (beanDefinitions.isEmpty()) {
                //如果没有扫描到任何文件发出警告
                logger.warn("No MyBatis mapper was found in '" + Arrays.toString(basePackages) +
"' package. Please check your configuration.");
            } else {
                for (BeanDefinitionHolder holder : beanDefinitions) {
                    GenericBeanDefinition definition = (GenericBeanDefinition) holder.getBeanDefinition();

                    if (logger.isDebugEnabled()) {
                        logger.debug("Creating MapperFactoryBean with name '" + holder.getBeanName()
                                + "' and '" + definition.getBeanClassName() + "' mapperInterface");
                    }

                    //开始构造 MapperFactoryBean 类型的 bean.
                    definition.getPropertyValues().add("mapperInterface", definition.getBeanClassName());
                    definition.setBeanClass(MapperFactoryBean.class);

                    definition.getPropertyValues().add("addToConfig", this.addToConfig);

                    boolean explicitFactoryUsed = false;
                    if (StringUtils.hasText(this.sqlSessionFactoryBeanName)) {
                        definition.getPropertyValues().add("sqlSessionFactory", new RuntimeBeanReference
(this.sqlSessionFactoryBeanName));
                        explicitFactoryUsed = true;
                    } else if (this.sqlSessionFactory != null) {
                        definition.getPropertyValues().add("sqlSessionFactory", this.sqlSessionFactory);
                        explicitFactoryUsed = true;
                    }

                    if (StringUtils.hasText(this.sqlSessionTemplateBeanName)) {
                        if (explicitFactoryUsed) {
                            logger.warn("Cannot use both: sqlSessionTemplate and sqlSessionFactory
together. sqlSessionFactory is ignored.");
```

```
                }
                definition.getPropertyValues().add("sqlSessionTemplate", new RuntimeBeanReference
(this.sqlSessionTemplateBeanName));
                explicitFactoryUsed = true;
            } else if (this.sqlSessionTemplate != null) {
                if (explicitFactoryUsed) {
                    logger.warn("Cannot use both: sqlSessionTemplate and sqlSessionFactory
together. sqlSessionFactory is ignored.");
                }
                definition.getPropertyValues().add("sqlSessionTemplate", this.sqlSessionTemplate);
                explicitFactoryUsed = true;
            }

            if (!explicitFactoryUsed) {
                if (logger.isDebugEnabled()) {
                    logger.debug("Enabling autowire by type for MapperFactoryBean with name '"
+ holder.getBeanName() + "'.");
                }
                definition.setAutowireMode(AbstractBeanDefinition.AUTOWIRE_BY_TYPE);
            }
        }
    }
```

此时，虽然还没有完成介绍到扫描的过程，但是我们也应该理解了 Spring 中对于自动扫描的注册，声明 MapperScannerConfigurer 类型的 bean 目的是不需要我们对于每个接口都注册一个 MapperFactoryBean 类型的对应的 bean，但是，不在配置文件中注册并不代表这个 bean 不存在，而是在扫描的过程中通过编码的方式动态注册。实现过程我们在上面的函数中可以看得非常清楚。

```
    protected Set<BeanDefinitionHolder> doScan(String... basePackages) {
        Assert.notEmpty(basePackages, "At least one base package must be specified");
        Set<BeanDefinitionHolder> beanDefinitions = new LinkedHashSet <BeanDefinition
Holder>();
        for (String basePackage : basePackages) {
            //扫描 basePackage 路径下 java 文件
            Set<BeanDefinition> candidates = findCandidateComponents(basePackage);
            for (BeanDefinition candidate : candidates) {
                //解析 scope 属性
                ScopeMetadata scopeMetadata = this.scopeMetadataResolver.resolveScope
Metadata (candidate);

                candidate.setScope(scopeMetadata.getScopeName());
                String beanName = this.beanNameGenerator.generateBeanName(candidate,
this.registry);
                if (candidate instanceof AbstractBeanDefinition) {
                    postProcessBeanDefinition((AbstractBeanDefinition) candidate,
beanName);
                }
                if (candidate instanceof AnnotatedBeanDefinition) {
                    //如果是 AnnotatedBeanDefinition 类型的 bean,需要检测下常用注解如：
Primary、Lazy 等
                    AnnotationConfigUtils.processCommonDefinitionAnnotations
((AnnotatedBeanDefinition) candidate);
                }
```

```java
                    //检测当前 bean 是否已经注册
                    if (checkCandidate(beanName, candidate)) {
                        BeanDefinitionHolder definitionHolder = new BeanDefinitionHolder(candidate, beanName);
                        //如果当前 bean 是用于生成代理的 bean 那么需要进一步处理
                        definitionHolder = AnnotationConfigUtils.applyScopedProxyMode(scopeMetadata, definitionHolder, this.registry);
                        beanDefinitions.add(definitionHolder);
                        registerBeanDefinition(definitionHolder, this.registry);
                    }
                }
            }
        return beanDefinitions;
    }
    public Set<BeanDefinition> findCandidateComponents(String basePackage) {
        Set<BeanDefinition> candidates = new LinkedHashSet<BeanDefinition>();
        try {
            String packageSearchPath = ResourcePatternResolver.CLASSPATH_ALL_URL_PREFIX +
                    resolveBasePackage(basePackage) + "/" + this.resourcePattern;
            Resource[] resources = this.resourcePatternResolver.getResources(packageSearchPath);
            boolean traceEnabled = logger.isTraceEnabled();
            boolean debugEnabled = logger.isDebugEnabled();
            for (Resource resource : resources) {
                if (traceEnabled) {
                    logger.trace("Scanning " + resource);
                }
                if (resource.isReadable()) {
                    try {
                        MetadataReader metadataReader = this.metadataReaderFactory.getMetadataReader(resource);
                        if (isCandidateComponent(metadataReader)) {
                            ScannedGenericBeanDefinition sbd = new ScannedGenericBeanDefinition(metadataReader);
                            sbd.setResource(resource);
                            sbd.setSource(resource);
                            if (isCandidateComponent(sbd)) {
                                if (debugEnabled) {
                                    logger.debug("Identified candidate component class: " + resource);
                                }
                                candidates.add(sbd);
                            }
                            else {
                                if (debugEnabled) {
                                    logger.debug("Ignored because not a concrete top-level class: " + resource);
                                }
                            }
                        }
                        else {
                            if (traceEnabled) {
                                logger.trace("Ignored because not matching any filter: " + resource);
```

```
                    }
                }
            }
            catch (Throwable ex) {
                throw new BeanDefinitionStoreException(
                        "Failed to read candidate component class: " + resource, ex);
            }
        }
        else {
            if (traceEnabled) {
                logger.trace("Ignored because not readable: " + resource);
            }
        }
    }
}
catch (IOException ex) {
    throw new BeanDefinitionStoreException("I/O failure during classpath scanning", ex);
}
return candidates;
```

findCandidateComponents 方法根据传入的包路径信息并结合类文件路径拼接成文件的绝对路径,同时完成了文件的扫描过程并且根据对应的文件生成了对应的 bean,使用 ScannedGenericBeanDefinition 类型的 bean 承载信息,bean 中只记录了 resource 和 source 信息。这里,我们更感兴趣的是 isCandidateComponent(metadataReader),此句代码用于判断当前扫描的文件是否符合要求,而我们之前注册的一些过滤器信息也正是在此时派上用场的。

```
protected boolean isCandidateComponent(MetadataReader metadataReader) throws IOException {
    for (TypeFilter tf : this.excludeFilters) {
        if (tf.match(metadataReader, this.metadataReaderFactory)) {
            return false;
        }
    }
    for (TypeFilter tf : this.includeFilters) {
        if (tf.match(metadataReader, this.metadataReaderFactory)) {
            AnnotationMetadata metadata = metadataReader.getAnnotationMetadata();
            if (!metadata.isAnnotated(Profile.class.getName())) {
                return true;
            }
            AnnotationAttributes profile = MetadataUtils.attributesFor(metadata, Profile.class);
            return this.environment.acceptsProfiles(profile.getStringArray("value"));
        }
    }
    return false;
}
```

我们看到了之前加入过滤器的两个属性 excludeFilters、includeFilters,并且知道对应的文件是否符合要求是根据过滤器中的 match 方法所返回的信息来判断的,当然用户可以实现并注册满足自己业务逻辑的过滤器来控制扫描的结果,metadataReader 中有你过滤所需要的全部文件信息。至此,我们完成了文件的扫描过程的分析。

第 10 章 事务

Spring 声明式事务让我们从复杂的事务处理中得到解脱，使我们再也不需要去处理获得连接、关闭连接、事务提交和回滚等操作，再也不需要在与事务相关的方法中处理大量的 try...catch...finally 代码。Spring 中事务的使用虽然已经相对简单得多，但是，还是有很多的使用及配置规则，有兴趣的读者可以自己查阅相关资料进行深入研究，这里只列举出最常用的使用方法。

同样，我们还是以最简单的示例来进行直观地介绍。

10.1 JDBC 方式下的事务使用示例

1．创建数据表结构

```
CREATE TABLE 'user' (
  'id' int(11) NOT NULL auto_increment,
  'name' varchar(255) default NULL,
  'age' int(11) default NULL,
  'sex' varchar(255) default NULL,
  PRIMARY KEY  ('id')
) ENGINE=InnoDB DEFAULT CHARSET=utf8;
```

2．创建对应数据表的 PO

```
public class User {

    private int id;
    private String name;
    private int age;
private String sex;

//省略 set/get 方法
}
```

3．创建表与实体间的映射

```
public class UserRowMapper implements RowMapper {

    @Override
```

```java
    public Object mapRow(ResultSet set, int index) throws SQLException {
        User person = new User(set.getInt("id"), set.getString("name"), set
                .getInt("age"), set.getString("sex"));
        return person;
    }
}
```

4. 创建数据操作接口

```java
@Transactional(propagation=Propagation.REQUIRED)
public interface UserService {

    public  void save(User user) throws Exception;

}
```

5. 创建数据操作接口实现类

```java
public class UserServiceImpl implements UserService {

    private JdbcTemplate jdbcTemplate;

    // 设置数据源
    public void setDataSource(DataSource dataSource) {
        this.jdbcTemplate = new JdbcTemplate(dataSource);
    }

    public void save(User user) throws Exception {
        jdbcTemplate.update("insert into user(name,age,sex)values(?,?,?)",
                new Object[] { user.getName(), user.getAge(),
                        user.getSex() }, new int[] { java.sql.Types.VARCHAR,
                        java.sql.Types.INTEGER, java.sql.Types.VARCHAR });

            //事务测试，加上这句代码则数据不会保存到数据库中
            throw new RuntimeException("aa");
    }
}
```

6. 创建 Spring 配置文件

```xml
<?xml version="1.0" encoding="UTF-8"?>
<beans xmlns="http://www.Springframework.org/schema/beans"
    xmlns:xsi="http://www.w3.org/2001/XMLSchema-instance"
    xmlns:tx="http://www.Springframework.org/schema/tx"
    xmlns:context="http://www.Springframework.org/schema/context"
    xsi:schemaLocation="
            http://www.Springframework.org/schema/beans http://www.Springframework.org/schema/beans/Spring-beans-2.5.xsd
            http://www.Springframework.org/schema/context http://www.Springframework.org/schema/context/Spring-context-2.5.xsd
            http://www.Springframework.org/schema/tx http://www.Springframework.org/schema/tx/Spring-tx-2.5.xsd
    ">


    <tx:annotation-driven transaction-manager="transactionManager" />
```

```xml
<bean id="transactionManager"
    class="org.Springframework.jdbc.datasource.DataSourceTransactionManager">
    <property name="dataSource" ref="dataSource" />
</bean>

<!--配置数据源 -->
<bean id="dataSource" class="org.apache.commons.dbcp.BasicDataSource"
    destroy-method="close">
    <property name="driverClassName" value="com.mysql.jdbc.Driver" />
    <property name="url" value="jdbc:mysql://localhost:3306/lexueba" />
    <property name="username" value="root" />
    <property name="password" value="haojia0421xixi" />
    <!-- 连接池启动时的初始值 -->
    <property name="initialSize" value="1" />
    <!-- 连接池的最大值 -->
    <property name="maxActive" value="300" />
    <!-- 最大空闲值.当经过一个高峰时间后,连接池可以慢慢将已经用不到的连接慢慢释放一部分,一直减
    少到maxIdle为止 -->
    <property name="maxIdle" value="2" />
    <!-- 最小空闲值.当空闲的连接数少于阀值时,连接池就会预申请去一些连接,以免洪峰来时来不及申请 -->
    <property name="minIdle" value="1" />
</bean>

<!-- 配置业务bean：PersonServiceBean -->
<bean id="userService" class="service.UserServiceImpl">
    <!-- 向属性dataSource注入数据源 -->
    <property name="dataSource" ref="dataSource"></property>
</bean>
</beans>
```

7．测试

```java
public static void main(String[] args) throws Exception {
    ApplicationContext act = new ClassPathXmlApplicationContext("bean.xml");

    UserService userService = (UserService) act.getBean("userService");
    User user = new User();
    user.setName("张三ccc");
    user.setAge(20);
    user.setSex("男");
    // 保存一条记录
    userService.save(user);

}
}
```

在上面的测试示例中，UserServiceImpl类对接口UserService中的save函数的实现最后加入了一句抛出异常的代码：throw new RuntimeException("aa")。当注掉这段代码执行测试类，那么会看到数据被成功的保存到了数据库中，但是如果加入这段代码时再次运行测试类，发现此处的操作并不会将数据保存到数据库中。

注意 默认情况下Spring中的事务处理只对RuntimeException方法进行回滚，所以，如果此处将Runtime Exception替换成普通的Exception不会产生回滚效果。

10.2 事务自定义标签

对于 Spring 中事务功能的代码分析,我们首先从配置文件开始入手,在配置文件中有这样一个配置:<tx:annotation-driven />。可以说此处配置是事务的开关,如果没有此处配置,那么 Spring 中将不存在事务的功能。那么我们就从这个配置开始分析。

根据之前的分析,我们因此可以判断,在自定义标签中的解析过程中一定是做了一些辅助操作,于是我们先从自定义标签入手进行分析。

使用 Eclipse 搜索全局代码,关键字 annotation-drive,最终锁定类 TxNamespaceHandler,在 TxNamespaceHandler 中的 init 方法中:

```
public void init() {
    registerBeanDefinitionParser("advice", new TxAdviceBeanDefinitionParser());
    registerBeanDefinitionParser("annotation-driven", new AnnotationDrivenBeanDefinitionParser());
    registerBeanDefinitionParser("jta-transaction-manager", new JtaTransactionManagerBeanDefinitionParser());
}
```

根据自定义标签的使用规则以及上面的代码,可以知道,在遇到诸如<tx:annotation-driven 为开头的配置后,Spring 都会使用 AnnotationDrivenBeanDefinitionParser 类的 parse 方法进行解析。

```
public BeanDefinition parse(Element element, ParserContext parserContext) {
    String mode = element.getAttribute("mode");
    if ("aspectj".equals(mode)) {
        // mode="aspectj"
        registerTransactionAspect(element, parserContext);
    }else {
        // mode="proxy"
        AopAutoProxyConfigurer.configureAutoProxyCreator(element, parserContext);
    }
    return null;
}
```

在解析中存在对于 mode 属性的判断,根据代码,如果我们需要使用 AspectJ 的方式进行事务切入(Spring 中的事务是以 AOP 为基础的),那么可以使用这样的配置:

```
<tx:annotation-driven transaction-manager="transactionManager" mode="aspectj" />
```

10.2.1 注册 InfrastructureAdvisorAutoProxyCreator

我们以默认配置为例子进行分析,进入 AopAutoProxyConfigurer 类的 configureAutoProxyCreator:

```
public static void configureAutoProxyCreator(Element element, ParserContext parserContext) {
    AopNamespaceUtils.registerAutoProxyCreatorIfNecessary(parserContext, element);
    //TRANSACTION_ADVISOR_BEAN_NAME ="org.Springframework.transaction.config.internal TransactionAdvisor";
    String txAdvisorBeanName = TransactionManagementConfigUtils.TRANSACTION_ADVISOR_BEAN_NAME;
    if (!parserContext.getRegistry().containsBeanDefinition (txAdvisorBean Name)) {
        Object eleSource = parserContext.extractSource(element);
```

```
            //创建 TransactionAttributeSource 的 bean
            RootBeanDefinition sourceDef = new RootBeanDefinition (Annotation
TransactionAttributeSource.class);
            sourceDef.setSource(eleSource);
            sourceDef.setRole(BeanDefinition.ROLE_INFRASTRUCTURE);
                //注册 bean,并使用 Spring 中的定义规则生成 beanname
            String sourceName = parserContext.getReaderContext(). RegisterWith
GeneratedName (sourceDef);

            //创建 TransactionInterceptor 的 bean
            RootBeanDefinition interceptorDef = new RootBeanDefinition
(TransactionInterceptor.class);
            interceptorDef.setSource(eleSource);
            interceptorDef.setRole(BeanDefinition.ROLE_INFRASTRUCTURE);
            registerTransactionManager(element, interceptorDef);
            interceptorDef.getPropertyValues().add("transactionAttributeSource",
new RuntimeBeanReference(sourceName));
                //注册 bean,并使用 Spring 中的定义规则生成 beanname
            String interceptorName = parserContext.getReaderContext(). Register
WithGeneratedName(interceptorDef);

            //创建 TransactionAttributeSourceAdvisor 的 bean
            RootBeanDefinition advisorDef = new RootBeanDefinition (BeanFactory
TransactionAttributeSourceAdvisor.class);
            advisorDef.setSource(eleSource);
            advisorDef.setRole(BeanDefinition.ROLE_INFRASTRUCTURE);
        //将 sourceName 的 bean 注入 advisorDef 的 transactionAttributeSource 属性中
            advisorDef.getPropertyValues().add("transactionAttributeSource", new
RuntimeBeanReference(sourceName));
        //将 interceptorName 的 bean 注入 advisorDef 的 adviceBeanName 属性中
            advisorDef.getPropertyValues().add("adviceBeanName", interceptorName);
            //如果配置了 order 属性,则加入到 bean 中
            if (element.hasAttribute("order")) {
                advisorDef.getPropertyValues().add("order", element.getAttribute
("order"));
            }
            parserContext.getRegistry().registerBeanDefinition(txAdvisorBeanName,
advisorDef);

            //创建 CompositeComponentDefinition
            CompositeComponentDefinition compositeDef = new CompositeComponent
Definition(element.getTagName(), eleSource);
            compositeDef.addNestedComponent(new BeanComponentDefinition (sourceDef,
sourceName));
            compositeDef.addNestedComponent(new BeanComponentDefinition(interceptorDef,
 interceptorName));
            compositeDef.addNestedComponent(new BeanComponentDefinition(advisorDef,
txAdvisorBeanName));
            parserContext.registerComponent(compositeDef);
        }
    }
```

10.2 事务自定义标签

上面的代码注册了代理类及 3 个 bean，很多读者会直接略过，认为只是注册 3 个 bean 而已，确实，这里只注册了 3 个 bean，但是这 3 个 bean 支撑了整个的事务功能，那么这 3 个 bean 是怎么组织起来的呢？

首先，其中的两个 bean 被注册到了一个名为 advisorDef 的 bean 中，advisorDef 使用 BeanFactoryTransactionAttributeSourceAdvisor 作为其 class 属性。也就是说 BeanFactoryTransactionAttributeSourceAdvisor 代表着当前 bean，如图 10-1 所示，具体代码如下：

```
advisorDef.getPropertyValues().add("adviceBeanName", interceptorName);
```

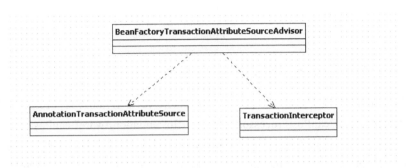

图 10-1　BeanFactoryTransactionAttributeSourceAdvisor 的组装

那么如此组装的目的是什么呢？我们暂且留下一个悬念，接着分析代码。上面函数 configureAutoProxyCreator 中的第一句貌似很简单但却是很重要的代码：

```
AopNamespaceUtils.registerAutoProxyCreatorIfNecessary(parserContext, element);
```

进入这个函数：

```
public static void registerAutoProxyCreatorIfNecessary(
        ParserContext parserContext, Element sourceElement) {

    BeanDefinition beanDefinition = AopConfigUtils.registerAutoProxyCreatorIfNecessary(
            parserContext.getRegistry(), parserContext.extractSource (source
Element));
    useClassProxyingIfNecessary(parserContext.getRegistry(), sourceElement);
    registerComponentIfNecessary(beanDefinition, parserContext);
}

public static BeanDefinition registerAutoProxyCreatorIfNecessary(BeanDefinitionRegistry registry, Object source) {
        return registerOrEscalateApcAsRequired(InfrastructureAdvisorAutoProxyCreator.class, registry, source);
}
```

对于解析来的代码流程 AOP 中已经有所分析，上面的两个函数主要目的是注册了 InfrastructureAdvisorAutoProxyCreator 类型的 bean，那么注册这个类的目的是什么呢？查看这个类的层次，如图 10-2 所示。

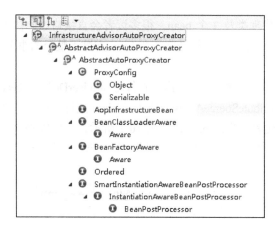

图 10-2　InfrastructureAdvisorAutoProxyCreator 类的层次结构图

从上面的层次结构中可以看到，InfrastructureAdvisorAutoProxyCreator 间接实现了 SmartInstantiationAwareBeanPostProcessor，而 SmartInstantiationAwareBeanPostProcessor 又继承自 InstantiationAwareBeanPostProcessor，也就是说在 Spring 中，所有 bean 实例化时 Spring 都会保证调用其 postProcessAfterInitialization 方法，其实现是在父类 AbstractAutoProxyCreator 类中实现。

以之前的示例为例，当实例化 userService 的 bean 时便会调用此方法，方法如下：

```
public Object postProcessAfterInitialization(Object bean, String beanName) throws BeansException {
        if (bean != null) {
            //根据给定的 bean 的 class 和 name 构建出个 key, beanClassName_beanName
            Object cacheKey = getCacheKey(bean.getClass(), beanName);
            //是否是由于避免循环依赖而创建的 bean 代理
            if (!this.earlyProxyReferences.contains(cacheKey)) {
                return wrapIfNecessary(bean, beanName, cacheKey);
            }
        }
        return bean;
    }
```

这里实现的主要目的是对指定 bean 进行封装，当然首先要确定是否需要封装，检测及封装的工作都委托给了 wrapIfNecessary 函数进行。

```
protected Object wrapIfNecessary(Object bean, String beanName, Object cacheKey) {
        //如果已经处理过
        if (this.targetSourcedBeans.contains(beanName)) {
            return bean;
        }
        if (this.nonAdvisedBeans.contains(cacheKey)) {
            return bean;
        }
        //给定的 bean 类是否代表一个基础设施类，不应代理，或者配置了指定 bean 不需要自动代理
        if (isInfrastructureClass(bean.getClass()) || shouldSkip(bean.getClass(), beanName)) {
```

```
            this.nonAdvisedBeans.add(cacheKey);
            return bean;
        }

        // Create proxy if we have advice.
        Object[] specificInterceptors = getAdvicesAndAdvisorsForBean(bean.getClass(),
beanName, null);
        if (specificInterceptors != DO_NOT_PROXY) {
            this.advisedBeans.add(cacheKey);
            Object proxy = createProxy(bean.getClass(), beanName, specificInterceptors,
new SingletonTargetSource(bean));
            this.proxyTypes.put(cacheKey, proxy.getClass());
            return proxy;
        }

        this.nonAdvisedBeans.add(cacheKey);
        return bean;
    }
```

wrapIfNecessary 函数功能实现起来很复杂，但是逻辑上理解起来还是相对简单的，在 wrapIfNecessary 函数中主要的工作如下。

- 找出指定 bean 对应的增强器。
- 根据找出的增强器创建代理。

听起来似乎简单的逻辑，Spring 中又做了哪些复杂的工作呢？对于创建代理的部分，通过之前的分析相信大家已经很熟悉了，但是对于增强器的获取，Spring 又是怎么做的呢？

10.2.2 获取对应 class/method 的增强器

获取指定 bean 对应的增强器，其中包含两个关键字：增强器与对应。也就是说在 getAdvicesAndAdvisorsForBean 函数中，不但要找出增强器，而且还需要判断增强器是否满足要求。

```
    protected Object[] getAdvicesAndAdvisorsForBean(Class beanClass, String beanName,
TargetSource targetSource) {
        List advisors = findEligibleAdvisors(beanClass, beanName);
        if (advisors.isEmpty()) {
            return DO_NOT_PROXY;
        }
        return advisors.toArray();
    }
    protected List<Advisor> findEligibleAdvisors(Class beanClass, String beanName) {
        List<Advisor> candidateAdvisors = findCandidateAdvisors();
        List<Advisor> eligibleAdvisors = findAdvisorsThatCanApply(candidateAdvisors,
beanClass, beanName);
        extendAdvisors(eligibleAdvisors);
        if (!eligibleAdvisors.isEmpty()) {
            eligibleAdvisors = sortAdvisors(eligibleAdvisors);
        }
        return eligibleAdvisors;
    }
```

其实我们也渐渐地体会到了 Spring 中代码的优秀，即使是一个很复杂的逻辑，在 Spring 中也会被拆分成若干个小的逻辑，然后在每个函数中实现，使得每个函数的逻辑简单到我们能快速地理解，而不会像有些人开发的那样，将一大堆的逻辑都罗列在一个函数中，给后期维护人员造成巨大的困扰。

同样，通过上面的函数，Spring 又将任务进行了拆分，分成了获取所有增强器与增强器是否匹配两个功能点。

1. 寻找候选增强器

在 findCandidateAdvisors 函数中完成的就是获取增强器的功能。

```java
protected List<Advisor> findCandidateAdvisors() {
    return this.advisorRetrievalHelper.findAdvisorBeans();
}

public List<Advisor> findAdvisorBeans() {
    // Determine list of advisor bean names, if not cached already.
    String[] advisorNames = null;
    synchronized (this) {
        advisorNames = this.cachedAdvisorBeanNames;
        if (advisorNames == null) {
            advisorNames = BeanFactoryUtils.beanNamesForTypeIncludingAncestors(
                    this.beanFactory, Advisor.class, true, false);
            this.cachedAdvisorBeanNames = advisorNames;
        }
    }
    if (advisorNames.length == 0) {
        return new LinkedList<Advisor>();
    }

    List<Advisor> advisors = new LinkedList<Advisor>();
    for (String name : advisorNames) {
        if (isEligibleBean(name) && !this.beanFactory.isCurrentlyInCreation(name)) {
            try {
                advisors.add(this.beanFactory.getBean(name, Advisor.class));
            }
            catch (BeanCreationException ex) {
                Throwable rootCause = ex.getMostSpecificCause();
                if (rootCause instanceof BeanCurrentlyInCreationException) {
                    BeanCreationException bce = (BeanCreationException) rootCause;
                    if (this.beanFactory.isCurrentlyInCreation(bce.getBeanName())) {
                        if (logger.isDebugEnabled()) {
                            logger.debug("Ignoring currently created advisor '" +
name + "': " + ex.getMessage());
                        }
                        continue;
                    }
                }
                throw ex;
            }
        }
```

```
            }
            return advisors;
}
```

对于上面的函数，你看懂其中的奥妙了吗？首先是通过 BeanFactoryUtils 类提供的工具方法获取所有对应 Advisor.class 的类，获取办法无非是使用 ListableBeanFactory 中提供的方法：

```
String[] getBeanNamesForType(Class<?> type, boolean includeNonSingletons, boolean allowEagerInit);
```

而当我们知道增强器在容器中的 beanName 时，获取增强器已经不是问题了，在 BeanFactory 中提供了这样的方法，可以帮助我们快速定位对应的 bean 实例。

```
<T> T getBean(String name, Class<T> requiredType) throws BeansException;
```

或许你已经忘了之前留下的悬念，在我们讲解自定义标签时曾经注册了一个类型为 BeanFactoryTransactionAttributeSourceAdvisor 的 bean，而在此 bean 中我们又注入了另外两个 Bean，那么此时这个 Bean 就会被开始使用了。因为 BeanFactoryTransactionAttribute Source Advisor 同样也实现了 Advisor 接口，那么在获取所有增强器时自然也会将此 bean 提取出来，并随着其他增强器一起在后续的步骤中被织入代理。

2. 候选增强器中寻找到匹配项

当找出对应的增强器后，接来的任务就是看这些增强器是否与对应的 class 匹配了，当然不只是 class，class 内部的方法如果匹配也可以通过验证。

```
public static List<Advisor> findAdvisorsThatCanApply(List<Advisor> candidateAdvisors, Class<?> clazz) {
            if (candidateAdvisors.isEmpty()) {
                return candidateAdvisors;
            }
            List<Advisor> eligibleAdvisors = new LinkedList<Advisor>();
            //首先处理引介增强
            for (Advisor candidate : candidateAdvisors) {
                if (candidate instanceof IntroductionAdvisor && canApply(candidate, clazz)) {
                    eligibleAdvisors.add(candidate);
                }
            }
            boolean hasIntroductions = !eligibleAdvisors.isEmpty();
            for (Advisor candidate : candidateAdvisors) {
                //引介增强已经处理
                if (candidate instanceof IntroductionAdvisor) {
                    continue;
                }
                //对于普通 bean 的处理
                if (canApply(candidate, clazz, hasIntroductions)) {
                    eligibleAdvisors.add(candidate);
                }
            }
            return eligibleAdvisors;
}

    public static boolean canApply(Advisor advisor, Class<?> targetClass, boolean hasIntroductions) {
```

第 10 章 事务

```
            if (advisor instanceof IntroductionAdvisor) {
                return ((IntroductionAdvisor) advisor).getClassFilter().Matches (targetClass);
            }else if (advisor instanceof PointcutAdvisor) {
                PointcutAdvisor pca = (PointcutAdvisor) advisor;
                return canApply(pca.getPointcut(), targetClass, hasIntroductions);
            }else {
                return true;
            }
        }
```

当前我们分析的是对于 UserService 是否适用于此增强方法，那么当前的 advisor 就是之前查找出来的类型为 BeanFactoryTransactionAttributeSourceAdvisor 的 bean 实例，而通过类的层次结构我们又知道：BeanFactoryTransactionAttributeSourceAdvisor 间接实现了 PointcutAdvisor。因此，在 canApply 函数中的第二个 if 判断时就会通过判断，会将 BeanFactory Transaction AttributeSourceAdvisor 中的 getPointcut()方法返回值作为参数继续调用 canApply 方法，而 getPoint()方法返回的是 TransactionAttributeSourcePointcut 类型的实例。对于 transactionAttribute Source 这个属性大家还有印象吗？这是在解析自定义标签时注入进去的。

```
    private final TransactionAttributeSourcePointcut pointcut = new TransactionAttribute
SourcePointcut() {
            @Override
            protected TransactionAttributeSource getTransactionAttributeSource() {
                return transactionAttributeSource;
            }
        };
```

那么，使用 ransactionAttributeSourcePointcut 类型的实例作为函数参数继续跟踪 canApply。

```
    public static boolean canApply(Pointcut pc, Class<?> targetClass, boolean hasIntroductions) {
            Assert.notNull(pc, "Pointcut must not be null");
            if (!pc.getClassFilter().matches(targetClass)) {
                return false;
            }

            //此时的 pc 表示 TransactionAttributeSourcePointcut
            //pc.getMethodMatcher()返回的正是自身(this)。
            MethodMatcher methodMatcher = pc.getMethodMatcher();
            IntroductionAwareMethodMatcher introductionAwareMethodMatcher = null;
            if (methodMatcher instanceof IntroductionAwareMethodMatcher) {
                introductionAwareMethodMatcher = (IntroductionAwareMethodMatcher) methodMatcher;
            }

            Set<Class> classes = new HashSet<Class>(ClassUtils. GetAllInterfaces
ForClassAsSet(targetClass));
            classes.add(targetClass);
            //classes:[interface test.IITestBean, class test.TestBean]
            for (Class<?> clazz : classes) {
                Method[] methods = clazz.getMethods();
                for (Method method : methods) {
                    if ((introductionAwareMethodMatcher != null &&
                            introductionAwareMethodMatcher.matches(method, targetClass, hasIntroductions)) ||

                            methodMatcher.matches(method, targetClass)) {
```

```
                    return true;
                }
            }
        }

        return false;
    }
```

通过上面函数大致可以理清大体脉络，首先获取对应类的所有接口并连同类本身一起遍历，遍历过程中又对类中的方法再次遍历，一旦匹配成功便认为这个类适用于当前增强器。

到这里我们不禁会有疑问，对于事务的配置不仅仅局限于在函数上配置，我们都知道，在类活接口上的配置可以延续到类中的每个函数，那么，如果针对每个函数进行检测，在类本身上配置的事务属性岂不是检测不到了吗？带着这个疑问，我们继续探求 matcher 方法。

做匹配的时候 methodMatcher.matches(method, targetClass)会使用 TransactionAttributeSource Pointcut 类的 matches 方法。

```
public boolean matches(Method method, Class targetClass) {
    //自定义标签解析时注入
    TransactionAttributeSource tas = getTransactionAttributeSource();
    return (tas == null || tas.getTransactionAttribute(method, targetClass) != null);
}
```

此时的 tas 表示 AnnotationTransactionAttributeSource 类型，而 AnnotationTransactionAttribute Source 类型的 getTransactionAttribute 方法如下：

```
public TransactionAttribute getTransactionAttribute(Method method, Class<?> targetClass) {
        Object cacheKey = getCacheKey(method, targetClass);
        Object cached = this.attributeCache.get(cacheKey);
        if (cached != null) {
            if (cached == NULL_TRANSACTION_ATTRIBUTE) {
                return null;
            }
            else {
                return (TransactionAttribute) cached;
            }
        }
        else {
            TransactionAttribute txAtt = computeTransactionAttribute(method, targetClass);
            // Put it in the cache.
            if (txAtt == null) {
                this.attributeCache.put(cacheKey, NULL_TRANSACTION_ATTRIBUTE);
            }
            else {
                if (logger.isDebugEnabled()) {
                    logger.debug("Adding transactional method '" + method.getName() + "' with attribute: " + txAtt);
                }
                this.attributeCache.put(cacheKey, txAtt);
            }
            return txAtt;
        }
    }
```

很遗憾，在 getTransactionAttribute 函数中并没有找到我们想要的代码，这里是指常规的一贯的套路。尝试从缓存加载，如果对应信息没有被缓存的话，工作又委托给了 computeTransactionAttribute 函数，在 computeTransactionAttribute 函数中终于的我们看到了事务标签的提取过程。

3. 提取事务标签

```
    private TransactionAttribute computeTransactionAttribute(Method method, Class<?> targetClass) {
        // Don't allow no-public methods as required.
        if (allowPublicMethodsOnly() && !Modifier.isPublic(method.getModifiers())) {
            return null;
        }

        // Ignore CGLIB subclasses - introspect the actual user class.
        Class<?> userClass = ClassUtils.getUserClass(targetClass);
        //method 代表接口中的方法，specificMethod 代表实现类中的方法
        Method specificMethod = ClassUtils.getMostSpecificMethod(method, userClass);
        // If we are dealing with method with generic parameters, find the original method.
        specificMethod = BridgeMethodResolver.findBridgedMethod(specificMethod);

        //查看方法中是否存在事务声明
        TransactionAttribute txAtt = findTransactionAttribute(specificMethod);
        if (txAtt != null) {
            return txAtt;
        }

        //查看方法所在类中是否存在事务声明
        txAtt = findTransactionAttribute(specificMethod.getDeclaringClass());
        if (txAtt != null) {
            return txAtt;
        }
        //如果存在接口，则到接口中去寻找
        if (specificMethod != method) {
            //查找接口方法
            txAtt = findTransactionAttribute(method);
            if (txAtt != null) {
                return txAtt;
            }
            //到接口中的类中去寻找
            return findTransactionAttribute(method.getDeclaringClass());
        }
        return null;
    }
```

对于事务属性的获取规则相信大家都已经很清楚，如果方法中存在事务属性，则使用方法上的属性，否则使用方法所在的类上的属性，如果方法所在类的属性上还是没有搜寻到对应的事务属性，那么再搜寻接口中的方法，再没有的话，最后尝试搜寻接口的类上面的声明。对于函数 computeTransactionAttribute 中的逻辑与我们所认识的规则并无差别，但是上面函数中并没有真正的去做搜寻事务属性的逻辑，而是搭建了个执行框架，将搜寻事务属性的任务委托给了

10.2 事务自定义标签

findTransactionAttribute 方法去执行。

```java
protected TransactionAttribute findTransactionAttribute(Method method) {
    return determineTransactionAttribute(method);
}

protected TransactionAttribute determineTransactionAttribute(AnnotatedElement ae) {
    for (TransactionAnnotationParser annotationParser : this.annotationParsers) {
        TransactionAttribute attr = annotationParser.parseTransaction Annotation (ae);
        if (attr != null) {
            return attr;
        }
    }
    return null;
}
```

this.annotationParsers 是在当前类 AnnotationTransactionAttributeSource 初始化的时候初始化的，其中的值被加入了 SpringTransactionAnnotationParser，也就是当进行属性获取的时候其实是使用 SpringTransactionAnnotationParser 类的 parseTransactionAnnotation 方法进行解析的。

```java
public TransactionAttribute parseTransactionAnnotation(AnnotatedElement ae) {
    Transactional ann = AnnotationUtils.getAnnotation(ae, Transactional.class);
    if (ann != null) {
        return parseTransactionAnnotation(ann);
    }
    else {
        return null;
    }
}
```

至此，我们终于看到了想看到的获取注解标记的代码。首先会判断当前的类是否含有 Transactional 注解，这是事务属性的基础，当然如果有的话会继续调用 parseTransactionAnnotation 方法解析详细的属性。

```java
public TransactionAttribute parseTransactionAnnotation(Transactional ann) {
    RuleBasedTransactionAttribute rbta = new RuleBasedTransactionAttribute();
    //解析 propagation
    rbta.setPropagationBehavior(ann.propagation().value());
    //解析 isolation
    rbta.setIsolationLevel(ann.isolation().value());
    //解析 timeout
    rbta.setTimeout(ann.timeout());
    //解析 readOnly
    rbta.setReadOnly(ann.readOnly());
    //解析 value
    rbta.setQualifier(ann.value());
    ArrayList<RollbackRuleAttribute> rollBackRules = new ArrayList<RollbackRuleAttribute>();
    //解析 rollbackFor
    Class[] rbf = ann.rollbackFor();
    for (Class rbRule : rbf) {
        RollbackRuleAttribute rule = new RollbackRuleAttribute(rbRule);
        rollBackRules.add(rule);
```

```
            }
            //解析rollbackForClassName
            String[] rbfc = ann.rollbackForClassName();
            for (String rbRule : rbfc) {
                RollbackRuleAttribute rule = new RollbackRuleAttribute(rbRule);
                rollBackRules.add(rule);
            }
            //解析noRollbackFor
            Class[] nrbf = ann.noRollbackFor();
            for (Class rbRule : nrbf) {
                NoRollbackRuleAttribute rule = new NoRollbackRuleAttribute(rbRule);
                rollBackRules.add(rule);
            }
            //解析noRollbackForClassName
            String[] nrbfc = ann.noRollbackForClassName();
            for (String rbRule : nrbfc) {
                NoRollbackRuleAttribute rule = new NoRollbackRuleAttribute(rbRule);
                rollBackRules.add(rule);
            }
            rbta.getRollbackRules().addAll(rollBackRules);
            return rbta;
}
```

上面方法中实现了对对应类或者方法的事务属性解析，你会在这个类中看到任何你常用或者不常用的属性提取。

至此，我们终于完成了事务标签的解析。我们是不是分析的太远了，似乎已经忘了从哪里开始了。再回顾一下，我们的现在的任务是找出某个增强器是否适合于对应的类，而是否匹配的关键则在于是否从指定的类或类中的方法中找到对应的事务属性，现在，我们以 UserServiceImpl 为例，已经在它的接口 UserService 中找到了事务属性，所以，它是与事务增强器匹配的，也就是它会被事务功能修饰。

至此，事务功能的初始化工作便结束了，当判断某个 bean 适用于事务增强时，也就是适用于增强器 BeanFactoryTransactionAttributeSourceAdvisor，没错，还是这个类，所以说，在自定义标签解析时，注入的类成为了整个事务功能的基础。

BeanFactoryTransactionAttributeSourceAdvisor 作为 Advisor 的实现类，自然要遵从 Advisor 的处理方式，当代理被调用时会调用这个类的增强方法，也就是此 bean 的 Advise，又因为在解析事务定义标签时我们把 TransactionInterceptor 类型的 bean 注入到了 BeanFactoryTransactionAttributeSourceAdvisor 中，所以，在调用事务增强器增强的代理类时会首先执行 TransactionInterceptor 进行增强，同时，也就是在 TransactionInterceptor 类中的 invoke 方法中完成了整个事务的逻辑。

10.3 事务增强器

TransactionInterceptor 支撑着整个事务功能的架构，逻辑还是相对复杂的，那么现在我们切入正题来分析此拦截器是如何实现事务特性的。TransactionInterceptor 类继承自 MethodInterceptor，所

10.3 事务增强器

所以调用该类是从其 invoke 方法开始的,首先预览下这个方法:

```java
public Object invoke(final MethodInvocation invocation) throws Throwable {
    Class<?> targetClass = (invocation.getThis() != null ? AopUtils.getTargetClass
(invocation.getThis()) : null);
        //获取对应事务属性
        final TransactionAttribute txAttr =
            getTransactionAttributeSource().getTransactionAttribute (invocation.
getMethod(), targetClass);
    //获取 beanFactory 中的 transactionManager
        final PlatformTransactionManager tm = determineTransactionManager(txAttr);
    //构造方法唯一标识(类.方法,如 service.UserServiceImpl.save)
        final String joinpointIdentification = methodIdentification (invocation.
getMethod(), targetClass);

        //声明式事务处理
        if (txAttr == null || !(tm instanceof CallbackPreferringPlatformTransaction
Manager)) {
            //创建 TransactionInfo
            TransactionInfo txInfo = createTransactionIfNecessary(tm, txAttr, joinpoint
Identification);
            Object retVal = null;
            try {
                //执行被增强方法
                retVal = invocation.proceed();
            }
            catch (Throwable ex) {
                //异常回滚
                completeTransactionAfterThrowing(txInfo, ex);
                throw ex;
            }
            finally {
                //清除信息
                cleanupTransactionInfo(txInfo);
            }
            //提交事务
            commitTransactionAfterReturning(txInfo);
            return retVal;
        }

        else {
            //编程式事务处理
            try {
                Object result = ((CallbackPreferringPlatformTransactionManager) tm).
execute(txAttr,
                    new TransactionCallback<Object>() {
                        public Object doInTransaction(TransactionStatus status) {
                            TransactionInfo txInfo = prepareTransactionInfo(tm,
txAttr, joinpointIdentification, status);
                            try {
                                return invocation.proceed();
                            }
                            catch (Throwable ex) {
```

```
                    if (txAttr.rollbackOn(ex)) {
                        // A RuntimeException: will lead to a rollback.
                        if (ex instanceof RuntimeException) {
                            throw (RuntimeException) ex;
                        }
                        else {
                            throw new ThrowableHolderException(ex);
                        }
                    }
                    else {
                        // A normal return value: will lead to a commit.
                        return new ThrowableHolder(ex);
                    }
                }
                finally {
                    cleanupTransactionInfo(txInfo);
                }
            }
        });

        // Check result: It might indicate a Throwable to rethrow.
        if (result instanceof ThrowableHolder) {
            throw ((ThrowableHolder) result).getThrowable();
        }
        else {
            return result;
        }
    }
    catch (ThrowableHolderException ex) {
        throw ex.getCause();
    }
}
```

从上面的函数中，我们尝试整理下事务处理的脉络，在 Spring 中支持两种事务处理的方式，分别是声明式事务处理与编程式事务处理，两者相对于开发人员来讲差别很大，但是对于 Spring 中的实现来讲，大同小异。在 invoke 中我们也可以看到这两种方式的实现。考虑到对事务的应用比声明式的事务处理使用起来方便，也相对流行些，我们就以此种方式进行分析。对于声明式的事务处理主要有以下几个步骤。

1. 获取事务的属性。

对于事务处理来说，最基础或者说最首要的工作便是获取事务属性了，这是支撑整个事务功能的基石，如果没有事务属性，其他功能也无从谈起，在分析事务准备阶段时我们已经分析了事务属性提取的功能，大家应该有所了解。

2. 加载配置中配置的 TransactionManager。

3. 不同的事务处理方式使用不同的逻辑。

对于声明式事务的处理与编程式事务的处理，第一点区别在于事务属性上，因为编程式的事务处理是不需要有事务属性的，第二点区别就是在 TransactionManager 上，CallbackPreferring PlatformTransactionManager 实现 PlatformTransactionManager 接口，暴露出一个方法用于执行事务处理中的回调。所以，这两种方式都可以用作事务处理方式的判断。

4. 在目标方法执行前获取事务并收集事务信息。

事务信息与事务属性并不相同，也就是 TransactionInfo 与 TransactionAttribute 并不相同，TransactionInfo 中包含 TransactionAttribute 信息，但是，除了 TransactionAttribute 外还有其他事务信息，例如 PlatformTransactionManager 以及 TransactionStatus 相关信息。

5. 执行目标方法。

6. 一旦出现异常，尝试异常处理。

并不是所有异常，Spring 都会将其回滚，默认只对 RuntimeException 回滚。

7. 提交事务前的事务信息清除。

8. 提交事务。

上面的步骤分析旨在让大家对事务功能与步骤有个大致的了解，具体的功能还需要详细地分析。

10.3.1 创建事务

我们先分析事务创建的过程。

```java
protected TransactionInfo createTransactionIfNecessary(
            PlatformTransactionManager tm, TransactionAttribute txAttr, final String joinpointIdentification) {

        // If no name specified, apply method identification as transaction name.
        //如果没有名称指定则使用方法唯一标识，并使用DelegatingTransactionAttribute封装txAttr
        if (txAttr != null && txAttr.getName() == null) {
            txAttr = new DelegatingTransactionAttribute(txAttr) {
                @Override
                public String getName() {
                    return joinpointIdentification;
                }
            };
        }

        TransactionStatus status = null;
        if (txAttr != null) {
            if (tm != null) {
                //获取TransactionStatus
                status = tm.getTransaction(txAttr);
            }
            else {
                if (logger.isDebugEnabled()) {
                    logger.debug("Skipping transactional joinpoint [" + joinpointIdentification +
                            "] because no transaction manager has been configured");
                }
            }
        }
        //根据指定的属性与status准备一个TransactionInfo
```

```
            return prepareTransactionInfo(tm, txAttr, joinpointIdentification, status);
}
```

对于 createTransactionIfNecessar 函数主要做了这样几件事情。

1. 使用 DelegatingTransactionAttribute 封装传入的 TransactionAttribute 实例。

对于传入的 TransactionAttribute 类型的参数 txAttr，当前的实际类型是 RuleBasedTransactionAttribute，是由获取事务属性时生成，主要用于数据承载，而这里之所以使用 DelegatingTransactionAttribute 进行封装，当然是提供了更多的功能。

2. 获取事务。

事务处理当然是以事务为核心，那么获取事务就是最重要的事情。

3. 构建事务信息。

根据之前几个步骤获取的信息构建 TransactionInfo 并返回。

我们分别对以上步骤进行详细的解析。

1. 获取事务

Spring 中使用 getTransaction 来处理事务的准备工作，包括事务获取以及信息的构建。

```
public final TransactionStatus getTransaction(TransactionDefinition definition) throws TransactionException {
            Object transaction = doGetTransaction();

            // Cache debug flag to avoid repeated checks.
            boolean debugEnabled = logger.isDebugEnabled();

            if (definition == null) {
                // Use defaults if no transaction definition given.
                definition = new DefaultTransactionDefinition();
            }
//判断当前线程是否存在事务，判读依据为当前线程记录的连接不为空且连接中(connectionHolder)中的
//transactionActive 属性不为空
            if (isExistingTransaction(transaction)) {
                //当前线程已经存在事务
                return handleExistingTransaction(definition, transaction, debugEnabled);
            }

            //事务超时设置验证
            if (definition.getTimeout() < TransactionDefinition.TIMEOUT_DEFAULT) {
                throw new InvalidTimeoutException("Invalid transaction timeout", definition.getTimeout());
            }
            //如果当前线程不存在事务，但是propagationBehavior 却被声明为 PROPAGATION_MANDATORY 抛
            //出异常
            if (definition.getPropagationBehavior() == TransactionDefinition. PROPAGATION_MANDATORY) {
                throw new IllegalTransactionStateException(
                        "No existing transaction found for transaction marked with propagation 'mandatory'");
            }else if (definition.getPropagationBehavior() == TransactionDefinition.PROPAGATION_REQUIRED ||
```

10.3 事务增强器

```
                    definition.getPropagationBehavior() == TransactionDefinition. PROPAGATION
_REQUIRES_NEW ||
                    definition.getPropagationBehavior() == TransactionDefinition.PROPAGATION_NESTED) {
                //PROPAGATION_REQUIRED、PROPAGATION_REQUIRES_NEW、PROPAGATION_NESTED 都需要
                //新建事务

                //空挂起
                SuspendedResourcesHolder suspendedResources = suspend(null);
                if (debugEnabled) {
                    logger.debug("Creating new transaction with name [" + definition.
getName() + "]: " + definition);
                }
                try {
                    boolean newSynchronization = (getTransactionSynchronization() !=
SYNCHRONIZATION_NEVER);
                    DefaultTransactionStatus status = newTransactionStatus(
                            definition, transaction, true, newSynchronization, debugEnabled,
 suspendedResources);
                    /*
                     * 构造 transaction,包括设置 ConnectionHolder、隔离级别、timout
                     * 如果是新连接,绑定到当前线程
                     */
                    doBegin(transaction, definition);
                    //新同步事务的设置,针对于当前线程的设置
                    prepareSynchronization(status, definition);
                    return status;
                }
                catch (RuntimeException ex) {
                    resume(null, suspendedResources);
                    throw ex;
                }
                catch (Error err) {
                    resume(null, suspendedResources);
                    throw err;
                }
            }
            else {
                // Create "empty" transaction: no actual transaction, but potentially
 synchronization.
                boolean newSynchronization = (getTransactionSynchronization() ==
SYNCHRONIZATION_ALWAYS);
                return prepareTransactionStatus(definition, null, true, newSynchronization,
 debugEnabled, null);
            }
        }
```

当然,在 Spring 中每个复杂的功能实现,并不是一次完成的,而是会通过入口函数进行一个框架的搭建,初步构建完整的逻辑,而将实现细节分摊给不同的函数。那么,让我们看看事务的准备工作都包括哪些。

1. 获取事务。

创建对应的事务实例,这里使用的是 DataSourceTransactionManager 中的 doGetTransaction 方法,创建基于 JDBC 的事务实例。如果当前线程中存在关于 dataSource 的连接,那么直接使

用。这里有一个对保存点的设置,是否开启允许保存点取决于是否设置了允许嵌入式事务。

```
protected Object doGetTransaction() {
    DataSourceTransactionObject txObject = new DataSourceTransactionObject();
    txObject.setSavepointAllowed(isNestedTransactionAllowed());
    //如果当前线程已经记录数据库连接则使用原有连接
    ConnectionHolder conHolder =
        (ConnectionHolder) TransactionSynchronizationManager.getResource(this.dataSource);
        //false 表示非新创建连接
    txObject.setConnectionHolder(conHolder, false);
    return txObject;
}
```

2. 如果当先线程存在事务,则转向嵌套事务的处理。

3. 事务超时设置验证。

4. 事务 propagationBehavior 属性的设置验证。

5. 构建 DefaultTransactionStatus。

6. 完善 transaction,包括设置 ConnectionHolder、隔离级别、timeout,如果是新连接,则绑定到当前线程。

对于一些隔离级别、timeout 等功能的设置并不是由 Spring 来完成的,而是委托给底层的数据库连接去做的,而对于数据库连接的设置就是在 doBegin 函数中处理的。

```
/**
 * 构造 transaction,包括设置 ConnectionHolder、隔离级别、timeout
 * 如果是新连接,绑定到当前线程
 */
@Override
protected void doBegin(Object transaction, TransactionDefinition definition) {
    DataSourceTransactionObject txObject = (DataSourceTransactionObject) transaction;
    Connection con = null;

    try {
        if (txObject.getConnectionHolder() == null ||
                txObject.getConnectionHolder().isSynchronizedWithTransaction()) {
            Connection newCon = this.dataSource.getConnection();
            if (logger.isDebugEnabled()) {
                logger.debug("Acquired Connection [" + newCon + "] for JDBC transaction");
            }
            txObject.setConnectionHolder(new ConnectionHolder(newCon), true);
        }

        txObject.getConnectionHolder().setSynchronizedWithTransaction(true);
        con = txObject.getConnectionHolder().getConnection();

        //设置隔离级别
        Integer previousIsolationLevel = DataSourceUtils.prepareConnectionForTransaction(con, definition);
        txObject.setPreviousIsolationLevel(previousIsolationLevel);

        //更改自动提交设置,由 Spring 控制提交
        if (con.getAutoCommit()) {
```

```
                    txObject.setMustRestoreAutoCommit(true);
                    if (logger.isDebugEnabled()) {
                        logger.debug("Switching JDBC Connection [" + con + "] to manual commit");
                    }
                    con.setAutoCommit(false);
                }
                //设置判断当前线程是否存在事务的依据
                txObject.getConnectionHolder().setTransactionActive(true);

                int timeout = determineTimeout(definition);
                if (timeout != TransactionDefinition.TIMEOUT_DEFAULT) {
                    txObject.getConnectionHolder().setTimeoutInSeconds(timeout);
                }

                // Bind the session holder to the thread.
                if (txObject.isNewConnectionHolder()) {
                    //将当前获取到的连接绑定到当前线程
                    TransactionSynchronizationManager.bindResource(getDataSource(),
txObject.getConnectionHolder());
                }
            }

            catch (Exception ex) {
                DataSourceUtils.releaseConnection(con, this.dataSource);
                throw new CannotCreateTransactionException("Could not open JDBC Connection
for transaction", ex);
            }
        }
```

可以说事务是从这个函数开始的,因为在这个函数中已经开始尝试了对数据库连接的获取,当然,在获取数据库连接的同时,一些必要的设置也是需要同步设置的。

1. 尝试获取连接。

当然并不是每次都会获取新的连接,如果当前线程中的 connectionHolder 已经存在,则没有必要再次获取,或者,对于事务同步表示设置为 true 的需要重新获取连接。

2. 设置隔离级别以及只读标识。

你是否有过这样的错觉?事务中的只读配置是 Spring 中做了一些处理呢? Spring 中确实是针对只读操作做了一些处理,但是核心的实现是设置 connection 上的 readOnly 属性。同样,对于隔离级别的控制也是交由 connection 去控制的。

3. 更改默认的提交设置。

如果事务属性是自动提交,那么需要改变这种设置,而将提交操作委托给 Spring 来处理。

4. 设置标志位,标识当前连接已经被事务激活。

5. 设置过期时间。

6. 将 connectionHolder 绑定到当前线程。

设置隔离级别的 prepareConnectionForTransaction 函数用于负责对底层数据库连接的设置,当然,只是包含只读标识和隔离级别的设置。由于强大的日志及异常处理,显得函数代码量比较大,但是单从业务角度去看,关键代码其实是不多的。

```java
public static Integer prepareConnectionForTransaction(Connection con, TransactionDefinition definition)
        throws SQLException {

    Assert.notNull(con, "No Connection specified");

    //设置数据连接的只读标识
    if (definition != null && definition.isReadOnly()) {
        try {
            if (logger.isDebugEnabled()) {
                logger.debug("Setting JDBC Connection [" + con + "] read-only");
            }
            con.setReadOnly(true);
        }
        catch (SQLException ex) {
            Throwable exToCheck = ex;
            while (exToCheck != null) {
                if (exToCheck.getClass().getSimpleName().contains("Timeout")) {
                    // Assume it's a connection timeout that would otherwise get lost: e.g. from JDBC 4.0
                    throw ex;
                }
                exToCheck = exToCheck.getCause();
            }
            logger.debug("Could not set JDBC Connection read-only", ex);
        }
        catch (RuntimeException ex) {
            Throwable exToCheck = ex;
            while (exToCheck != null) {
                if (exToCheck.getClass().getSimpleName().contains("Timeout")) {
                    throw ex;
                }
                exToCheck = exToCheck.getCause();
            }
            logger.debug("Could not set JDBC Connection read-only", ex);
        }
    }

    //设置数据库连接的隔离级别
    Integer previousIsolationLevel = null;
    if (definition != null && definition.getIsolationLevel() != TransactionDefinition.ISOLATION_DEFAULT) {
        if (logger.isDebugEnabled()) {
            logger.debug("Changing isolation level of JDBC Connection [" + con + "] to " +
                    definition.getIsolationLevel());
        }
        int currentIsolation = con.getTransactionIsolation();
        if (currentIsolation != definition.getIsolationLevel()) {
            previousIsolationLevel = currentIsolation;
            con.setTransactionIsolation(definition.getIsolationLevel());
        }
    }

    return previousIsolationLevel;
}
```

7. 将事务信息记录在当前线程中。

```
protected void prepareSynchronization(DefaultTransactionStatus status, Transaction Definition definition) {
            if (status.isNewSynchronization()) {
                TransactionSynchronizationManager.setActualTransactionActive(status.hasTransaction());
                TransactionSynchronizationManager.setCurrentTransactionIsolationLevel(
                        (definition.getIsolationLevel() != TransactionDefinition.ISOLATION_DEFAULT) ?
                                definition.getIsolationLevel() : null);
                TransactionSynchronizationManager.setCurrentTransactionReadOnly(definition.isReadOnly());
                TransactionSynchronizationManager.setCurrentTransactionName(definition.getName());
                TransactionSynchronizationManager.initSynchronization();
            }
    }
```

2. 处理已经存在的事务

之前讲述了普通事务建立的过程，但是 Spring 中支持多种事务的传播规则，比如 PROPAGATION_NESTED、PROPAGATION_REQUIRES_NEW 等，这些都是在已经存在事务的基础上进行进一步的处理，那么，对于已经存在的事务，准备操作是如何进行的呢？

```
    private TransactionStatus handleExistingTransaction(
            TransactionDefinition definition, Object transaction, boolean debugEnabled)
            throws TransactionException {

        if (definition.getPropagationBehavior() == TransactionDefinition.PROPAGATION_NEVER) {
            throw new IllegalTransactionStateException(
                    "Existing transaction found for transaction marked with propagation 'never'");
        }

        if (definition.getPropagationBehavior() == TransactionDefinition.PROPAGATION_NOT_SUPPORTED) {
            if (debugEnabled) {
                logger.debug("Suspending current transaction");
            }
            Object suspendedResources = suspend(transaction);
            boolean newSynchronization = (getTransactionSynchronization() == SYNCHRONIZATION_ALWAYS);
            return prepareTransactionStatus(
                    definition, null, false, newSynchronization, debugEnabled, suspendedResources);
        }

        if (definition.getPropagationBehavior() == TransactionDefinition.PROPAGATION_REQUIRES_NEW) {
            if (debugEnabled) {
                logger.debug("Suspending current transaction, creating new transaction with name [" +
```

```java
                    definition.getName() + "]");
        }
        //新事务的建立
        SuspendedResourcesHolder suspendedResources = suspend(transaction);
        try {
            boolean newSynchronization = (getTransactionSynchronization() != SYNCHRONIZATION_NEVER);
            DefaultTransactionStatus status = newTransactionStatus(
                    definition, transaction, true, newSynchronization, debugEnabled, suspendedResources);
            doBegin(transaction, definition);
            prepareSynchronization(status, definition);
            return status;
        }
        catch (RuntimeException beginEx) {
            resumeAfterBeginException(transaction, suspendedResources, beginEx);
            throw beginEx;
        }
        catch (Error beginErr) {
            resumeAfterBeginException(transaction, suspendedResources, beginErr);
            throw beginErr;
        }
    }
    //嵌入式事务的处理
    if (definition.getPropagationBehavior() == TransactionDefinition.PROPAGATION_NESTED) {
        if (!isNestedTransactionAllowed()) {
            throw new NestedTransactionNotSupportedException(
                    "Transaction manager does not allow nested transactions by default - " +
                    "specify 'nestedTransactionAllowed' property with value 'true'");
        }
        if (debugEnabled) {
            logger.debug("Creating nested transaction with name [" + definition.getName() + "]");
        }
        if (useSavepointForNestedTransaction()) {
            //如果没有可以使用保存点的方式控制事务回滚，那么在嵌入式事务的建立初始建立保存点
            DefaultTransactionStatus status =
                    prepareTransactionStatus(definition, transaction, false, false, debugEnabled, null);
            status.createAndHoldSavepoint();
            return status;
        }else {
            //有些情况是不能使用保存点操作，比如JTA，那么建立新事务
            boolean newSynchronization = (getTransactionSynchronization() != SYNCHRONIZATION_NEVER);
            DefaultTransactionStatus status = newTransactionStatus(
                    definition, transaction, true, newSynchronization, debugEnabled, null);
            doBegin(transaction, definition);
            prepareSynchronization(status, definition);
            return status;
        }
```

10.3 事务增强器

```java
            }
            if (debugEnabled) {
                logger.debug("Participating in existing transaction");
            }
            if (isValidateExistingTransaction()) {
                if (definition.getIsolationLevel() != TransactionDefinition.PROPAGATION_REQUIRES_NEW) {
                    Integer currentIsolationLevel = TransactionSynchronizationManager.getCurrentTransactionIsolationLevel();
                    if (currentIsolationLevel == null || currentIsolationLevel != definition.getIsolationLevel()) {
                        Constants isoConstants = DefaultTransactionDefinition.constants;
                        throw new IllegalTransactionStateException("Participating transaction with definition [" +
                                definition + "] specifies isolation level which is incompatible with existing transaction: " +
                                (currentIsolationLevel != null ?
                                        isoConstants.toCode(currentIsolationLevel, DefaultTransactionDefinition.PREFIX_ISOLATION) :
                                        "(unknown)"));
                    }
                }
                if (!definition.isReadOnly()) {
                    if (TransactionSynchronizationManager.isCurrentTransactionReadOnly()) {
                        throw new IllegalTransactionStateException("Participating transaction with definition [" +
                                definition + "] is not marked as read-only but existing transaction is");
                    }
                }
            }
            boolean newSynchronization = (getTransactionSynchronization() != SYNCHRONIZATION_NEVER);
            return prepareTransactionStatus(definition, transaction, false, newSynchronization, debugEnabled, null);
        }
```

对于已经存在事务的处理过程中，我们看到了很多熟悉的操作，但是，也有些不同的地方，函数中对已经存在的事务处理考虑两种情况。

- PROPAGATION_REQUIRES_NEW 表示当前方法必须在它自己的事务里运行，一个新的事务将被启动，而如果有一个事务正在运行的话，则在这个方法运行期间被挂起。而 Spring 中对于此种传播方式的处理与新事务建立最大的不同点在于使用 suspend 方法将原事务挂起。将信息挂起的目的当然是为了在当前事务执行完毕后在将原事务还原。
- PROPAGATION_NESTED 表示如果当前正有一个事务在运行中，则该方法应该运行在一个嵌套的事务中，被嵌套的事务可以独立于封装事务进行提交或者回滚，如果封装事务不存在，行为就像 PROPAGATION_REQUIRES_NEW。对于嵌入式事务的处理，Spring 中主要考虑了两种方式的处理。
  - ◆ Spring 中允许嵌入事务的时候，则首选设置保存点的方式作为异常处理的回滚。

- 对于其他方式，比如 JTA 无法使用保存点的方式，那么处理方式与 PROPAGATION_REQUIRES_NEW 相同，而一旦出现异常，则由 Spring 的事务异常处理机制去完成后续操作。

对于挂起操作的主要目的是记录原有事务的状态，以便于后续操作对事务的恢复：

```
protected final SuspendedResourcesHolder suspend(Object transaction) throws Transaction Exception {
    if (TransactionSynchronizationManager.isSynchronizationActive()) {
        List<TransactionSynchronization> suspendedSynchronizations = doSuspendSynchronization();
        try {
            Object suspendedResources = null;
            if (transaction != null) {
                suspendedResources = doSuspend(transaction);
            }
            String name = TransactionSynchronizationManager. GetCurrent Transaction Name();
            TransactionSynchronizationManager.setCurrentTransactionName(null);
            boolean readOnly = TransactionSynchronizationManager.isCurrentTransaction ReadOnly();
            TransactionSynchronizationManager.setCurrentTransactionReadOnly(false);
            Integer isolationLevel = TransactionSynchronizationManager.getCurrent TransactionIsolationLevel();
            TransactionSynchronizationManager.setCurrentTransactionIsolation Level(null);
            boolean wasActive = TransactionSynchronizationManager.isActual TransactionActive();
            TransactionSynchronizationManager.setActualTransactionActive(false);
            return new SuspendedResourcesHolder(
                    suspendedResources, suspendedSynchronizations, name, readOnly, isolationLevel, wasActive);
        }
        catch (RuntimeException ex) {
            // doSuspend failed - original transaction is still active...
            doResumeSynchronization(suspendedSynchronizations);
            throw ex;
        }
        catch (Error err) {
            doResumeSynchronization(suspendedSynchronizations);
            throw err;
        }
    }
    else if (transaction != null) {
        Object suspendedResources = doSuspend(transaction);
        return new SuspendedResourcesHolder(suspendedResources);
    }
    else {
        return null;
    }
}
```

3. 准备事务信息

当已经建立事务连接并完成了事务信息的提取后，我们需要将所有的事务信息统一记录在 TransactionInfo 类型的实例中，这个实例包含了目标方法开始前的所有状态信息，一旦事务执行失败，Spring 会通过 TransactionInfo 类型的实例中的信息来进行回滚等后续工作。

```java
protected TransactionInfo prepareTransactionInfo(PlatformTransactionManager tm,
            TransactionAttribute txAttr, String joinpointIdentification, Transaction
Status status) {

            TransactionInfo txInfo = new TransactionInfo(tm, txAttr, joinpointIdentification);
            if (txAttr != null) {
                // We need a transaction for this method
                if (logger.isTraceEnabled()) {
                    logger.trace("Getting transaction for [" + txInfo.getJoinpoint
Identification() + "]");
                }
                //记录事务状态
                txInfo.newTransactionStatus(status);
            }
            else {
                if (logger.isTraceEnabled())
                    logger.trace("Don't need to create transaction for [" + joinpoint
Identification +
                        "]: This method isn't transactional.");
            }
            txInfo.bindToThread();
            return txInfo;
    }
```

10.3.2 回滚处理

之前已经完成了目标方法运行前的事务准备工作，而这些准备工作最大的目的无非是对于程序没有按照我们期待的那样进行，也就是出现特定的错误，那么，当出现错误的时候，Spring 是怎么对数据进行恢复的呢？

```java
protected void completeTransactionAfterThrowing(TransactionInfo txInfo, Throwable ex) {
            //当抛出异常时首先判断当前是否存在事务，这是基础依据
            if (txInfo != null && txInfo.hasTransaction()) {
                if (logger.isTraceEnabled()) {
                    logger.trace("Completing transaction for [" + txInfo.getJoinpoint
Identification() +
                        "] after exception: " + ex);
                }
                //这里判断是否回滚默认的依据是抛出的异常是否是 RuntimeException 或者是 Error 的类型
                if (txInfo.transactionAttribute.rollbackOn(ex)) {
                    try {
                        //根据 TransactionStatus 信息进行回滚处理
                        txInfo.getTransactionManager().rollback(txInfo.GetTransaction Status());
                    }
                    catch (TransactionSystemException ex2) {
```

```
                    logger.error("Application exception overridden by rollback
exception", ex);
                    ex2.initApplicationException(ex);
                    throw ex2;
                }
                catch (RuntimeException ex2) {
                    logger.error("Application exception overridden by rollback
exception", ex);
                    throw ex2;
                }
                catch (Error err) {
                    logger.error("Application exception overridden by rollback error", ex);
                    throw err;
                }
            }else {
                //如果不满足回滚条件即使抛出异常也同样会提交
                try {
                    txInfo.getTransactionManager().commit(txInfo.getTransactionStatus());
                }
                catch (TransactionSystemException ex2) {
                    logger.error("Application exception overridden by commit exception", ex);
                    ex2.initApplicationException(ex);
                    throw ex2;
                }
                catch (RuntimeException ex2) {
                    logger.error("Application exception overridden by commit exception", ex);
                    throw ex2;
                }
                catch (Error err) {
                    logger.error("Application exception overridden by commit error", ex);
                    throw err;
                }
            }
        }
    }
```

在对目标方法的执行过程中,一旦出现 Throwable 就会被引导至此方法处理,但是并不代表所有的 Throwable 都会被回滚处理,比如我们常用的 Exception,默认是不会被处理的。默认情况下,即使出现异常,数据也会被正常提交,而这个关键的地方就是在 txInfo.transactionAttribute.rollbackOn(ex)这个函数。

1. 回滚条件

```
public boolean rollbackOn(Throwable ex) {
    return (ex instanceof RuntimeException || ex instanceof Error);
}
```

看到了吗?默认情况下 Spring 中的事务异常处理机制只对 RuntimeException 和 Error 两种情况感兴趣,当然你可以通过扩展来改变,不过,我们最常用的还是使用事务提供的属性设置,利用注解方式的使用,例如:

```
@Transactional(propagation=Propagation.REQUIRED,rollbackFor=Exception.class)
```

2. 回滚处理

当然，一旦符合回滚条件，那么 Spring 就会将程序引导至回滚处理函数中。

```
public final void rollback(TransactionStatus status) throws TransactionException {
            //如果事务已经完成，那么再次回滚会抛出异常
            if (status.isCompleted()) {
                throw new IllegalTransactionStateException(
                        "Transaction is already completed - do not call commit or rollback more than once per transaction");
            }

            DefaultTransactionStatus defStatus = (DefaultTransactionStatus) status;
            processRollback(defStatus);
}
    private void processRollback(DefaultTransactionStatus status) {
            try {
                try {
                    //激活所有 TransactionSynchronization 中对应的方法
                    triggerBeforeCompletion(status);
                    if (status.hasSavepoint()) {
                        if (status.isDebug()) {
                            logger.debug("Rolling back transaction to savepoint");
                        }
                        //如果有保存点，也就是当前事务为单独的线程则会退到保存点
                        status.rollbackToHeldSavepoint();
                    }
                    else if (status.isNewTransaction()) {
                        if (status.isDebug()) {
                            logger.debug("Initiating transaction rollback");
                        }
                        //如果当前事务为独立的新事务，则直接回退
                        doRollback(status);
                    }
                    else if (status.hasTransaction()) {
                        if (status.isLocalRollbackOnly() || isGlobalRollbackOnParticipationFailure()) {
                            if (status.isDebug()) {
                                logger.debug("Participating transaction failed - marking existing transaction as rollback-only");
                            }
                            //如果当前事务不是独立的事务，那么只能标记状态，等到事务链执行完毕后统一回滚
                            doSetRollbackOnly(status);
                        }
                        else {
                            if (status.isDebug()) {
                                logger.debug("Participating transaction failed - letting transaction originator decide on rollback");
                            }
                        }
                    }
                    else {
                        logger.debug("Should roll back transaction but cannot - no transaction available");
                    }
```

```
            catch (RuntimeException ex) {
                triggerAfterCompletion(status, TransactionSynchronization.STATUS_UNKNOWN);
                throw ex;
            }
            catch (Error err) {
                triggerAfterCompletion(status, TransactionSynchronization.STATUS_UNKNOWN);
                throw err;
            }
            //激活所有TransactionSynchronization中对应的方法
            triggerAfterCompletion(status, TransactionSynchronization.STATUS_ROLLED_BACK);
        }
        finally {
            //清空记录的资源并将挂起的资源恢复
            cleanupAfterCompletion(status);
        }
    }
```

同样，对于在 Spring 中的复杂的逻辑处理过程，在入口函数一般都会给出个整体的处理脉络，而把实现细节委托给其他函数去执行。我们尝试总结下 Spring 中对于回滚处理的大致脉络如下。

1. 首先是自定义触发器的调用，包括在回滚前、完成回滚后的调用，当然完成回滚包括正常回滚与回滚过程中出现异常，自定义的触发器会根据这些信息作进一步处理，而对于触发器的注册，常见是在回调过程中通过 TransactionSynchronizationManager 类中的静态方法直接注册：

```
public static void registerSynchronization(TransactionSynchronization synchronization)
```

2. 除了触发监听函数外，就是真正的回滚逻辑处理了。

- 当之前已经保存的事务信息中有保存点信息的时候，使用保存点信息进行回滚。常用于嵌入式事务，对于嵌入式的事务的处理，内嵌的事务异常并不会引起外部事务的回滚。根据保存点回滚的实现方式其实是根据底层的数据库连接进行的。

```
public void rollbackToHeldSavepoint() throws TransactionException {
        if (!hasSavepoint()) {
            throw new TransactionUsageException("No savepoint associated with current transaction");
        }
        getSavepointManager().rollbackToSavepoint(getSavepoint());
        setSavepoint(null);
    }
```

这里使用的是 JDBC 的方式进行数据库连接，那么 getSavepointManager() 函数返回的是 JdbcTransactionObjectSupport，也就是说上面函数会调用 JdbcTransactionObjectSupport 中的 rollbackToSavepoint 方法。

```
public void rollbackToSavepoint(Object savepoint) throws TransactionException {
        try {
            getConnectionHolderForSavepoint().getConnection().rollback((Savepoint) savepoint);
        }
        catch (Throwable ex) {
            throw new TransactionSystemException("Could not roll back to JDBC savepoint", ex);
        }
    }
```

- 当之前已经保存的事务信息中的事务为新事务,那么直接回滚。常用于单独事务的处理。对于没有保存点的回滚,Spring 同样是使用底层数据库连接提供的 API 来操作的。由于我们使用的是 DataSourceTransactionManager,那么 doRollback 函数会使用此类中的实现:

```java
protected void doRollback(DefaultTransactionStatus status) {
    DataSourceTransactionObject txObject = (DataSourceTransactionObject) status.getTransaction();
    Connection con = txObject.getConnectionHolder().getConnection();
    if (status.isDebug()) {
        logger.debug("Rolling back JDBC transaction on Connection [" + con + "]");
    }
    try {
        con.rollback();
    }
    catch (SQLException ex) {
        throw new TransactionSystemException("Could not roll back JDBC transaction", ex);
    }
}
```

- 当前事务信息中表明是存在事务的,又不属于以上两种情况,多数用于 JTA,只做回滚标识,等到提交的时候统一不提交。

3. 回滚后的信息清除

对于回滚逻辑执行结束后,无论回滚是否成功,都必须要做的事情就是事务结束后的收尾工作。

```java
private void cleanupAfterCompletion(DefaultTransactionStatus status) {
    //设置完成状态
    status.setCompleted();
    if (status.isNewSynchronization()) {
        TransactionSynchronizationManager.clear();
    }
    if (status.isNewTransaction()) {
        doCleanupAfterCompletion(status.getTransaction());
    }
    if (status.getSuspendedResources() != null) {
        if (status.isDebug()) {
            logger.debug("Resuming suspended transaction after completion of inner transaction");
        }
        //结束之前事务的挂起状态
        resume(status.getTransaction(), (SuspendedResourcesHolder) status.getSuspendedResources());
    }
}
```

从函数中得知,事务处理的收尾处理工作包括如下内容。

- 设置状态是对事务信息作完成标识以避免重复调用。
- 如果当前事务是新的同步状态,需要将绑定到当前线程的事务信息清除。
- 如果是新事务需要做些清除资源的工作。

```java
protected void doCleanupAfterCompletion(Object transaction) {
    DataSourceTransactionObject txObject = (DataSourceTransactionObject) transaction;
```

```java
            if (txObject.isNewConnectionHolder()) {
                //将数据库连接从当前线程中解除绑定
                TransactionSynchronizationManager.unbindResource(this.dataSource);
            }

            //释放链接
            Connection con = txObject.getConnectionHolder().getConnection();
            try {
                if (txObject.isMustRestoreAutoCommit()) {
                    //恢复数据库连接的自动提交属性
                    con.setAutoCommit(true);
                }
                //重置数据库连接
                DataSourceUtils.resetConnectionAfterTransaction(con, txObject.getPrevious
IsolationLevel());
            }
            catch (Throwable ex) {
                logger.debug("Could not reset JDBC Connection after transaction", ex);
            }

            if (txObject.isNewConnectionHolder()) {
                if (logger.isDebugEnabled()) {
                    logger.debug("Releasing JDBC Connection [" + con + "] after transaction");
                }
                //如果当前事务时独立的新创建的事务则在事务完成时释放数据库连接
                DataSourceUtils.releaseConnection(con, this.dataSource);
            }

            txObject.getConnectionHolder().clear();
    }
```

- 如果在事务执行前有事务挂起，那么当前事务执行结束后需要将挂起事务恢复。

```java
    protected final void resume(Object transaction, SuspendedResourcesHolder resourcesHolder)
            throws TransactionException {

        if (resourcesHolder != null) {
            Object suspendedResources = resourcesHolder.suspendedResources;
            if (suspendedResources != null) {
                doResume(transaction, suspendedResources);
            }
            List<TransactionSynchronization> suspendedSynchronizations = resourcesHolder.
suspendedSynchronizations;
            if (suspendedSynchronizations != null) {
                TransactionSynchronizationManager.setActualTransactionActive (resources
Holder.wasActive);
                TransactionSynchronizationManager.setCurrentTransactionIsolationLevel
(resourcesHolder.isolationLevel);
                TransactionSynchronizationManager.setCurrentTransactionReadOnly
(resourcesHolder.readOnly);
                TransactionSynchronizationManager.setCurrentTransactionName
(resourcesHolder.name);
                doResumeSynchronization(suspendedSynchronizations);
            }
        }
    }
```

10.3.3 事务提交

之前我们分析了 Spring 的事务异常处理机制，那么事务的执行并没有出现任何的异常，也就意味着事务可以走正常事务提交的流程了。

```
protected void commitTransactionAfterReturning(TransactionInfo txInfo) {
    if (txInfo != null && txInfo.hasTransaction()) {
        if (logger.isTraceEnabled()) {
            logger.trace("Completing transaction for [" + txInfo.getJoinpoint
Identification() + "]");
        }
        txInfo.getTransactionManager().commit(txInfo.getTransactionStatus());
    }
}
```

在真正的数据提交之前，还需要做个判断。不知道大家还有没有印象，在我们分析事务异常处理规则的时候，当某个事务既没有保存点又不是新事务，Spring 对它的处理方式只是设置一个回滚标识。这个回滚标识在这里就会派上用场了，主要的应用场景如下。

某个事务是另一个事务的嵌入事务，但是，这些事务又不在 Spring 的管理范围内，或者无法设置保存点，那么 Spring 会通过设置回滚标识的方式来禁止提交。首先当某个嵌入事务发生回滚的时候会设置回滚标识，而等到外部事务提交时，一旦判断出当前事务流被设置了回滚标识，则由外部事务来统一进行整体事务的回滚。

所以，当事务没有被异常捕获的时候也并不意味着一定会执行提交的过程。

```
public final void commit(TransactionStatus status) throws TransactionException {
    if (status.isCompleted()) {
        throw new IllegalTransactionStateException(
                "Transaction is already completed - do not call commit or rollback
more than once per transaction");
    }

    DefaultTransactionStatus defStatus = (DefaultTransactionStatus) status;
    //如果在事务链中已经被标记回滚，那么不会尝试提交事务，直接回滚
    if (defStatus.isLocalRollbackOnly()) {
        if (defStatus.isDebug()) {
            logger.debug("Transactional code has requested rollback");
        }
        processRollback(defStatus);
        return;
    }
    if (!shouldCommitOnGlobalRollbackOnly() && defStatus.isGlobalRollbackOnly()) {
        if (defStatus.isDebug()) {
            logger.debug("Global transaction is marked as rollback-only but
transactional code requested commit");
        }
        processRollback(defStatus);
        if (status.isNewTransaction() || isFailEarlyOnGlobalRollbackOnly()) {
            throw new UnexpectedRollbackException(
                    "Transaction rolled back because it has been marked as rollback-only");
        }
        return;
```

```
            }
            //处理事务提交
            processCommit(defStatus);
}
```

而当事务执行一切都正常的时候,便可以真正地进入提交流程了。

```
private void processCommit(DefaultTransactionStatus status) throws TransactionException {
    try {
        boolean beforeCompletionInvoked = false;
        try {
            //预留
            prepareForCommit(status);
            //添加的 TransactionSynchronization 中的对应方法的调用
            triggerBeforeCommit(status);
            //添加的 TransactionSynchronization 中的对应方法的调用
            triggerBeforeCompletion(status);
            beforeCompletionInvoked = true;
            boolean globalRollbackOnly = false;
            if (status.isNewTransaction() || isFailEarlyOnGlobalRollbackOnly()) {
                globalRollbackOnly = status.isGlobalRollbackOnly();
            }
            if (status.hasSavepoint()) {
                if (status.isDebug()) {
                    logger.debug("Releasing transaction savepoint");
                }
                //如果存在保存点则清除保存点信息
                status.releaseHeldSavepoint();
            }
            else if (status.isNewTransaction()) {
                if (status.isDebug()) {
                    logger.debug("Initiating transaction commit");
                }
                //如果是独立的事务则直接提交
                doCommit(status);
            }
            if (globalRollbackOnly) {
                throw new UnexpectedRollbackException(
                        "Transaction silently rolled back because it has been marked as rollback-only");
            }
        }
        catch (UnexpectedRollbackException ex) {
            triggerAfterCompletion(status, TransactionSynchronization.STATUS_ROLLED_BACK);
            throw ex;
        }
        catch (TransactionException ex) {
            if (isRollbackOnCommitFailure()) {
                doRollbackOnCommitException(status, ex);
            }
            else {
                triggerAfterCompletion(status,TransactionSynchronization.STATUS_UNKNOWN);
            }
            throw ex;
        }
        catch (RuntimeException ex) {
```

```
                if (!beforeCompletionInvoked) {
                    triggerBeforeCompletion(status);
                }
                doRollbackOnCommitException(status, ex);
                throw ex;
            }
            catch (Error err) {
                if (!beforeCompletionInvoked) {
                    //添加的 TransactionSynchronization 中的对应方法的调用
                    triggerBeforeCompletion(status);
                }
                //提交过程中出现异常则回滚
                doRollbackOnCommitException(status, err);
                throw err;
            }
            try {
                //添加的 TransactionSynchronization 中的对应方法的调用
                triggerAfterCommit(status);
            }
            finally {
                triggerAfterCompletion(status, TransactionSynchronization.STATUS_COMMITTED);
            }

        }
        finally {
            cleanupAfterCompletion(status);
        }
    }
```

在提交过程中也并不是直接提交的,而是考虑了诸多的方面,符合提交的条件如下。

- 当事务状态中有保存点信息的话便不会去提交事务。
- 当事务非新事务的时候也不会去执行提交事务操作。

此条件主要考虑内嵌事务的情况,对于内嵌事务,在 Spring 中正常的处理方式是将内嵌事务开始之前设置保存点,一旦内嵌事务出现异常便根据保存点信息进行回滚,但是如果没有出现异常,内嵌事务并不会单独提交,而是根据事务流由最外层事务负责提交,所以如果当前存在保存点信息便不是最外层事务,不做保存操作,对于是否是新事务的判断也是基于此考虑。

如果程序流通过了事务的层层把关,最后顺利地进入了提交流程,那么同样,Spring 会将事务提交的操作引导至底层数据库连接的 API,进行事务提交。

```
protected void doCommit(DefaultTransactionStatus status) {
    DataSourceTransactionObject txObject = (DataSourceTransactionObject) status.getTransaction();
    Connection con = txObject.getConnectionHolder().getConnection();
    if (status.isDebug()) {
        logger.debug("Committing JDBC transaction on Connection [" + con + "]");
    }
    try {
        con.commit();
    }
    catch (SQLException ex) {
        throw new TransactionSystemException("Could not commit JDBC transaction", ex);
    }
}
```

第 11 章 SpringMVC

Spring 框架提供了构建 Web 应用程序的全功能 MVC 模块。通过策略接口，Spring 框架是高度可配置的，而且支持多种视图技术，例如 JavaServer Pages（JSP）、Velocity、Tiles、iText 和 POI。SpringMVC 框架并不知道使用的视图，所以不会强迫您只使用 JSP 技术。SpringMVC 分离了控制器、模型对象、分派器以及处理程序对象的角色，这种分离让它们更容易进行定制。

Spring 的 MVC 是基于 Servlet 功能实现的，通过实现 Servlet 接口的 DispatcherServlet 来封装其核心功能实现，通过将请求分派给处理程序，同时带有可配置的处理程序映射、视图解析、本地语言、主题解析以及上载文件支持。默认的处理程序是非常简单的 Controller 接口，只有一个方法 ModelAndView handleRequest(request, response)。Spring 提供了一个控制器层次结构，可以派生子类。如果应用程序需要处理用户输入表单，那么可以继承 AbstractFormController。如果需要把多页输入处理到一个表单，那么可以继承 AbstractWizardFormController。

对 SpringMVC 或者其他比较成熟的 MVC 框架而言，解决的问题无外乎以下几点。
- 将 Web 页面的请求传给服务器。
- 根据不同的请求处理不同的逻辑单元。
- 返回处理结果数据并跳转至响应的页面。

我们首先通过一个简单示例来快速回顾 SpringMVC 的使用。

11.1 SpringMVC 快速体验

1．配置 web.xml

一个 Web 中可以没有 web.xml 文件，也就是说，web.xml 文件并不是 Web 工程必需的。web.xml 文件用来初始化配置信息，比如 Welcome 页面、servlet、servlet-mapping、filter、listener、启动加载级别等。但是，SpringMVC 的实现原理是通过 servlet 拦截所有 URL 来达到控制的目的，所以 web.xml 的配置是必需的。

```xml
<?xml version="1.0" encoding="UTF-8"?>
<web-app id="WebApp_ID" version="2.5" xmlns="http://java.sun.com/xml/ns/javaee" xmlns:xsi="http://www.w3.org/2001/XMLSchema-instance" xsi:schemaLocation="http://java.sun.com/xml/ns/javaee http://java.sun.com/xml/ns/javaee/web-app_2_5.xsd">
    <display-name>Springmvc</display-name>

    <!-- 使用 ContextLoaderListener 配置时，需要告诉它 Spring 配置文件的位置 -->
    <context-param>
        <param-name>contextConfigLocation</param-name>
        <param-value>classpath:applicationContext.xml</param-value>
    </context-param>

    <!-- SpringMVC 的前端控制器 -->
    <!-- 当 DispatcherServlet 载入后，它将从一个 XML 文件中载入 Spring 的应用上下文，该 XML 文件的名
    字取决于<servlet-name> -->
    <!-- 这里 DispatcherServlet 将试图从一个叫作 Springmvc-servlet.xml 的文件中载入应用上下文，
    其默认位于 WEB-INF 目录下 -->
    <servlet>
        <servlet-name>Springmvc</servlet-name>
        <servlet-class>org.Springframework.web.servlet.DispatcherServlet</servlet-class>
        <load-on-startup>1</load-on-startup>
    </servlet>
    <servlet-mapping>
        <servlet-name>Springmvc</servlet-name>
        <url-pattern>*.htm</url-pattern>
    </servlet-mapping>

    <!-- 配置上下文载入器 -->
    <!-- 上下文载入器载入除 DispatcherServlet 载入的配置文件之外的其他上下文配置文件 -->
    <!-- 最常用的上下文载入器是一个 Servlet 监听器，其名称为 ContextLoaderListener -->
    <listener>
        <listener-class>org.Springframework.web.context.ContextLoaderListener</listener-class>
    </listener>
</web-app>
```

Spring 的 MVC 之所以必须要配置 web.xml，其实最关键的是要配置两个地方。

- contextConfigLocation：Spring 的核心就是配置文件，可以说配置文件是 Spring 中必不可少的东西，而这个参数就是使 Web 与 Spring 的配置文件相结合的一个关键配置。
- DispatcherServlet：包含了 SpringMVC 的请求逻辑，Spring 使用此类拦截 Web 请求并进行相应的逻辑处理。

2. 创建 Spring 配置文件 applicationContext.xml

```xml
<?xml version="1.0" encoding="UTF-8"?>
<beans xmlns="http://www.Springframework.org/schema/beans"
    xmlns:xsi="http://www.w3.org/2001/XMLSchema-instance"
    xmlns:tx="http://www.Springframework.org/schema/tx"
    xsi:schemaLocation="http://www.Springframework.org/schema/beans
    http://www.Springframework.org/schema/beans/Spring-beans-2.5.xsd
    http://www.Springframework.org/schema/tx
    http://www.Springframework.org/schema/tx/Spring-tx-2.5.xsd">
```

```xml
<bean id="viewResolver" class="org.Springframework.web.servlet.view.InternalResourceViewResolver">
    <property name="prefix" value="/WEB-INF/jsp/"/>
    <property name="suffix" value=".jsp"/>
</bean>
</beans>
```

InternalResourceViewResolver 是一个辅助 bean，会在 ModelAndView 返回的视图名前加上 prefix 指定的前缀，再在最后加上 suffix 指定的后缀，例如：由于 XXController 返回的 ModelAndView 中的视图名是 testview，故该视图解析器将在/WEB-INF/jsp/testview.jsp 处查找视图。

3. 创建 model

模型对于 SpringMVC 来说并不是必不可少，如果处理程序非常简单，完全可以忽略。模型创建主要的目的就是承载数据，使数据传输更加方便。

```java
public class User {
    private String username;
    private Integer age;
    public String getUsername() {
        return username;
    }
    public void setUsername(String username) {
        this.username = username;
    }
    public Integer getAge() {
        return age;
    }
    public void setAge(Integer age) {
        this.age = age;
    }
}
```

4. 创建 controller

控制器用于处理 Web 请求，每个控制器都对应着一个逻辑处理。

```java
public class UserController extends AbstractController {
    @Override
    protected ModelAndView handleRequestInternal(HttpServletRequest arg0, HttpServletResponse arg1) throws Exception {
        List<User> userList = new ArrayList<User>();
        User userA = new User();
        User userB = new User();
        userA.setUsername("张三");
        userA.setAge(27);
        userB.setUsername("李四");
        userB.setAge(37);
        userList.add(userA);
        userList.add(userB);

        return new ModelAndView("userlist", "users", userList);
    }
}
```

在请求的最后返回了 ModelAndView 类型的实例。ModelAndView 类在 SpringMVC 中占有

很重要的地位，控制器执行方法都必须返回一个 ModelAndView，ModelAndView 对象保存了视图以及视图显示的模型数据，例如其中的参数如下。

- 第 1 个参数 userlist：视图组件的逻辑名称。这里视图的逻辑名称就是 userlist，视图解析器会使用该名称查找实际的 View 对象。
- 第 2 个参数 users：传递给视图的模型对象的名称。
- 第 3 个参数 userList：传递给视图的模型对象的值。

5. 创建视图文件 userlist.jsp

```
<%@ page language="java" pageEncoding="UTF-8"%>
<%@ taglib prefix="c" uri="http://java.sun.com/jsp/jstl/core"%>
<h2>This is  SpringMVC demo page</h2>
<c:forEach items="${users}" var="user">
    <c:out value="${user.username}"/><br/>
    <c:out value="${user.age}"/><br/>
</c:forEach>
```

视图文件用于展现请求处理结果，通过对 JSTL 的支持，可以很方便地展现在控制器中放入 ModelAndView 中的处理结果数据。

6. 创建 Servlet 配置文件 Spring-servlet.xml

```
<?xml version="1.0" encoding="UTF-8"?>
<beans xmlns="http://www.Springframework.org/schema/beans"
    xmlns:xsi="http://www.w3.org/2001/XMLSchema-instance" xmlns:tx="http://www. Spring framework. org/schema/tx"
    xsi:schemaLocation="http://www.Springframework.org/schema/beans
    http://www.Springframework.org/schema/beans/Spring-beans-2.5.xsd
    http://www.Springframework.org/schema/tx
    http://www.Springframework.org/schema/tx/Spring-tx-2.5.xsd">

  <bean id="simpleUrlMapping"
      class="org.Springframework.web.servlet.handler.SimpleUrlHandlerMapping">
      <property name="mappings">
          <props>
              <prop key="/userlist.htm">userController</prop>
          </props>
      </property>
  </bean>

        <!-- 这里的 id="userController"对应的是<bean id="simpleUrlMapping">中的<prop>里面的 value -->
        <bean id="userController" class="test.controller.UserController" />

</beans>
```

因为 SpringMVC 是基于 Servlet 的实现，所以在 Web 启动的时候，服务器会首先尝试加载对应于 Servlet 的配置文件，而为了让项目更加模块化，通常我们将 Web 部分的配置都存放于此配置文件中。

至此，已经完成了 SpringMVC 的搭建，启动服务器，输入网址 http://localhost:8080/Springmvc/userlist.htm。

看到了服务器返回界面，如图 11-1 所示。

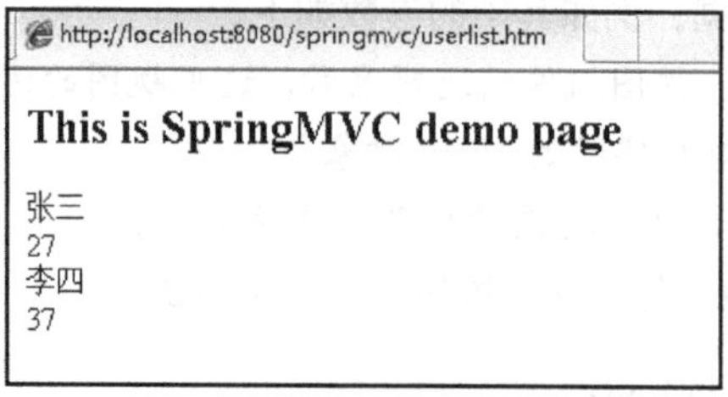

图 11-1　SpringMVC 快速体验

11.2　ContextLoaderListener

对于 SpringMVC 功能实现的分析，我们首先从 web.xml 开始，在 web.xml 文件中我们首先配置的就是 ContextLoaderListener，那么它所提供的功能有哪些，又是如何实现的呢？

当使用编程方式的时候我们可以直接将 Spring 配置信息作为参数传入 Spring 容器中，如
`ApplicationContext ac=new ClassPathXmlApplicationContext("applicationContext.xml");`
但是在 Web 下，我们需要更多的是与 Web 环境相互结合，通常的办法是将路径以 context-param 的方式注册并使用 ContextLoaderListener 进行监听读取。

ContextLoaderListener 的作用就是启动 Web 容器时，自动装配 ApplicationContext 的配置信息。因为它实现了 ServletContextListener 这个接口，在 web.xml 配置这个监听器，启动容器时，就会默认执行它实现的方法，使用 ServletContextListener 接口，开发者能够在为客户端请求提供服务之前向 ServletContext 中添加任意的对象。这个对象在 ServletContext 启动的时候被初始化，然后在 ServletContext 整个运行期间都是可见的。

每一个 Web 应用都有一个 ServletContext 与之相关联。ServletContext 对象在应用启动时被创建，在应用关闭的时候被销毁。ServletContext 在全局范围内有效，类似于应用中的一个全局变量。

在 ServletContextListener 中的核心逻辑便是初始化 WebApplicationContext 实例并存放至 ServletContext 中。

11.2.1　ServletContextListener 的使用

正式分析代码前我们同样还是首先了解 ServletContextListener 的使用。

1．创建自定义 ServletContextListener

首先我们创建 ServletContextListener，目标是在系统启动时添加自定义的属性，以便于在全局范围内可以随时调用。系统启动的时候会调用 ServletContextListener 实现类的 contextInitialized 方法，所以需要在这个方法中实现我们的初始化逻辑。

```
public class MyDataContextListener implements ServletContextListener {
```

```
    private ServletContext context = null;

    public MyDataContextListener () {

    }

    //该方法在ServletContext启动之后被调用，并准备好处理客户端请求
    public void contextInitialized(ServletContextEvent event)   {
this.context = event.getServletContext();

//通过你可以实现自己的逻辑并将结果记录在属性中
    context = setAttribute("myData","this is myData");
    }

    //这个方法在ServletContext将要关闭的时候调用
    public void contextDestroyed(ServletContextEvent event){
      this.context = null;
    }
}
```

2. 注册监听器

在 web.xml 文件中需要注册自定义的监听器。
```
<listener>
com.test.MyDataContextListener
</listener>
```

3. 测试

一旦 Web 应用启动的时候，我们就能在任意的 Servlet 或者 JSP 中通过下面的方式获取我们初始化的参数，如下：
```
String myData = (String) getServletContext().getAttribute("myData");
```

11.2.2 Spring 中的 ContextLoaderListener

分析了 ServletContextListener 的使用方式后再来分析 Spring 中的 ContextLoaderListener 的实现就容易理解得多，虽然 ContextLoaderListener 实现的逻辑要复杂得多，但是大致的套路还是万变不离其宗。

ServletContext 启动之后会调用 ServletContextListener 的 contextInitialized 方法，那么，我们就从这个函数开始进行分析。
```
    public void contextInitialized(ServletContextEvent event) {
        this.contextLoader = createContextLoader();
        if (this.contextLoader == null) {
            this.contextLoader = this;
        }
//初始化WebApplicationContext
        this.contextLoader.initWebApplicationContext(event.getServletContext());
}
```
这里涉及了一个常用类 WebApplicationContext：在 Web 应用中，我们会用到 WebApplication

Context,WebApplicationContext 继承自 ApplicationContext,在 ApplicationContext 的基础上又追加了一些特定于 Web 的操作及属性,非常类似于我们通过编程方式使用 Spring 时使用的 ClassPathXmlApplicationContext 类提供的功能。继续跟踪代码:

```java
public WebApplicationContext initWebApplicationContext(ServletContext servletContext) {
    if (servletContext.getAttribute(WebApplicationContext.ROOT_WEB_APPLICATION_CONTEXT_ATTRIBUTE) != null) {
        //web.xml 中存在多次 ContextLoader 定义
        throw new IllegalStateException(
                "Cannot initialize context because there is already a root application context present - " +
                "check whether you have multiple ContextLoader* definitions in your web.xml!");
    }

    Log logger = LogFactory.getLog(ContextLoader.class);
    servletContext.log("Initializing Spring root WebApplicationContext");
    if (logger.isInfoEnabled()) {
        logger.info("Root WebApplicationContext: initialization started");
    }
    long startTime = System.currentTimeMillis();

    try {
        // Store context in local instance variable, to guarantee that
        // it is available on ServletContext shutdown.
        if (this.context == null) {
            //初始化 context
            this.context = createWebApplicationContext(servletContext);
        }
        if (this.context instanceof ConfigurableWebApplicationContext) {
            configureAndRefreshWebApplicationContext((ConfigurableWebApplicationContext)this.context, servletContext);
        }
        //记录在 servletContext 中
        servletContext.setAttribute(WebApplicationContext.ROOT_WEB_APPLICATION_CONTEXT_ATTRIBUTE, this.context);

        ClassLoader ccl = Thread.currentThread().getContextClassLoader();
        if (ccl == ContextLoader.class.getClassLoader()) {
            currentContext = this.context;
        }
        else if (ccl != null) {
            currentContextPerThread.put(ccl, this.context);
        }

        if (logger.isDebugEnabled()) {
            logger.debug("Published root WebApplicationContext as ServletContext attribute with name [" +
                    WebApplicationContext.ROOT_WEB_APPLICATION_CONTEXT_ATTRIBUTE+"]");
        }
        if (logger.isInfoEnabled()) {
            long elapsedTime = System.currentTimeMillis() - startTime;
            logger.info("Root WebApplicationContext: initialization completed in " + elapsedTime + " ms");
        }
```

11.2 ContextLoaderListener

```
            return this.context;
        }
        catch (RuntimeException ex) {
            logger.error("Context initialization failed", ex);
            servletContext.setAttribute(WebApplicationContext.ROOT_WEB_APPLICATION_
CONTEXT_ATTRIBUTE, ex);
            throw ex;
        }
        catch (Error err) {
            logger.error("Context initialization failed", err);
            servletContext.setAttribute(WebApplicationContext.ROOT_WEB_APPLICATION_
CONTEXT_ATTRIBUTE, err);
            throw err;
        }
    }
```

initWebApplicationContext 函数主要是体现了创建 WebApplicationContext 实例的一个功能架构, 从函数中我们看到了初始化的大致步骤。

1. WebApplicationContext 存在性的验证。

在配置中只允许声明一次 ServletContextListener, 多次声明会扰乱 Spring 的执行逻辑, 所以这里首先做的就是对此验证, 在 Spring 中如果创建 WebApplicationContext 实例会记录在 ServletContext 中以方便全局调用, 而使用的 key 就是 WebApplicationContext.ROOT_WEB_APPLICATION_CONTEXT_ATTRIBUTE, 所以验证的方式就是查看 ServletContext 实例中是否有对应 key 的属性。

2. 创建 WebApplicationContext 实例。

如果通过验证, 则 Spring 将创建 WebApplicationContext 实例的工作委托给了 createWebApplicationContext 函数。

```
    protected WebApplicationContext createWebApplicationContext(ServletContext sc) {
        Class<?> contextClass = determineContextClass(sc);
        if (!ConfigurableWebApplicationContext.class.isAssignableFrom(contextClass)) {
            throw new ApplicationContextException("Custom context class [" + context
Class.getName() +
                "] is not of type [" + ConfigurableWebApplicationContext.class.getName
() + "]");
        }
        ConfigurableWebApplicationContext wac =
                (ConfigurableWebApplicationContext) BeanUtils.instantiateClass(
contextClass);
        return wac;
    }


    protected Class<?> determineContextClass(ServletContext servletContext) {
//CONTEXT_CLASS_PARAM = "contextClass";
        String contextClassName = servletContext.getInitParameter(CONTEXT_CLASS_PARAM);
        if (contextClassName != null) {
            try {
                return ClassUtils.forName(contextClassName, ClassUtils.getDefaultClass
Loader());
            }
```

```
                catch (ClassNotFoundException ex) {
                    throw new ApplicationContextException(
                         "Failed to load custom context class [" + contextClassName + "]",ex);
                }
            }
            else {
                contextClassName = defaultStrategies.getProperty(WebApplicationContext.class.getName());
                try {
                    return ClassUtils.forName(contextClassName, ContextLoader.class.getClassLoader());
                }
                catch (ClassNotFoundException ex) {
                    throw new ApplicationContextException(
                         "Failed to load default context class [" + contextClassName + "]",ex);
                }
            }
        }
```

其中，在 ContextLoader 类中有这样的静态代码块：

```
    static {
        // Load default strategy implementations from properties file.
        // This is currently strictly internal and not meant to be customized
        // by application developers.
        try {
            //DEFAULT_STRATEGIES_PATH = "ContextLoader.properties"
            ClassPathResource resource = new ClassPathResource(DEFAULT_STRATEGIES_PATH, ContextLoader.class);
            defaultStrategies = PropertiesLoaderUtils.loadProperties(resource);
        }
        catch (IOException ex) {
            throw new IllegalStateException("Could not load 'ContextLoader.properties': " + ex.getMessage());
        }
    }
```

根据以上静态代码块的内容，我们推断在当前类 ContextLoader 同样目录下必定会存在属性文件 ContextLoader.properties，查看后果然存在，内容如下：

```
org.Springframework.web.context.WebApplicationContext=org.Springframework.web.context.support.XmlWebApplicationContext
```

综合以上代码分析，在初始化的过程中，程序首先会读取 ContextLoader 类的同目录下的属性文件 ContextLoader.properties，并根据其中的配置提取将要实现 WebApplicationContext 接口的实现类，并根据这个实现类通过反射的方式进行实例的创建。

3. 将实例记录在 servletContext 中。
4. 映射当前的类加载器与创建的实例到全局变量 currentContextPerThread 中。

11.3 DispatcherServlet

在 Spring 中，ContextLoaderListener 只是辅助功能，用于创建 WebApplicationContext 类型实例，而真正的逻辑实现其实是在 DispatcherServlet 中进行的，DispatcherServlet 是实现 servlet

接口的实现类。

servlet 是一个 Java 编写的程序，此程序是基于 HTTP 协议的，在服务器端运行的（如 Tomcat），是按照 servlet 规范编写的一个 Java 类。主要是处理客户端的请求并将其结果发送到客户端。servlet 的生命周期是由 servlet 的容器来控制的，它可以分为 3 个阶段：初始化、运行和销毁。

1. 初始化阶段

- servlet 容器加载 servlet 类，把 servlet 类的.class 文件中的数据读到内存中。
- servlet 容器创建一个 ServletConfig 对象。ServletConfig 对象包含了 servlet 的初始化配置信息。
- servlet 容器创建一个 servlet 对象。
- servlet 容器调用 servlet 对象的 init 方法进行初始化。

2. 运行阶段

当 servlet 容器接收到一个请求时，servlet 容器会针对这个请求创建 servletRequest 和 servletResponse 对象，然后调用 service 方法。并把这两个参数传递给 service 方法。service 方法通过 servletRequest 对象获得请求的信息。并处理该请求。再通过 servletResponse 对象生成这个请求的响应结果。然后销毁 servletRequest 和 servletResponse 对象。我们不管这个请求是 post 提交的还是 get 提交的，最终这个请求都会由 service 方法来处理。

3. 销毁阶段

当 Web 应用被终止时，servlet 容器会先调用 servlet 对象的 destroy 方法，然后再销毁 servlet 对象，同时也会销毁与 servlet 对象相关联的 servletConfig 对象。我们可以在 destroy 方法的实现中，释放 servlet 所占用的资源，如关闭数据库连接，关闭文件输入输出流等。

servlet 的框架是由两个 Java 包组成：javax.servlet 和 javax.servlet.http。在 javax.servlet 包中定义了所有的 servlet 类都必须实现或扩展的通用接口和类，在 javax.servlet.http 包中定义了采用 HTTP 通信协议的 HttpServlet 类。

servlet 被设计成请求驱动，servlet 的请求可能包含多个数据项，当 Web 容器接收到某个 servlet 请求时，servlet 把请求封装成一个 HttpServletRequest 对象，然后把对象传给 servlet 的对应的服务方法。

HTTP 的请求方式包括 delete、get、options、post、put 和 trace，在 HttpServlet 类中分别提供了相应的服务方法，它们是 doDelete()、doGet()、doOptions()、doPost()、doPut()和 doTrace()。

11.3.1 servlet 的使用

我们同样还是以最简单的 servlet 来快速体验其用法。

1. 建立 servlet

```
public class MyServlet extends HttpServlet{
    public void init(){
        System.out.println("this is init method");
    }

    public void doGet(HttpServletRequest request, HttpServletResponse response){
        handleLogic(request,response);
    }
    public void doPost(HttpServletRequest request, HttpServletResponse response){
        handleLogic(request,response);
    }

    private void handleLogic(HttpServletRequest request, HttpServletResponse response){
        System.out.println("handle myLogic");
        ServletContext sc = getServletContext();

        RequestDispatcher rd = null;

        rd = sc.getRequestDispatcher("/index.jsp");         //定向的页面
        try {
            rd.forward(request, response);
        } catch (ServletException | IOException e) {
            e.printStackTrace();
        }
    }
}
```

麻雀虽小，五脏俱全。实例中包含了对 init 方法和 get/post 方法的处理，init 方法保证在 servlet 加载的时候能做一些逻辑操作，而 HttpServlet 类则会帮助我们根据方法类型的不同而将逻辑引入不同的函数。在子类中我们只需要重写对应的函数逻辑便可，如以上代码重写了 doGet 和 doPost 方法并将逻辑处理部分引导至 handleLogic 函数中，最后，又将页面跳转至 index.jsp。

2. 添加配置

为了使 servlet 能够正常使用，需要在 web.xml 文件中添加以下配置：

```xml
<servlet>
    <servlet-name>myservlet</servlet-name>
    <servlet-class>test.servlet.MyServlet</servlet-class>
    <load-on-startup>1</load-on-startup>
</servlet>

<servlet-mapping>
    <servlet-name>myservlet</servlet-name>
    <url-pattern>*.htm</url-pattern>
</servlet-mapping>
```

配置后便可以根据对应的配置访问相应的路径了。

11.3.2 DispatcherServlet 的初始化

通过上面的实例我们了解到，在 servlet 初始化阶段会调用其 init 方法，所以我们首先要查

11.3 DispatcherServlet

看在 DispatcherServlet 中是否重写了 init 方法。我们在其父类 HttpServletBean 中找到了该方法。

```
public final void init() throws ServletException {
    if (logger.isDebugEnabled()) {
        logger.debug("Initializing servlet '" + getServletName() + "'");
    }

    try {
        //解析 init-param 并封装至 pvs 中
        PropertyValues pvs = new ServletConfigPropertyValues(getServletConfig(), this.requiredProperties);
        //将当前的这个 servlet 类转化为一个 BeanWrapper,从而能够以 Spring 的方式来对 init-param
        //的值进行注入
        BeanWrapper bw = PropertyAccessorFactory.forBeanPropertyAccess(this);
        ResourceLoader resourceLoader = new ServletContextResourceLoader(getServletContext());
        //注册自定义属性编辑器,一旦遇到 Resource 类型的属性将会使用 ResourceEditor 进行解析
        bw.registerCustomEditor(Resource.class, new ResourceEditor(resourceLoader, this.environment));
        //空实现,留给子类覆盖
        initBeanWrapper(bw);
        //属性注入
        bw.setPropertyValues(pvs, true);
    }
    catch (BeansException ex) {
        logger.error("Failed to set bean properties on servlet '" + getServletName() + "'", ex);
        throw ex;
    }

    //留给子类扩展
    initServletBean();

    if (logger.isDebugEnabled()) {
        logger.debug("Servlet '" + getServletName() + "' configured successfully");
    }
}
```

DipatcherServlet 的初始化过程主要是通过将当前的 servlet 类型实例转换为 BeanWrapper 类型实例,以便使用 Spring 中提供的注入功能进行对应属性的注入。这些属性如 contextAttribute、contextClass、nameSpace、contextConfigLocation 等,都可以在 web.xml 文件中以初始化参数的方式配置在 servlet 的声明中。DispatcherServlet 继承自 FrameworkServlet,FrameworkServlet 类上包含对应的同名属性,Spring 会保证这些参数被注入到对应的值中。属性注入主要包含以下几个步骤。

1. 封装及验证初始化参数

ServletConfigPropertyValues 除了封装属性外还有对属性验证的功能。

```
public ServletConfigPropertyValues(ServletConfig config, Set<String> requiredProperties)
        throws ServletException {

    Set<String> missingProps = (requiredProperties != null && !requiredProperties.isEmpty()) ?
            new HashSet<String>(requiredProperties) : null;
```

```
                Enumeration en = config.getInitParameterNames();
                while (en.hasMoreElements()) {
                    String property = (String) en.nextElement();
                    Object value = config.getInitParameter(property);
                    addPropertyValue(new PropertyValue(property, value));
                    if (missingProps != null) {
                        missingProps.remove(property);
                    }
                }

                // Fail if we are still missing properties.
                if (missingProps != null && missingProps.size() > 0) {
                    throw new ServletException(
                        "Initialization from ServletConfig for servlet '" + config.getServlet
Name() +
                        "' failed; the following required properties were missing: " +
                        StringUtils.collectionToDelimitedString(missingProps, ", "));
                }
            }
```

从代码中得知，封装属性主要是对初始化的参数进行封装，也就是 servlet 中配置的 <init-param>中配置的封装。当然，用户可以通过对 requiredProperties 参数的初始化来强制验证某些属性的必要性，这样，在属性封装的过程中，一旦检测到 requiredProperties 中的属性没有指定初始值，就会抛出异常。

2. 将当前 servlet 实例转化成 BeanWrapper 实例

PropertyAccessorFactory.forBeanPropertyAccess 是 Spring 中提供的工具方法，主要用于将指定实例转化为 Spring 中可以处理的 BeanWrapper 类型的实例。

3. 注册相对于 Resource 的属性编辑器

属性编辑器，我们在上文中已经介绍并且分析过其原理，这里使用属性编辑器的目的是在对当前实例（DispatcherServlet）属性注入过程中一旦遇到 Resource 类型的属性就会使用 ResourceEditor 去解析。

4. 属性注入

BeanWrapper 为 Spring 中的方法，支持 Spring 的自动注入。其实我们最常用的属性注入无非是 contextAttribute、contextClass、nameSpace、contextConfigLocation 等。

5. servletBean 的初始化

在 ContextLoaderListener 加载的时候已经创建了 WebApplicationContext 实例，而在这个函数中最重要的就是对这个实例进行进一步的补充初始化。

继续查看 initServletBean()。父类 FrameworkServlet 覆盖了 HttpServletBean 中的 initServletBean

函数，如下：
```
protected final void initServletBean() throws ServletException {
    getServletContext().log("Initializing Spring FrameworkServlet '" + getServlet Name()+"'");
    if (this.logger.isInfoEnabled()) {
        this.logger.info("FrameworkServlet '" + getServletName()+ "': initialization started");
    }
    long startTime = System.currentTimeMillis();

    try {
        this.webApplicationContext = initWebApplicationContext();
        //设计为子类覆盖
        initFrameworkServlet();
    }
    catch (ServletException ex) {
        this.logger.error("Context initialization failed", ex);
        throw ex;
    }
    catch (RuntimeException ex) {
        this.logger.error("Context initialization failed", ex);
        throw ex;
    }

    if (this.logger.isInfoEnabled()) {
        long elapsedTime = System.currentTimeMillis() - startTime;
        this.logger.info("FrameworkServlet '" + getServletName() + "': initialization completed in " +
            elapsedTime + " ms");
    }
}
```

上面的函数设计了计时器来统计初始化的执行时间，而且提供了一个扩展方法 initFrameworkServlet()用于子类的覆盖操作，而作为关键的初始化逻辑实现委托给了 initWebApplicationContext()。

11.3.3 WebApplicationContext 的初始化

initWebApplicationContext 函数的主要工作就是创建或刷新 WebApplicationContext 实例并对 servlet 功能所使用的变量进行初始化。

```
Protected WebApplicationContext initWebApplicationContext() {
    WebApplicationContext rootContext =
        WebApplicationContextUtils.getWebApplicationContext(getServletContext());
    WebApplicationContext wac = null;

    if (this.webApplicationContext != null) {
    //context 实例在构造函数中被注入
        wac = this.webApplicationContext;
        if (wac instanceof ConfigurableWebApplicationContext) {
            ConfigurableWebApplicationContext cwac = (ConfigurableWebApplication Context) wac;
            if (!cwac.isActive()) {
```

```java
            if (cwac.getParent() == null) {
                cwac.setParent(rootContext);
            }
            //刷新上下文环境
            configureAndRefreshWebApplicationContext(cwac);
        }
    }
}
if (wac == null) {
    //根据 contextAttribute 属性加载 WebApplicationContext
    wac = findWebApplicationContext();
}
if (wac == null) {
    // No context instance is defined for this servlet -> create a local one
    wac = createWebApplicationContext(rootContext);
}

if (!this.refreshEventReceived) {
    // Either the context is not a ConfigurableApplicationContext with refresh

    // support or the context injected at construction time had already been
    // refreshed -> trigger initial onRefresh manually here.
    onRefresh(wac);
}

if (this.publishContext) {
    // Publish the context as a servlet context attribute.
    String attrName = getServletContextAttributeName();
    getServletContext().setAttribute(attrName, wac);
    if (this.logger.isDebugEnabled()) {
        this.logger.debug("Published WebApplicationContext of servlet '" + getServletName() +
            "' as ServletContext attribute with name [" + attrName + "]");
    }
}

return wac;
}
```

对于本函数中的初始化主要包含几个部分。

1. 寻找或创建对应的 WebApplicationContext 实例

WebApplicationContext 的寻找及创建包括以下几个步骤。

1. 通过构造函数的注入进行初始化。

当进入 initWebApplicationContext 函数后通过判断 this.webApplicationContext != null 后，便可以确定 this.webApplicationContext 是否是通过构造函数来初始化的。可是有读者可能会有疑问，在 initServletBean 函数中明明是把创建好的实例记录在了 this.webApplicationContext 中：

```java
this.webApplicationContext= initWebApplicationContext();
```

何以判定这个参数是通过构造函数初始化,而不是通过上一次的函数返回值初始化呢?如果存在这个问题,那么就是读者忽略一个问题了:在 Web 中包含 SpringWeb 的核心逻辑的 DispatcherServlet 只可以被声明为一次,在 Spring 中已经存在验证,所以这就确保了如果 this.webApplicationContext != null,则可以直接判定 this.webApplicationContext 已经通过构造函数初始化。

2. 通过 contextAttribute 进行初始化。

通过在 web.xml 文件中配置的 servlet 参数 contextAttribute 来查找 ServletContext 中对应的属性,默认为 WebApplicationContext.class.getName() + ".ROOT",也就是在 ContextLoaderListener 加载时会创建 WebApplicationContext 实例,并将实例以 WebApplicationContext.class.getName() + ".ROOT" 为 key 放入 ServletContext 中,当然读者可以重写初始化逻辑使用自己创建的 WebApplicationContext,并在 servlet 的配置中通过初始化参数 contextAttribute 指定 key。

```
protected WebApplicationContext findWebApplicationContext() {
    String attrName = getContextAttribute();
    if (attrName == null) {
        return null;
    }
    WebApplicationContext wac =
            WebApplicationContextUtils.getWebApplicationContext(getServletContext(), attrName);
    if (wac == null) {
        throw new IllegalStateException("No WebApplicationContext found: initializer not registered?");
    }
    return wac;
}
```

3. 重新创建 WebApplicationContext 实例。

如果通过以上两种方式并没有找到任何突破,那就没办法了,只能在这里重新创建新的实例了。

```
protected WebApplicationContext createWebApplicationContext(WebApplicationContext parent) {
    return createWebApplicationContext((ApplicationContext) parent);
}

protected WebApplicationContext createWebApplicationContext(ApplicationContext parent) {
    //获取servlet的初始化参数contextClass,如果没有配置默认为XmlWebApplicationContext.class
    Class<?> contextClass = getContextClass();
    if (this.logger.isDebugEnabled()) {
        this.logger.debug("Servlet with name '" + getServletName() +
                "' will try to create custom WebApplicationContext context of class '" +
                contextClass.getName() + "'" + ", using parent context [" + parent + "]");
    }
    if (!ConfigurableWebApplicationContext.class.isAssignableFrom(contextClass)) {
        throw new ApplicationContextException(
                "Fatal initialization error in servlet with name '" + getServletName() +
                "': custom WebApplicationContext class [" + contextClass.getName() +
                "] is not of type ConfigurableWebApplicationContext");
    }
```

```java
//通过反射方式实例化contextClass
            ConfigurableWebApplicationContext wac =
                (ConfigurableWebApplicationContext) BeanUtils.instantiateClass
(contextClass);
    //parent为在ContextLoaderListener中创建的实例
            //在ContextLoaderListener加载的时候初始化的WebApplicationContext类型实例
    wac.setParent(parent);
            //获取contextConfigLocation属性，配置在servlet初始化参数中
    wac.setConfigLocation(getContextConfigLocation());
    //初始化Spring环境包括加载配置文件等
            configureAndRefreshWebApplicationContext(wac);

            return wac;
    }
```

2. configureAndRefreshWebApplicationContext

无论是通过构造函数注入还是单独创建，都会调用configureAndRefreshWebApplicationContext方法来对已经创建的WebApplicationContext实例进行配置及刷新，那么这个步骤又做了哪些工作呢？

```java
    protected void configureAndRefreshWebApplicationContext(ConfigurableWebApplication Context wac, ServletContext sc) {
            if (ObjectUtils.identityToString(wac).equals(wac.getId())) {
                // The application context id is still set to its original default value
                // -> assign a more useful id based on available information
                String idParam = sc.getInitParameter(CONTEXT_ID_PARAM);
                if (idParam != null) {
                    wac.setId(idParam);
                }
                else {
                    // Generate default id...
                    if (sc.getMajorVersion() == 2 && sc.getMinorVersion() < 5) {
                        // Servlet <= 2.4: resort to name specified in web.xml, if any.
                        wac.setId(ConfigurableWebApplicationContext.APPLICATION_ CONTEXT_
ID_PREFIX +
                            ObjectUtils.getDisplayString(sc.getServletContextName()));
                    }
                    else {
                        wac.setId(ConfigurableWebApplicationContext.APPLICATION_CONTEXT_ ID
_PREFIX +
                            ObjectUtils.getDisplayString(sc.getContextPath()));
                    }
                }
            }

            // Determine parent for root web application context, if any.
            ApplicationContext parent = loadParentContext(sc);

            wac.setParent(parent);
            wac.setServletContext(sc);
            String initParameter = sc.getInitParameter(CONFIG_LOCATION_PARAM);
            if (initParameter != null) {
```

```
            wac.setConfigLocation(initParameter);
        }
        customizeContext(sc, wac);
//加载配置文件及整合 parent 到 wac
        wac.refresh();
}
```

无论调用方式如何变化，只要是使用 AlicationContext 所提供的功能最后都免不了使用公共父类 AbstractApplicationContext 提供的 refresh()进行配置文件加载。

3. 刷新

onRefresh 是 FrameworkServlet 类中提供的模板方法，在其子类 DispatcherServlet 中进行了重写，主要用于刷新 Spring 在 Web 功能实现中所必须使用的全局变量。下面我们会介绍它们的初始化过程以及使用场景，而至于具体的使用细节会在稍后的章节中再做详细介绍。

```
protected void onRefresh(ApplicationContext context) {
    initStrategies(context);
}

protected void initStrategies(ApplicationContext context) {
    //(1)初始化 MultipartResolver
        initMultipartResolver(context);
    //(2)初始化 LocaleResolver
        initLocaleResolver(context);
    //(3)初始化 ThemeResolver
        initThemeResolver(context);
    //(4)初始化 HandlerMappings
        initHandlerMappings(context);
    //(5)初始化 HandlerAdapters
        initHandlerAdapters(context);
    //(6)初始化 HandlerExceptionResolvers
        initHandlerExceptionResolvers(context);
    //(7)初始化 RequestToViewNameTranslator
        initRequestToViewNameTranslator(context);
    //(8)初始化 ViewResolvers
        initViewResolvers(context);
    //(9)初始化 FlashMapManager
        initFlashMapManager(context);
}
```

1. 初始化 MultipartResolver。

在 Spring 中，MultipartResolver 主要用来处理文件上传。默认情况下，Spring 是没有 multipart 处理的，因为一些开发者想要自己处理它们。如果想使用 Spring 的 multipart，则需要在 Web 应用的上下文中添加 multipart 解析器。这样，每个请求就会被检查是否包含 multipart。然而，如果请求中包含 multipart，那么上下文中定义的 MultipartResolver 就会解析它，这样请求中的 multipart 属性就会像其他属性一样被处理。常用配置如下：

```
<bean id="multipartResolver" class="org.Springframework.web.multipart.commons. Commons MultipartResolver">
    <!-- 该属性用来配置可上传文件的最大字节数 -->
    <property name="maximumFileSize"><value>100000</value></property>
</bean>
```

当然，CommonsMultipartResolver 还提供了其他功能用于帮助用户完成上传功能，有兴趣的读者可以进一步查看。

那么 MultipartResolver 就是在 initMultipartResolver 中被加入到 DispatcherServlet 中的。

```
private void initMultipartResolver(ApplicationContext context) {
    try {
        this.multipartResolver = context.getBean(MULTIPART_RESOLVER_BEAN_NAME,
MultipartResolver.class);
        if (logger.isDebugEnabled()) {
            logger.debug("Using MultipartResolver [" + this.multipartResolver + "]");
        }
    }
    catch (NoSuchBeanDefinitionException ex) {
        // Default is no multipart resolver.
        this.multipartResolver = null;
        if (logger.isDebugEnabled()) {
            logger.debug("Unable to locate MultipartResolver with name '" +
MULTIPART_RESOLVER_BEAN_NAME +
                    "': no multipart request handling provided");
        }
    }
}
```

因为之前的步骤已经完成了 Spring 中配置文件的解析，所以在这里只要在配置文件注册过都可以通过 ApplicationContext 提供的 getBean 方法来直接获取对应 bean，进而初始化 MultipartResolver 中的 multipartResolver 变量。

2. 初始化 LocaleResolver。

在 Spring 的国际化配置中一共有 3 种使用方式。

- 基于 URL 参数的配置。

通过 URL 参数来控制国际化，比如你在页面上加一句`<a href="?locale=zh_CN">`简体中文`</a>`来控制项目中使用的国际化参数。而提供这个功能的就是 AcceptHeaderLocaleResolver，默认的参数名为 locale，注意大小写。里面放的就是你的提交参数，比如 en_US、zh_CN 之类的，具体配置如下：

```
<bean id="localeResolver" class="org.Springframework.web.servlet.i18n.AcceptHeaderLocaleResolver"/>
```

- 基于 session 的配置。

它通过检验用户会话中预置的属性来解析区域。最常用的是根据用户本次会话过程中的语言设定决定语言种类（例如，用户登录时选择语言种类，则此次登录周期内统一使用此语言设定），如果该会话属性不存在，它会根据 accept-language HTTP 头部确定默认区域。

```
<bean id="localeResolver" class="org.Springframework.web.servlet.i18n.SessionLocaleResolver"/>
```

- 基于 cookie 的国际化配置。

CookieLocaleResolver 用于通过浏览器的 cookie 设置取得 Locale 对象。这种策略在应用程序不支持会话或者状态必须保存在客户端时有用，配置如下：

```
<bean id="localeResolver" class="org.Springframework.web.servlet.i18n.CookieLocaleResolver"/>
```

这 3 种方式都可以解决国际化的问题，但是，对于 LocalResolver 的使用基础是在

11.3 DispatcherServlet

DispatcherServlet 中的初始化。

```
    private void initLocaleResolver(ApplicationContext context) {
        try {
            this.localeResolver = context.getBean(LOCALE_RESOLVER_BEAN_NAME, LocaleResolver.class);
            if (logger.isDebugEnabled()) {
                logger.debug("Using LocaleResolver [" + this.localeResolver + "]");
            }
        }
        catch (NoSuchBeanDefinitionException ex) {
            // We need to use the default.
            this.localeResolver = getDefaultStrategy(context, LocaleResolver.class);
            if (logger.isDebugEnabled()) {
                logger.debug("Unable to locate LocaleResolver with name '" + LOCALE_RESOLVER_BEAN_NAME +
                        "': using default [" + this.localeResolver + "]");
            }
        }
    }
```

提取配置文件中设置的 LocaleResolver 来初始化 DispatcherServlet 中的 localeResolver 属性。

3. 初始化 ThemeResolver。

在 Web 开发中经常会遇到通过主题 Theme 来控制网页风格，这将进一步改善用户体验。简单地说，一个主题就是一组静态资源（比如样式表和图片），它们可以影响应用程序的视觉效果。Spring 中的主题功能和国际化功能非常类似。Spring 主题功能的构成主要包括如下内容。

- 主题资源。

org.Springframework.ui.context.ThemeSource 是 Spring 中主题资源的接口，Spring 的主题需要通过 ThemeSource 接口来实现存放主题信息的资源。

org.Springframework.ui.context.support.ResourceBundleThemeSource 是 ThemeSource 接口默认实现类（也就是通过 ResourceBundle 资源的方式定义主题），在 Spring 中的配置如下：

```
<bean id="themeSource" class="org.Springframework.ui.context.support.ResourceBundleThemeSource">
    <property name="basenamePrefix" value="com.test. "></property>
</bean>
```

默认状态下是在类路径根目录下查找相应的资源文件，也可以通过 basenamePrefix 来制定。这样，DispatcherServlet 就会在 com.test 包下查找资源文件。

- 主题解析器。

ThemeSource 定义了一些主题资源，那么不同的用户使用什么主题资源由谁定义呢？org.Springframework.web.servlet.ThemeResolver 是主题解析器的接口，主题解析的工作便由它的子类来完成。

对于主题解析器的子类主要有 3 个比较常用的实现。以主题文件 summer.properties 为例。

① **FixedThemeResolver** 用于选择一个固定的主题。

```
<bean id="themeResolver" class="org.Springframework.web.servlet.theme.FixedTheme Resolver">
<property name="defaultThemeName" value="summer"/>
</bean>
```

以上配置的作用是设置主题文件为 summer.properties，在整个项目内固定不变。

② **CookieThemeResolver** 用于实现用户所选的主题，以 cookie 的形式存放在客户端的机器上，配置如下：

```
<bean id="themeResolver" class="org.Springframework.web.servlet.theme.CookieThemeResolver">
<property name="defaultThemeName" value="summer"/>
</bean>
```

③ **SessionThemeResolver** 用于主题保存在用户的 HTTP Session 中。

```
<bean id="themeResolver" class="org.Springframework.web.servlet.theme.SessionThemeResolver">
<property name="defaultThemeName" value="summer"/>
</bean>
```

以上配置用于设置主题名称，并且将该名称保存在用户的 HttpSession 中。

④ **AbstractThemeResolver** 是一个抽象类被 SessionThemeResolver 和 FixedThemeResolver 继承，用户也可以继承它来自定义主题解析器。

- 拦截器。

如果需要根据用户请求来改变主题，那么 Spring 提供了一个已经实现的拦截器——ThemeChangeInterceptor 拦截器了，配置如下：

```
<bean id="themeChangeInterceptor" class="org.Springframework.web.servlet.theme. ThemeChangeInterceptor">
    <property name="paramName" value="themeName"></property>
</bean>
```

其中设置用户请求参数名为 themeName，即 URL 为?themeName=具体的主题名称。此外，还需要在 handlerMapping 中配置拦截器。当然需要在 HandleMapping 中添加拦截器。

```
<property name="interceptors">
    <list>
        <ref local="themeChangeInterceptor" />
    </list>
</property>
```

了解了主题文件的简单使用方式后，再来查看解析器的初始化工作，与其他变量的初始化工作相同，主题文件解析器的初始化工作并没有任何需要特别说明的地方。

```
        private void initThemeResolver(ApplicationContext context) {
            try {
                this.themeResolver = context.getBean(THEME_RESOLVER_BEAN_NAME, ThemeResolver.class);
                if (logger.isDebugEnabled()) {
                    logger.debug("Using ThemeResolver [" + this.themeResolver + "]");
                }
            }
            catch (NoSuchBeanDefinitionException ex) {
                // We need to use the default.
                this.themeResolver = getDefaultStrategy(context, ThemeResolver.class);
                if (logger.isDebugEnabled()) {
                    logger.debug(
                            "Unable to locate ThemeResolver with name '" + THEME_ RESOLVER_ BEAN_NAME + "': using default [" +
                                    this.themeResolver + "]");
                }
```

    }
}
4. 初始化 HandlerMappings。

当客户端发出 Request 时 DispatcherServlet 会将 Request 提交给 HandlerMapping, 然后 HanlerMapping 根据 Web Application Context 的配置来回传给 DispatcherServlet 相应的 Controller。

在基于 SpringMVC 的 Web 应用程序中, 我们可以为 DispatcherServlet 提供多个 Handler Mapping 供其使用。DispatcherServlet 在选用 HandlerMapping 的过程中, 将根据我们所指定的一系列 HandlerMapping 的优先级进行排序, 然后优先使用优先级在前的 HandlerMapping。如果当前的 HandlerMapping 能够返回可用的 Handler, DispatcherServlet 则使用当前返回的 Handler 进行 Web 请求的处理, 而不再继续询问其他的 HandlerMapping。否则, DispatcherServlet 将继续按照各个 HandlerMapping 的优先级进行询问, 直到获取一个可用的 Handler 为止。初始化配置如下:

```
private void initHandlerMappings(ApplicationContext context) {
    this.handlerMappings = null;

    if (this.detectAllHandlerMappings) {

        Map<String, HandlerMapping> matchingBeans =
            BeanFactoryUtils.beansOfTypeIncludingAncestors(context,
HandlerMapping. class, true, false);
        if (!matchingBeans.isEmpty()) {
            this.handlerMappings = new ArrayList<HandlerMapping> (matchingBeans.
values());
            // We keep HandlerMappings in sorted order.
            OrderComparator.sort(this.handlerMappings);
        }
    }
    else {
        try {
            HandlerMapping hm = context.getBean(HANDLER_MAPPING_BEAN_NAME,
HandlerMapping.class);
            this.handlerMappings = Collections.singletonList(hm);
        }
        catch (NoSuchBeanDefinitionException ex) {
            // Ignore, we'll add a default HandlerMapping later.
        }
    }

    // Ensure we have at least one HandlerMapping, by registering
    // a default HandlerMapping if no other mappings are found.
    if (this.handlerMappings == null) {
        this.handlerMappings = getDefaultStrategies(context, HandlerMapping.class);
        if (logger.isDebugEnabled()) {
            logger.debug("No HandlerMappings found in servlet '" + getServletName() +
"': using default");
        }
    }
}
```

默认情况下，SpringMVC 将加载当前系统中所有实现了 HandlerMapping 接口的 bean。如果只期望 SpringMVC 加载指定的 handlermapping 时，可以修改 web.xml 中的 DispatcherServlet 的初始参数，将 detectAllHandlerMappings 的值设置为 false：

```xml
<init-param>
    <param-name>detectAllHandlerMappings</param-name>
    <param-value>false</param-value>
</init-param>
```

此时，SpringMVC 将查找名为 "handlerMapping" 的 bean，并作为当前系统中唯一的 handlermapping。如果没有定义 handlerMapping 的话，则 SpringMVC 将按照 org.Springframework.web.servlet.DispatcherServlet 所在目录下的 DispatcherServlet.properties 中所定义的 org.Springframework.web.servlet.HandlerMapping 的内容来加载默认的 handlerMapping（用户没有自定义 Strategies 的情况下）。

5. 初始化 HandlerAdapters。

从名字也能联想到这是一个典型的适配器模式的使用，在计算机编程中，适配器模式将一个类的接口适配成用户所期待的。使用适配器，可以使接口不兼容而无法在一起工作的类协同工作，做法是将类自己的接口包裹在一个已存在的类中。那么在处理 handler 时为什么会使用适配器模式呢？回答这个问题我们首先要分析它的初始化逻辑。

```java
private void initHandlerAdapters(ApplicationContext context) {
    this.handlerAdapters = null;

    if (this.detectAllHandlerAdapters) {
        // Find all HandlerAdapters in the ApplicationContext, including ancestor contexts.
        Map<String, HandlerAdapter> matchingBeans =
                BeanFactoryUtils.beansOfTypeIncludingAncestors(context, HandlerAdapter.class, true, false);
        if (!matchingBeans.isEmpty()) {
            this.handlerAdapters = new ArrayList<HandlerAdapter>(matchingBeans.values());
            // We keep HandlerAdapters in sorted order.
            OrderComparator.sort(this.handlerAdapters);
        }
    }
    else {
        try {
            HandlerAdapter ha = context.getBean(HANDLER_ADAPTER_BEAN_NAME, HandlerAdapter.class);
            this.handlerAdapters = Collections.singletonList(ha);
        }
        catch (NoSuchBeanDefinitionException ex) {
            // Ignore, we'll add a default HandlerAdapter later.
        }
    }

    // Ensure we have at least some HandlerAdapters, by registering
    // default HandlerAdapters if no other adapters are found.
    if (this.handlerAdapters == null) {
```

11.3 DispatcherServlet

```
                this.handlerAdapters = getDefaultStrategies(context, HandlerAdapter.class);
                if (logger.isDebugEnabled()) {
                    logger.debug("No HandlerAdapters found in servlet '" + getServletName
() + "': using default");
                }
            }
        }
```

同样在初始化的过程中涉及了一个变量 detectAllHandlerAdapters, detectAllHandlerAdapters 作用和 detectAllHandlerMappings 类似, 只不过作用对象为 handlerAdapter。亦可通过如下配置来强制系统只加载 bean name 为 "handlerAdapter" handlerAdapter。

```
        <init-param>
            <param-name>detectAllHandlerAdapters</param-name>
            <param-value>false</param-value>
        </init-param>
```

如果无法找到对应的 bean, 那么系统会尝试加载默认的适配器。

```
    protected <T> List<T> getDefaultStrategies(ApplicationContext context, Class<T>
strategyInterface) {
            String key = strategyInterface.getName();
            String value = defaultStrategies.getProperty(key);
            if (value != null) {
                String[] classNames = StringUtils.commaDelimitedListToStringArray(value);
                List<T> strategies = new ArrayList<T>(classNames.length);
                for (String className : classNames) {
                    try {
                        Class<?> clazz = ClassUtils.forName(className, DispatcherServlet.
class.getClassLoader());
                        Object strategy = createDefaultStrategy(context, clazz);
                        strategies.add((T) strategy);
                    }
                    catch (ClassNotFoundException ex) {
                        throw new BeanInitializationException(
                                "Could not find DispatcherServlet's default strategy class
[" + className +
                                "] for interface [" + key + "]", ex);
                    }
                    catch (LinkageError err) {
                        throw new BeanInitializationException(
                                "Error loading DispatcherServlet's default strategy class
[" + className +
                                "] for interface [" + key +"]: problem with class
file or dependent class", err);
                    }
                }
                return strategies;
            }
            else {
                return new LinkedList<T>();
            }
        }
```

在 getDefaultStrategies 函数中, Spring 会尝试从 defaultStrategies 中加载对应的 HandlerAdapter 的属性, 那么 defaultStrategies 是如何初始化的呢?

在当前类 DispatcherServlet 中存在这样一段初始化代码块:
```
static {
    try {
        // DEFAULT_STRATEGIES_PATH = "DispatcherServlet.properties";
        ClassPathResource resource = new ClassPathResource(DEFAULT_STRATEGIES_PATH, DispatcherServlet.class);
        defaultStrategies = PropertiesLoaderUtils.loadProperties(resource);
    }
    catch (IOException ex) {
        throw new IllegalStateException("Could not load 'Dispatcher Servlet.properties': " + ex.getMessage());
    }
}
```

在系统加载的时候，defaultStrategies 根据当前路径 DispatcherServlet.properties 来初始化本身，查看 DispatcherServlet.properties 中对应于 HandlerAdapter 的属性：

```
org.Springframework.web.servlet.HandlerAdapter=org.Springframework.web.servlet.mvc.HttpRequestHandlerAdapter,\
    org.Springframework.web.servlet.mvc.SimpleControllerHandlerAdapter,\
    org.Springframework.web.servlet.mvc.annotation.AnnotationMethodHandlerAdapter
```

由此得知，如果程序开发人员没有在配置文件中定义自己的适配器，那么 Spring 会默认加载配置文件中的 3 个适配器。

作为总控制器的派遣器 servlet 通过处理器映射得到处理器后，会轮询处理器适配器模块，查找能够处理当前 HTTP 请求的处理器适配器的实现，处理器适配器模块根据处理器映射返回的处理器类型，例如简单的控制器类型、注解控制器类型或者远程调用处理器类型，来选择某一个适当的处理器适配器的实现，从而适配当前的 HTTP 请求。

- HTTP 请求处理器适配器（HttpRequestHandlerAdapter）。

HTTP 请求处理器适配器仅仅支持对 HTTP 请求处理器的适配。它简单地将 HTTP 请求对象和响应对象传递给 HTTP 请求处理器的实现，它并不需要返回值。它主要应用在基于 HTTP 的远程调用的实现上。

- 简单控制器处理器适配器（SimpleControllerHandlerAdapter）。

这个实现类将 HTTP 请求适配到一个控制器的实现进行处理。这里控制器的实现是一个简单的控制器接口的实现。简单控制器处理器适配器被设计成一个框架类的实现，不需要被改写，客户化的业务逻辑通常是在控制器接口的实现类中实现的。

- 注解方法处理器适配器（AnnotationMethodHandlerAdapter）。

这个类的实现是基于注解的实现，它需要结合注解方法映射和注解方法处理器协同工作。它通过解析声明在注解控制器的请求映射信息来解析相应的处理器方法来处理当前的 HTTP 请求。在处理的过程中，它通过反射来发现探测处理器方法的参数，调用处理器方法，并且映射返回值到模型和控制器对象，最后返回模型和控制器对象给作为主控制器的派遣器 Servlet。

所以我们现在基本上可以回答之前的问题了，Spring 中所使用的 Handler 并没有任何特殊的联系，但是为了统一处理，Spring 提供了不同情况下的适配器。

6. 初始化 HandlerExceptionResolvers。

基于 HandlerExceptionResolver 接口的异常处理，使用这种方式只需要实现 resolveException 方法，该方法返回一个 ModelAndView 对象，在方法内部对异常的类型进行判断，然后尝试生成对应的 ModelAndView 对象，如果该方法返回了 null，则 Spring 会继续寻找其他的实现了 HandlerExceptionResolver 接口的 bean。换句话说，Spring 会搜索所有注册在其环境中的实现了 HandlerExceptionResolver 接口的 bean，逐个执行，直到返回了一个 ModelAndView 对象。

```java
import javax.servlet.http.HttpServletRequest;
import javax.servlet.http.HttpServletResponse;

import org.apache.commons.logging.Log;
import org.apache.commons.logging.LogFactory;
import org.Springframework.stereotype.Component;
import org.Springframework.web.servlet.HandlerExceptionResolver;
import org.Springframework.web.servlet.ModelAndView;

@Component
public class ExceptionHandler implements HandlerExceptionResolver
{
    private static final Log logs = LogFactory.getLog(ExceptionHandler.class);

    @Override
    public ModelAndView resolveException(HttpServletRequest request, HttpServletResponse response, Object obj,
        Exception exception)
    {
        request.setAttribute("exception", exception.toString());
        request.setAttribute("exceptionStack", exception);
        logs.error(exception.toString(), exception);
        return new ModelAndView("error/exception");
    }

}
```

这个类必须声明到 Spring 中去，让 Spring 管理它，在 Spring 的配置文件 applicationContext.xml 中增加以下内容：

```xml
<bean id="exceptionHandler" class="com.test.exception.MyExceptionHandler"/>
```

初始化代码如下：

```java
private void initHandlerExceptionResolvers(ApplicationContext context) {
        this.handlerExceptionResolvers = null;

        if (this.detectAllHandlerExceptionResolvers) {
            // Find all HandlerExceptionResolvers in the ApplicationContext, including ancestor contexts.
            Map<String, HandlerExceptionResolver> matchingBeans = BeanFactoryUtils
                    .beansOfTypeIncludingAncestors(context, HandlerExceptionResolver.class, true, false);
            if (!matchingBeans.isEmpty()) {
                this.handlerExceptionResolvers = new ArrayList<HandlerExceptionResolver>(matchingBeans.values());
                // We keep HandlerExceptionResolvers in sorted order.
```

```
                    OrderComparator.sort(this.handlerExceptionResolvers);
                }
            }
            else {
                try {
                    HandlerExceptionResolver her =
                            context.getBean(HANDLER_EXCEPTION_RESOLVER_BEAN_NAME, HandlerExceptionResolver.class);
                    this.handlerExceptionResolvers = Collections.singletonList(her);
                }
                catch (NoSuchBeanDefinitionException ex) {
                    // Ignore, no HandlerExceptionResolver is fine too.
                }
            }

            // Ensure we have at least some HandlerExceptionResolvers, by registering
            // default HandlerExceptionResolvers if no other resolvers are found.
            if (this.handlerExceptionResolvers == null) {
                this.handlerExceptionResolvers = getDefaultStrategies(context, HandlerExceptionResolver.class);
                if (logger.isDebugEnabled()) {
                    logger.debug("No HandlerExceptionResolvers found in servlet '" + getServletName() + "': using default");
                }
            }
        }
```

7. 初始化 RequestToViewNameTranslator。

当 Controller 处理器方法没有返回一个 View 对象或逻辑视图名称，并且在该方法中没有直接往 response 的输出流里面写数据的时候，Spring 就会采用约定好的方式提供一个逻辑视图名称。这个逻辑视图名称是通过 Spring 定义的 org.Springframework.web.servlet.RequestToViewNameTranslator 接口的 getViewName 方法来实现的，我们可以实现自己的 Request ToViewNameTranslator 接口来约定好没有返回视图名称的时候如何确定视图名称。Spring 已经给我们提供了一个它自己的实现，那就是 org.Springframework.web.servlet.view.DefaultRequest ToViewName Translator。

在介绍 DefaultRequestToViewNameTranslator 是如何约定视图名称之前，先来看一下它支持用户定义的属性。

- prefix：前缀，表示约定好的视图名称需要加上的前缀，默认是空串。
- suffix：后缀，表示约定好的视图名称需要加上的后缀，默认是空串。
- separator：分隔符，默认是斜杠 "/"。
- stripLeadingSlash：如果首字符是分隔符，是否要去除，默认是 true。
- stripTrailingSlash：如果最后一个字符是分隔符，是否要去除，默认是 true。
- stripExtension：如果请求路径包含扩展名是否要去除，默认是 true。
- urlDecode：是否需要对 URL 解码，默认是 true。它会采用 request 指定的编码或者 ISO-8859-1 编码对 URL 进行解码。

当我们没有在 SpringMVC 的配置文件中手动的定义一个名为 viewNameTranlator 的 Bean 的时候，Spring 就会为我们提供一个默认的 viewNameTranslator，即 DefaultRequestToViewNameTranslator。

接下来看一下，当 Controller 处理器方法没有返回逻辑视图名称时，DefaultRequestToViewNameTranslator 是如何约定视图名称的。DefaultRequestToViewNameTranslator 会获取到请求的 URI，然后根据提供的属性做一些改造，把改造之后的结果作为视图名称返回。这里以请求路径 http://localhost/app/test/index.html 为例，来说明一下 DefaultRequestToViewNameTranslator 是如何工作的。该请求路径对应的请求 URI 为/test/index.html，我们来看以下几种情况，它分别对应的逻辑视图名称是什么。

- prefix 和 suffix 如果都存在，其他为默认值，那么对应返回的逻辑视图名称应该是 prefixtest/indexsuffix。
- stripLeadingSlash 和 stripExtension 都为 false，其他默认，这时候对应的逻辑视图名称是/product/index.html。
- 都采用默认配置时，返回的逻辑视图名称应该是 product/index。

如果逻辑视图名称跟请求路径相同或者相关关系都是一样的，那么我们就可以采用 Spring 为我们事先约定好的逻辑视图名称返回，这可以大大简化我们的开发工作，而以上功能实现的关键属性 viewNameTranslator，则是在 initRequestToViewNameTranslator 中完成。

```
private void initRequestToViewNameTranslator(ApplicationContext context) {
    try {
        this.viewNameTranslator =
                context.getBean(REQUEST_TO_VIEW_NAME_TRANSLATOR_BEAN_NAME,
RequestToViewNameTranslator.class);
        if (logger.isDebugEnabled()) {
            logger.debug("Using RequestToViewNameTranslator [" + this.viewNameTranslator + "]");
        }
    }
    catch (NoSuchBeanDefinitionException ex) {
        // We need to use the default.
        this.viewNameTranslator = getDefaultStrategy(context, RequestToViewNameTranslator.class);
        if (logger.isDebugEnabled()) {
            logger.debug("Unable to locate RequestToViewNameTranslator with name '" +
                    REQUEST_TO_VIEW_NAME_TRANSLATOR_BEAN_NAME + "': using default
[" + this.viewNameTranslator +
                    "]");
        }
    }
}
```

8. 初始化 ViewResolvers。

在 SpringMVC 中，当 Controller 将请求处理结果放入到 ModelAndView 中以后，DispatcherServlet 会根据 ModelAndView 选择合适的视图进行渲染。那么在 SpringMVC 中是如何选择合适的 View 呢？View 对象是是如何创建的呢？答案就在 ViewResolver 中。ViewResolver

接口定义了 resolverViewName 方法，根据 viewName 创建合适类型的 View 实现。

那么如何配置 ViewResolver 呢？在 Spring 中，ViewResolver 作为 Spring Bean 存在，可以在 Spring 配置文件中进行配置，例如下面的代码，配置了 JSP 相关的 viewResolver。

```
<bean class="org.Springframework.web.servlet.view.InternalResourceViewResolver">
    <property name="prefix" value="/WEB-INF/views/"/>
    <property name="suffix" value=".jsp"/>
</bean>
```

viewResolvers 属性的初始化工作在 initViewResolvers 中完成。

```
private void initViewResolvers(ApplicationContext context) {
    this.viewResolvers = null;
    if (this.detectAllViewResolvers) {
        // Find all ViewResolvers in the ApplicationContext, including ancestor contexts.
        Map<String, ViewResolver> matchingBeans =
                BeanFactoryUtils.beansOfTypeIncludingAncestors(context, ViewResolver.class, true, false);
        if (!matchingBeans.isEmpty()) {
            this.viewResolvers = new ArrayList<ViewResolver>(matchingBeans.values());
            // We keep ViewResolvers in sorted order.
            OrderComparator.sort(this.viewResolvers);
        }
    }
    else {
        try {
            ViewResolver vr = context.getBean(VIEW_RESOLVER_BEAN_NAME, ViewResolver.class);
            this.viewResolvers = Collections.singletonList(vr);
        }
        catch (NoSuchBeanDefinitionException ex) {
            // Ignore, we'll add a default ViewResolver later.
        }
    }

    // Ensure we have at least one ViewResolver, by registering
    // a default ViewResolver if no other resolvers are found.
    if (this.viewResolvers == null) {
        this.viewResolvers = getDefaultStrategies(context, ViewResolver.class);
        if (logger.isDebugEnabled()) {
            logger.debug("No ViewResolvers found in servlet '" + getServletName() +
"': using default");
        }
    }
}
```

9. 初始化 FlashMapManager。

SpringMVC Flash attributes 提供了一个请求存储属性，可供其他请求使用。在使用重定向时候非常必要，例如 Post/Redirect/Get 模式。Flash attributes 在重定向之前暂存（就像存在 session 中）以便重定向之后还能使用，并立即删除。

SpringMVC 有两个主要的抽象来支持 flash attributes。FlashMap 用于保持 flash attributes，

而 FlashMapManager 用于存储、检索、管理 FlashMap 实例。

flash attribute 支持默认开启（"on"）并不需要显式启用，它永远不会导致 HTTP Session 的创建。这两个 FlashMap 实例都可以通过静态方法 RequestContextUtils 从 Spring MVC 的任何位置访问。

flashMapManager 的初始化在 initFlashMapManager 中完成。

```java
private void initFlashMapManager(ApplicationContext context) {
    try {
        this.flashMapManager =
                context.getBean(FLASH_MAP_MANAGER_BEAN_NAME, FlashMapManager.class);
        if (logger.isDebugEnabled()) {
            logger.debug("Using FlashMapManager [" + this.flashMapManager + "]");
        }
    }
    catch (NoSuchBeanDefinitionException ex) {
        // We need to use the default.
        this.flashMapManager = getDefaultStrategy(context, FlashMapManager.class);
        if (logger.isDebugEnabled()) {
            logger.debug("Unable to locate FlashMapManager with name '" +
                    FLASH_MAP_MANAGER_BEAN_NAME + "': using default [" + this.flashMapManager + "]");
        }
    }
}
```

11.4 DispatcherServlet 的逻辑处理

根据之前的示例，我们知道在 HttpServlet 类中分别提供了相应的服务方法，它们是 doDelete()、doGet()、doOptions()、doPost()、doPut()和 doTrace()，它会根据请求的不同形式将程序引导至对应的函数进行处理。这几个函数中最常用的函数无非就是 doGet()和 doPost()，那么我们就直接查看 DispatcherServlet 中对于这两个函数的逻辑实现。

```java
@Override
protected final void doGet(HttpServletRequest request, HttpServletResponse response)
        throws ServletException, IOException {

    processRequest(request, response);
}

@Override
protected final void doPost(HttpServletRequest request, HttpServletResponse response)
        throws ServletException, IOException {

    processRequest(request, response);
}
```

对于不同的方法，Spring 并没有做特殊处理，而是统一将程序再一次地引导至 processRequest(request, response)中。

```java
protected final void processRequest(HttpServletRequest request, HttpServletResponse response)
        throws ServletException, IOException {
```

//记录当前时间,用于计算 web 请求的处理时间
        long startTime = System.currentTimeMillis();
        Throwable failureCause = null;

        // Expose current LocaleResolver and request as LocaleContext.
        LocaleContext previousLocaleContext = LocaleContextHolder.getLocaleContext();
        LocaleContextHolder.setLocaleContext(buildLocaleContext(request), this.thread
ContextInheritable);

        // Expose current RequestAttributes to current thread.
        RequestAttributes previousRequestAttributes = RequestContextHolder. GetRequest
Attributes();
        ServletRequestAttributes requestAttributes = null;
        if (previousRequestAttributes == null || previousRequestAttributes. getClass().
equals(ServletRequestAttributes.class)) {
            requestAttributes = new ServletRequestAttributes(request);
            RequestContextHolder.setRequestAttributes(requestAttributes, this.thread
ContextInheritable);
        }

        if (logger.isTraceEnabled()) {
            logger.trace("Bound request context to thread: " + request);
        }

        try {
            doService(request, response);
        }
        catch (ServletException ex) {
            failureCause = ex;
            throw ex;
        }
        catch (IOException ex) {
            failureCause = ex;
            throw ex;
        }
        catch (Throwable ex) {
            failureCause = ex;
            throw new NestedServletException("Request processing failed", ex);
        }

        finally {
            // Clear request attributes and reset thread-bound context.
            LocaleContextHolder.setLocaleContext(previousLocaleContext,this.threadContext
Inheritable);
            if (requestAttributes != null) {
                RequestContextHolder.setRequestAttributes(previousRequestAttributes,
this.threadContextInheritable);
                requestAttributes.requestCompleted();
            }
            if (logger.isTraceEnabled()) {
                logger.trace("Cleared thread-bound request context: " + request);
            }

```java
            if (logger.isDebugEnabled()) {
                if (failureCause != null) {
                    this.logger.debug("Could not complete request", failureCause);
                }
                else {
                    this.logger.debug("Successfully completed request");
                }
            }
            if (this.publishEvents) {
                // Whether or not we succeeded, publish an event.
                long processingTime = System.currentTimeMillis() - startTime;
                this.webApplicationContext.publishEvent(
                        new ServletRequestHandledEvent(this,
                                request.getRequestURI(), request.getRemoteAddr(),
                                request.getMethod(), getServletConfig().getServletName(),
                                WebUtils.getSessionId(request), getUsernameForRequest(request),
                                processingTime, failureCause));
            }
        }
    }
```

函数中已经开始了对请求的处理，虽然把细节转移到了 doService 函数中实现，但是我们不难看出处理请求前后所做的准备与处理工作。

1. 为了保证当前线程的 LocaleContext 以及 RequestAttributes 可以在当前请求后还能恢复，提取当前线程的两个属性。
2. 根据当前 request 创建对应的 LocaleContext 和 RequestAttributes，并绑定到当前线程。
3. 委托给 doService 方法进一步处理。
4. 请求处理结束后恢复线程到原始状态。
5. 请求处理结束后无论成功与否发布事件通知。

继续查看 doService 方法。

```java
protected void doService(HttpServletRequest request, HttpServletResponse response) throws Exception {
            if (logger.isDebugEnabled()) {
                String requestUri = urlPathHelper.getRequestUri(request);
                logger.debug("DispatcherServlet with name '" + getServletName() + "' processing " + request.getMethod() +
                        " request for [" + requestUri + "]");
            }

            // Keep a snapshot of the request attributes in case of an include,
            // to be able to restore the original attributes after the include.
            Map<String, Object> attributesSnapshot = null;
            if (WebUtils.isIncludeRequest(request)) {
                logger.debug("Taking snapshot of request attributes before include");
                attributesSnapshot = new HashMap<String, Object>();
                Enumeration<?> attrNames = request.getAttributeNames();
                while (attrNames.hasMoreElements()) {
                    String attrName = (String) attrNames.nextElement();
                    if (this.cleanupAfterInclude || attrName.startsWith ("org.Springframework.web.servlet")) {
```

```
                    attributesSnapshot.put(attrName, request.getAttribute (attrName));
                }
            }
        }

        // Make framework objects available to handlers and view objects.
        request.setAttribute(WEB_APPLICATION_CONTEXT_ATTRIBUTE, getWebApplicationContext());
        request.setAttribute(LOCALE_RESOLVER_ATTRIBUTE, this.localeResolver);
        request.setAttribute(THEME_RESOLVER_ATTRIBUTE, this.themeResolver);
        request.setAttribute(THEME_SOURCE_ATTRIBUTE, getThemeSource());

        FlashMap inputFlashMap = this.flashMapManager.retrieveAndUpdate(request, response);
        if (inputFlashMap != null) {
            request.setAttribute(INPUT_FLASH_MAP_ATTRIBUTE, Collections.unmodifiableMap(inputFlashMap));
        }
        request.setAttribute(OUTPUT_FLASH_MAP_ATTRIBUTE, new FlashMap());
        request.setAttribute(FLASH_MAP_MANAGER_ATTRIBUTE, this.flashMapManager);

        try {
            doDispatch(request, response);
        }
        finally {
            // Restore the original attribute snapshot, in case of an include.
            if (attributesSnapshot != null) {
                restoreAttributesAfterInclude(request, attributesSnapshot);
            }
        }
    }
```

我们猜想对请求处理至少应该包括一些诸如寻找 Handler 并页面跳转之类的逻辑处理，但是，在 doService 中我们并没有看到想看到的逻辑，相反却同样是一些准备工作，但是这些准备工作却是必不可少的。Spring 将已经初始化的功能辅助工具变量，比如 localeResolver、themeResolver 等设置在 request 属性中，而这些属性会在接下来的处理中派上用场。

经过层层的准备工作，终于在 doDispatch 函数中看到了完整的请求处理过程。

```
protected void doDispatch(HttpServletRequest request, HttpServletResponse response) throws Exception {
    HttpServletRequest processedRequest = request;
    HandlerExecutionChain mappedHandler = null;
    int interceptorIndex = -1;

    try {
        ModelAndView mv;
        boolean errorView = false;

        try {
//如果是 MultipartContent 类型的 request 则转换 request 为 MultipartHttpServletRequest 类型的 request
            processedRequest = checkMultipart(request);

//根据 request 信息寻找对应的 Handler
            mappedHandler = getHandler(processedRequest, false);
```

11.4 DispatcherServlet 的逻辑处理

```
            if (mappedHandler == null || mappedHandler.getHandler() == null) {
                //如果没有找到对应的 handler 则通过 response 反馈错误信息
                noHandlerFound(processedRequest, response);
                return;
            }

            //根据当前的 handler 寻找对应的 HandlerAdapter
            HandlerAdapter ha = getHandlerAdapter(mappedHandler.getHandler());

            //如果当前 handler 支持 last-modified 头处理
            String method = request.getMethod();
            boolean isGet = "GET".equals(method);
            if (isGet || "HEAD".equals(method)) {
                long lastModified = ha.getLastModified(request, mappedHandler.getHandler());
                if (logger.isDebugEnabled()) {
                    String requestUri = urlPathHelper.getRequestUri(request);
                    logger.debug("Last-Modified value for [" + requestUri + "] is : " + lastModified);
                }
                if (new ServletWebRequest(request, response).checkNotModified(lastModified) && isGet) {
                    return;
                }
            }

            //拦截器的 preHandler 方法的调用
            HandlerInterceptor[] interceptors = mappedHandler.getInterceptors();
            if (interceptors != null) {
                for (int i = 0; i < interceptors.length; i++) {
                    HandlerInterceptor interceptor = interceptors[i];
                    if (!interceptor.preHandle(processedRequest, response, mappedHandler.getHandler())) {
                        triggerAfterCompletion(mappedHandler, interceptorIndex, processedRequest, response, null);
                        return;
                    }
                    interceptorIndex = i;
                }
            }

            //真正的激活 handler 并返回视图
            mv = ha.handle(processedRequest, response, mappedHandler.getHandler());

            //视图名称转换应用于需要添加前缀后缀的情况
            if (mv != null && !mv.hasView()) {
                mv.setViewName(getDefaultViewName(request));
            }

            //应用所有拦截器的 postHandle 方法
            if (interceptors != null) {
                for (int i = interceptors.length - 1; i >= 0; i--) {
                    HandlerInterceptor interceptor = interceptors[i];
                    interceptor.postHandle(processedRequest, response, mappedHandler.getHandler(), mv);
```

```java
                    }
                }
            }
            catch (ModelAndViewDefiningException ex) {
                logger.debug("ModelAndViewDefiningException encountered", ex);
                mv = ex.getModelAndView();
            }
            catch (Exception ex) {
                Object handler = (mappedHandler != null ? mappedHandler.getHandler() : null);
                mv = processHandlerException(processedRequest, response, handler, ex);
                errorView = (mv != null);
            }

            // Did the handler return a view to render?
            //如果在Handler实例的处理中返回了view，那么需要做页面的处理
            if (mv != null && !mv.wasCleared()) {
                //处理页面跳转
                render(mv, processedRequest, response);
                if (errorView) {
                    WebUtils.clearErrorRequestAttributes(request);
                }
            }
            else {
                if (logger.isDebugEnabled()) {
                    logger.debug("Null ModelAndView returned to DispatcherServlet with name '" + getServletName() +
                            "': assuming HandlerAdapter completed request handling");
                }
            }

            //完成处理激活触发器
            triggerAfterCompletion(mappedHandler, interceptorIndex, processedRequest, response, null);
        }
        catch (Exception ex) {
            // Trigger after-completion for thrown exception.
            triggerAfterCompletion(mappedHandler, interceptorIndex, processedRequest, response, ex);
            throw ex;
        }
        catch (Error err) {
            ServletException ex = new NestedServletException("Handler processing failed", err);
            // Trigger after-completion for thrown exception.
            triggerAfterCompletion(mappedHandler, interceptorIndex, processedRequest, response, ex);
            throw ex;
        }
        finally {
            // Clean up any resources used by a multipart request.
            if (processedRequest != request) {
```

```
            cleanupMultipart(processedRequest);
        }
    }
}
```

doDispatch 函数中展示了 Spring 请求处理所涉及的主要逻辑，而我们之前设置在 request 中的各种辅助属性也都有被派上了用场。下面回顾一下逻辑处理的全过程。

11.4.1　MultipartContent 类型的 request 处理

对于请求的处理，Spring 首先考虑的是对于 Multipart 的处理，如果是 MultipartContent 类型的 request，则转换 request 为 MultipartHttpServletRequest 类型的 request。

```
protected HttpServletRequest checkMultipart(HttpServletRequest request) throws MultipartException
{
        if (this.multipartResolver != null && this.multipartResolver.isMultipart(request)) {
            if (request instanceof MultipartHttpServletRequest) {
                logger.debug("Request is already a MultipartHttpServletRequest - if not in a forward, " +
                        "this typically results from an additional MultipartFilter in web.xml");
            }
            else {
                return this.multipartResolver.resolveMultipart(request);
            }
        }
        // If not returned before: return original request.
        return request;
}
```

11.4.2　根据 request 信息寻找对应的 Handler

在 Spring 中最简单的映射处理器配置如下：

```
<bean id="simpleUrlMapping"
        class="org.Springframework.web.servlet.handler.SimpleUrlHandlerMapping">
    <property name="mappings">
        <props>
            <prop key="/userlist.htm">userController</prop>
        </props>
    </property>
</bean>
```

在 Spring 加载的过程中，Spring 会将类型为 SimpleUrlHandlerMapping 的实例加载到 this.handlerMappings 中，按照常理推断，根据 request 提取对应的 Handler，无非就是提取当前实例中的 userController，但是 userController 为继承自 AbstractController 类型实例，与 Handler ExecutionChain 并无任何关联，那么这一步是如何封装的呢？

```
protected HandlerExecutionChain getHandler(HttpServletRequest request, boolean cache) throws Exception {
        return getHandler(request);
}
```

```
protected HandlerExecutionChain getHandler(HttpServletRequest request) throws Exception {
    for (HandlerMapping hm : this.handlerMappings) {
        if (logger.isTraceEnabled()) {
            logger.trace(
                "Testing handler map [" + hm + "] in DispatcherServlet with name '" + getServletName() + "'");
        }
        HandlerExecutionChain handler = hm.getHandler(request);
        if (handler != null) {
            return handler;
        }
    }
    return null;
}
```

在之前的内容我们提过，在系统启动时 Spring 会将所有的映射类型的 bean 注册到 this.handlerMappings 变量中，所以此函数的目的就是遍历所有的 HandlerMapping，并调用其 getHandler 方法进行封装处理。以 SimpleUrlHandlerMapping 为例查看其 getHandler 方法如下：

```
public final HandlerExecutionChain getHandler(HttpServletRequest request) throws Exception {
    //根据 request 获取对应的 handler
    Object handler = getHandlerInternal(request);
    if (handler == null) {
        //如果没有对应 request 的 handler 则使用默认的 handler
        handler = getDefaultHandler();
    }
    //如果也没有提供默认的 handler 则无法继续处理返回 null
    if (handler == null) {
        return null;
    }
    // Bean name or resolved handler?
    if (handler instanceof String) {
        String handlerName = (String) handler;
        handler = getApplicationContext().getBean(handlerName);
    }
    return getHandlerExecutionChain(handler, request);
}
```

函数中首先会使用 getHandlerInternal 方法根据 request 信息获取对应的 Handler，如果以 SimpleUrlHandlerMapping 为例分析，那么我们推断此步骤提供的功能很可能就是根据 URL 找到匹配的 Controller 并返回，当然如果没有找到对应的 Controller 处理器那么程序会尝试去查找配置中的默认处理器，当然，当查找的 controller 为 String 类型时，那就意味着返回的是配置的 bean 名称，需要根据 bean 名称查找对应的 bean，最后，还要通过 getHandlerExecutionChain 方法对返回的 Handler 进行封装，以保证满足返回类型的匹配。下面详细分析这个过程。

1. 根据 request 查找对应的 Handler

首先从根据 request 查找对应的 Handler 开始分析。

```
protected Object getHandlerInternal(HttpServletRequest request) throws Exception {
    //截取用于匹配的 url 有效路径
    String lookupPath = getUrlPathHelper().getLookupPathForRequest(request);
    //根据路径寻找 Handler
```

11.4　DispatcherServlet 的逻辑处理

```java
        Object handler = lookupHandler(lookupPath, request);
        if (handler == null) {
            Object rawHandler = null;
            if ("/".equals(lookupPath)) {
                //如果请求的路径仅仅是"/"，那么使用 RootHandler 进行处理
                rawHandler = getRootHandler();
            }
            if (rawHandler == null) {
                //无法找到 handler 则使用默认 handler
                rawHandler = getDefaultHandler();
            }
            if (rawHandler != null) {
                    //根据 beanName 获取对应的 bean
                if (rawHandler instanceof String) {
                    String handlerName = (String) rawHandler;
                    rawHandler = getApplicationContext().getBean(handlerName);
                }
                //模版方法
                validateHandler(rawHandler, request);
                handler = buildPathExposingHandler(rawHandler, lookupPath, lookupPath, null);
            }
        }
        if (handler != null && logger.isDebugEnabled()) {
            logger.debug("Mapping [" + lookupPath + "] to " + handler);
        }
        else if (handler == null && logger.isTraceEnabled()) {
            logger.trace("No handler mapping found for [" + lookupPath + "]");
        }
        return handler;
    }


protected Object lookupHandler(String urlPath, HttpServletRequest request) throws Exception {
    //直接匹配情况的处理
        Object handler = this.handlerMap.get(urlPath);
        if (handler != null) {
            // Bean name
            if (handler instanceof String) {
                String handlerName = (String) handler;
                handler = getApplicationContext().getBean(handlerName);
            }
            validateHandler(handler, request);
            return buildPathExposingHandler(handler, urlPath, urlPath, null);
        }
        // 通配符匹配的处理
        List<String> matchingPatterns = new ArrayList<String>();
        for (String registeredPattern : this.handlerMap.keySet()) {
            if (getPathMatcher().match(registeredPattern, urlPath)) {
                matchingPatterns.add(registeredPattern);
            }
        }
        String bestPatternMatch = null;
        Comparator<String> patternComparator = getPathMatcher().getPatternComparator(urlPath);
        if (!matchingPatterns.isEmpty()) {
```

```java
                    Collections.sort(matchingPatterns, patternComparator);
                    if (logger.isDebugEnabled()) {
                        logger.debug("Matching patterns for request [" + urlPath + "] are " +
matchingPatterns);
                    }
                    bestPatternMatch = matchingPatterns.get(0);
                }
                if (bestPatternMatch != null) {
                    handler = this.handlerMap.get(bestPatternMatch);
                    // Bean name or resolved handler?
                    if (handler instanceof String) {
                        String handlerName = (String) handler;
                        handler = getApplicationContext().getBean(handlerName);
                    }
                    validateHandler(handler, request);
                    String pathWithinMapping = getPathMatcher().extractPathWithinPattern
(bestPatternMatch, urlPath);

                    // There might be multiple 'best patterns', let's make sure we have the
correct URI template variables
                    // for all of them
                    Map<String, String> uriTemplateVariables = new LinkedHashMap<String, String>();
                    for (String matchingPattern : matchingPatterns) {
                        if (patternComparator.compare(bestPatternMatch, matchingPattern) == 0) {
                            uriTemplateVariables
                                    .putAll(getPathMatcher().extractUriTemplateVariables
(matchingPattern, urlPath));
                        }
                    }
                    if (logger.isDebugEnabled()) {
                        logger.debug("URI Template variables for request [" + urlPath + "] 
are " + uriTemplateVariables);
                    }
                    return buildPathExposingHandler(handler, bestPatternMatch, pathWithinMapping
, uriTemplateVariables);
                }
                // No handler found...
                return null;
        }
```

根据 URL 获取对应 Handler 的匹配规则代码实现起来虽然很长，但是并不难理解，考虑了直接匹配与通配符两种情况。其中要提及的是 buildPathExposingHandler 函数，它将 Handler 封装成了 HandlerExecutionChain 类型。

```java
        protected Object buildPathExposingHandler(Object rawHandler, String bestMatchingPattern,
                String pathWithinMapping, Map<String, String> uriTemplateVariables) {

            HandlerExecutionChain chain = new HandlerExecutionChain(rawHandler);
            chain.addInterceptor(new PathExposingHandlerInterceptor(bestMatchingPattern, 
pathWithinMapping));
            if (!CollectionUtils.isEmpty(uriTemplateVariables)) {
                chain.addInterceptor(new UriTemplateVariablesHandlerInterceptor
(uri TemplateVariables));
            }
```

```
            return chain;
    }
```

在函数中我们看到了通过将 Handler 以参数形式传入，并构建 HandlerExecutionChain 类型实例，加入了两个拦截器。此时我们似乎已经了解了 Spring 这样大费周折的目的。链处理机制，是 Spring 中非常常用的处理方式，是 AOP 中的重要组成部分，可以方便地对目标对象进行扩展及拦截，这是非常优秀的设计。

2．加入拦截器到执行链

getHandlerExecutionChain 函数最主要的目的是将配置中的对应拦截器加入到执行链中，以保证这些拦截器可以有效地作用于目标对象。

```
protected HandlerExecutionChain getHandlerExecutionChain(Object handler, HttpServletRequest request) {
        HandlerExecutionChain chain =
            (handler instanceof HandlerExecutionChain) ?
                (HandlerExecutionChain) handler : new HandlerExecutionChain(handler);

        chain.addInterceptors(getAdaptedInterceptors());

        String lookupPath = urlPathHelper.getLookupPathForRequest(request);
        for (MappedInterceptor mappedInterceptor : mappedInterceptors) {
            if (mappedInterceptor.matches(lookupPath, pathMatcher)) {
                chain.addInterceptor(mappedInterceptor.getInterceptor());
            }
        }

        return chain;
    }
```

11.4.3　没找到对应的 Handler 的错误处理

每个请求都应该对应着一 Handler，因为每个请求都会在后台有相应的逻辑对应，而逻辑的实现就是在 Handler 中，所以一旦遇到没有找到 Handler 的情况（正常情况下如果没有 URL 匹配的 Handler，开发人员可以设置默认的 Handler 来处理请求，但是如果默认请求也未设置就会出现 Handler 为空的情况），就只能通过 response 向用户返回错误信息。

```
    protected void noHandlerFound(HttpServletRequest request, HttpServletResponse response) throws Exception {
        if (pageNotFoundLogger.isWarnEnabled()) {
            String requestUri = urlPathHelper.getRequestUri(request);
            pageNotFoundLogger.warn("No mapping found for HTTP request with URI [" + requestUri +
                "] in DispatcherServlet with name '" + getServletName() + "'");
        }
        response.sendError(HttpServletResponse.SC_NOT_FOUND);
    }
```

11.4.4 根据当前 Handler 寻找对应的 HandlerAdapter

在 WebApplicationContext 的初始化过程中我们讨论了 HandlerAdapters 的初始化，了解了在默认情况下普通的 Web 请求会交给 SimpleControllerHandlerAdapter 去处理。下面我们以 SimpleControllerHandlerAdapter 为例来分析获取适配器的逻辑。

```
protected HandlerAdapter getHandlerAdapter(Object handler) throws ServletException {
    for (HandlerAdapter ha : this.handlerAdapters) {
        if (logger.isTraceEnabled()) {
            logger.trace("Testing handler adapter [" + ha + "]");
        }
        if (ha.supports(handler)) {
            return ha;
        }
    }
    throw new ServletException("No adapter for handler [" + handler +
        "]: Does your handler implement a supported interface like Controller?");
}
```

通过上面的函数我们了解到，对于获取适配器的逻辑无非就是遍历所有适配器来选择合适的适配器并返回它，而某个适配器是否适用于当前的 Handler 逻辑被封装在具体的适配器中。进一步查看 SimpleControllerHandlerAdapter 中的 supports 方法。

```
public boolean supports(Object handler) {
    return (handler instanceof Controller);
}
```

分析到这里，一切已经明了，SimpleControllerHandlerAdapter 就是用于处理普通的 Web 请求的，而且对于 SpringMVC 来说，我们会把逻辑封装至 Controller 的子类中，例如我们之前的引导示例 UserController 就是继承自 AbstractController，而 AbstractController 实现 Controller 接口。

11.4.5 缓存处理

在研究 Spring 对缓存处理的功能支持前，我们先了解一个概念：Last-Modified 缓存机制。

1. 在客户端第一次输入 URL 时，服务器端会返回内容和状态码 200，表示请求成功，同时会添加一个 "Last-Modified" 的响应头，表示此文件在服务器上的最后更新时间，例如，"Last-Modified:Wed, 14 Mar 2012 10:22:42 GMT" 表示最后更新时间为（2012-03-14 10:22）。

2. 客户端第二次请求此 URL 时，客户端会向服务器发送请求头 "If-Modified-Since"，询问服务器该时间之后当前请求内容是否有被修改过，如 "If-Modified-Since: Wed, 14 Mar 2012 10:22:42 GMT"，如果服务器端的内容没有变化，则自动返回 HTTP 304 状态码（只要响应头，内容为空，这样就节省了网络带宽）。

Spring 提供的对 Last-Modified 机制的支持，只需要实现 LastModified 接口，如下所示：

```
public class HelloWorldLastModifiedCacheController extends AbstractController implemen
ts LastModified {
```

```java
        private long lastModified;
        protected ModelAndView handleRequestInternal(HttpServletRequest req, Http
ServletResponse resp) throws Exception {
            //点击后再次请求当前页面
            resp.getWriter().write("<a href=''>this</a>");
            return null;
        }
        public long getLastModified(HttpServletRequest request) {
            if(lastModified == 0L) {
                //第一次或者逻辑有变化的时候，应该重新返回内容最新修改的时间戳
                lastModified = System.currentTimeMillis();
            }
            return lastModified;
        }
    }
```

HelloWorldLastModifiedCacheController 只需要实现 LastModified 接口的 getLastModified 方法，保证当内容发生改变时返回最新的修改时间即可。

Spring 判断是否过期，通过判断请求的 "If-Modified-Since" 是否大于等于当前的 getLastModified 方法的时间戳，如果是，则认为没有修改。上面的 controller 与普通的 controller 并无太大差别，声明如下：

```xml
<bean name="/helloLastModified" class="com.test.controller.HelloWorldLastModifiedCacheController"/>
```

11.4.6　HandlerInterceptor 的处理

Servlet API 定义的 servlet 过滤器可以在 servlet 处理每个 Web 请求的前后分别对它进行前置处理和后置处理。此外，有些时候，你可能只想处理由某些 SpringMVC 处理程序处理的 Web 请求，并在这些处理程序返回的模型属性被传递到视图之前，对它们进行一些操作。

SpringMVC 允许你通过处理拦截 Web 请求，进行前置处理和后置处理。处理拦截是在 Spring 的 Web 应用程序上下文中配置的，因此它们可以利用各种容器特性，并引用容器中声明的任何 bean。处理拦截是针对特殊的处理程序映射进行注册的，因此它只拦截通过这些处理程序映射的请求。每个处理拦截都必须实现 HandlerInterceptor 接口，它包含三个需要你实现的回调方法：preHandle()、postHandle() 和 afterCompletion()。第一个和第二个方法分别是在处理程序处理请求之前和之后被调用的。第二个方法还允许访问返回的 ModelAndView 对象，因此可以在它里面操作模型属性。最后一个方法是在所有请求处理完成之后被调用的（如视图呈现之后），以下是 HandlerInterceptor 的简单实现：

```java
public class MyTestInterceptor implements HandlerInterceptor{
    public boolean preHandle(HttpServletRequest request,
        HttpServletResponse response,Object handler)throws Exception{{
        long startTime = System.currentTimeMillis();
        request.setAttribute("startTime",startTime);
        return true;
    }
    public void postHandle(HttpServletRequest request,HttpServletResponse response,
```

```
        Object handler,ModelAndView modelAndView)throws Exception{
        long startTime = (Long)request.getAttribute("startTime");
        request.removeAttribute("startTime");
        long endTime = System.currentTimeMillis();
        modelAndView.addObject("handlingTime",endTime-startTime);
    }
    public void afterCompletion(HttpServletRequest request,
        HttpServletResponse response,Object handler,Exception ex)throws Exception{

    }
}
```

在这个拦截器的 preHandler()方法中，你记录了起始时间，并将它保存到请求属性中。这个方法应该返回 true，允许 DispatcherServlet 继续处理请求。否则，DispatcherServlet 会认为这个方法已经处理了请求，直接将响应返回给用户。然后，在 postHandler()方法中，从请求属性中加载起始时间，并将它与当前时间进行比较。你可以计算总的持续时间，然后把这个时间添加到模型中，传递给视图。最后，afterCompletion()方法无事可做，空着就可以了。

11.4.7 逻辑处理

对于逻辑处理其实是通过适配器中转调用 Handler 并返回视图的，对应代码如下：

```
mv = ha.handle(processedRequest, response, mappedHandler.getHandler());
```

同样，还是以引导示例为基础进行处理逻辑分析，之前分析过，对于普通的 Web 请求，Spring 默认使用 SimpleControllerHandlerAdapter 类进行处理，我们进入 SimpleControllerHandlerAdapter 类的 handle 方法如下：

```
    public ModelAndView handle(HttpServletRequest request, HttpServletResponse response,
Object handler)
            throws Exception {

        return ((Controller) handler).handleRequest(request, response);
    }
```

但是回顾引导示例中的 UserController，我们的逻辑是写在 handleRequestInternal 函数中而不是 handleRequest 函数，所以我们还需要进一步分析这期间所包含的处理流程。

```
    public ModelAndView handleRequest(HttpServletRequest request, HttpServletResponse response)
            throws Exception {

        // Delegate to WebContentGenerator for checking and preparing.
        checkAndPrepare(request, response, this instanceof LastModified);

        //如果需要session 内的同步执行
        if (this.synchronizeOnSession) {
            HttpSession session = request.getSession(false);
            if (session != null) {
                Object mutex = WebUtils.getSessionMutex(session);
                synchronized (mutex) {
                    //调用用户的逻辑
                    return handleRequestInternal(request, response);
                }
```

```
        }
    }
    //调用用户逻辑
    return handleRequestInternal(request, response);
}
```

11.4.8 异常视图的处理

有时候系统运行过程中出现异常,而我们并不希望就此中断对用户的服务,而是至少告知客户当前系统在处理逻辑的过程中出现了异常,甚至告知他们因为什么原因导致的。Spring 中的异常处理机制会帮我们完成这个工作。其实,这里 Spring 主要的工作就是将逻辑引导至 HandlerExceptionResolver 类的 resolveException 方法,而 HandlerExceptionResolver 的使用,我们在讲解 WebApplicationContext 的初始化的时候已经介绍过了。

```
protected ModelAndView processHandlerException(HttpServletRequest request, HttpServletResponse response,
            Object handler, Exception ex) throws Exception {

        // Check registered HandlerExceptionResolvers...
        ModelAndView exMv = null;
        for (HandlerExceptionResolver handlerExceptionResolver : this.handlerExceptionResolvers) {
            exMv = handlerExceptionResolver.resolveException(request, response, handler, ex);
            if (exMv != null) {
                break;
            }
        }
        if (exMv != null) {
            if (exMv.isEmpty()) {
                return null;
            }
            // We might still need view name translation for a plain error model...
            if (!exMv.hasView()) {
                exMv.setViewName(getDefaultViewName(request));
            }
            if (logger.isDebugEnabled()) {
                logger.debug("Handler execution resulted in exception - forwarding to resolved error view: " + exMv, ex);
            }
            WebUtils.exposeErrorRequestAttributes(request, ex, getServletName());
            return exMv;
        }

        throw ex;
    }
```

11.4.9 根据视图跳转页面

无论是一个系统还是一个站点,最重要的工作都是与用户进行交互,用户操作系统后无论下发的命令成功与否都需要给用户一个反馈,以便于用户进行下一步的判断。所以,在逻辑处理的最后一定会涉及一个页面跳转的问题。

```java
protected void render(ModelAndView mv, HttpServletRequest request, HttpServletResponse response) throws Exception {
    // Determine locale for request and apply it to the response.
    Locale locale = this.localeResolver.resolveLocale(request);
    response.setLocale(locale);

    View view;
    if (mv.isReference()) {
        // We need to resolve the view name.
        view = resolveViewName(mv.getViewName(), mv.getModelInternal(), locale, request);
        if (view == null) {
            throw new ServletException(
                    "Could not resolve view with name '" + mv.getViewName() + "' in servlet with name '" +
                            getServletName() + "'");
        }
    }
    else {
        // No need to lookup: the ModelAndView object contains the actual View object.
        view = mv.getView();
        if (view == null) {
            throw new ServletException("ModelAndView [" + mv + "] neither contains a view name nor a " +
                    "View object in servlet with name '" + getServletName() + "'");
        }
    }

    // Delegate to the View object for rendering.
    if (logger.isDebugEnabled()) {
        logger.debug("Rendering view [" + view + "] in DispatcherServlet with name '" + getServletName() + "'");
    }
    view.render(mv.getModelInternal(), request, response);
}
```

1. 解析视图名称

在上文中我们提到 DispatcherServlet 会根据 ModelAndView 选择合适的视图来进行渲染，而这一功能就是在 resolveViewName 函数中完成的。

```java
protected View resolveViewName(String viewName, Map<String, Object> model, Locale locale,
        HttpServletRequest request) throws Exception {
    for (ViewResolver viewResolver : this.viewResolvers) {
        View view = viewResolver.resolveViewName(viewName, locale);
        if (view != null) {
            return view;
        }
    }
    return null;
}
```

我们以 org.Springframework.web.servlet.view.InternalResourceViewResolver 为例来分析

ViewResolver 逻辑的解析过程,其中 resolveViewName 函数的实现是在其父类 AbstractCaching ViewResolver 中完成的。

```java
public View resolveViewName(String viewName, Locale locale) throws Exception {
    if (!isCache()) {
        //不存在缓存的情况下直接创建视图
        return createView(viewName, locale);
    }
    else {
        //直接从缓存中提取
        Object cacheKey = getCacheKey(viewName, locale);
        synchronized (this.viewCache) {
            View view = this.viewCache.get(cacheKey);
            if (view == null && (!this.cacheUnresolved || !this.viewCache.containsKey(cacheKey))) {
                // Ask the subclass to create the View object.
                view = createView(viewName, locale);
                if (view != null || this.cacheUnresolved) {
                    this.viewCache.put(cacheKey, view);
                    if (logger.isTraceEnabled()) {
                        logger.trace("Cached view [" + cacheKey + "]");
                    }
                }
            }
            return view;
        }
    }
}
```

在父类 UrlBasedViewResolver 中重写了 createView 函数。

```java
protected View createView(String viewName, Locale locale) throws Exception {
    //如果当前解析器不支持当前解析器如 viewName 为空等情况
    if (!canHandle(viewName, locale)) {
        return null;
    }

    //处理前缀为 redirect:xx 的情况
    if (viewName.startsWith(REDIRECT_URL_PREFIX)) {
        String redirectUrl = viewName.substring(REDIRECT_URL_PREFIX.length());
        RedirectView view = new RedirectView(redirectUrl, isRedirectContextRelative(), isRedirectHttp10Compatible());
        return applyLifecycleMethods(viewName, view);
    }

    //处理前缀为 forward:xx 的情况
    if (viewName.startsWith(FORWARD_URL_PREFIX)) {
        String forwardUrl = viewName.substring(FORWARD_URL_PREFIX.length());
        return new InternalResourceView(forwardUrl);
    }
    // Else fall back to superclass implementation: calling loadView.
    return super.createView(viewName, locale);
}

protected View createView(String viewName, Locale locale) throws Exception {
```

```
            return loadView(viewName, locale);
    }

    protected View loadView(String viewName, Locale locale) throws Exception {
        AbstractUrlBasedView view = buildView(viewName);
        View result = applyLifecycleMethods(viewName, view);
        return (view.checkResource(locale) ? result : null);
    }

    protected AbstractUrlBasedView buildView(String viewName) throws Exception {
        AbstractUrlBasedView view = (AbstractUrlBasedView) BeanUtils.instantiateClass
(getViewClass());
        //添加前缀以及后缀
        view.setUrl(getPrefix() + viewName + getSuffix());
        String contentType = getContentType();
        if (contentType != null) {
            //设置ContentType
            view.setContentType(contentType);
        }
        view.setRequestContextAttribute(getRequestContextAttribute());
        view.setAttributesMap(getAttributesMap());
        if (this.exposePathVariables != null) {
            view.setExposePathVariables(exposePathVariables);
        }
        return view;
    }
```

通读以上代码，我们发现对于 InternalResourceViewResolver 所提供的解析功能主要考虑到了几个方面的处理。

- 基于效率的考虑，提供了缓存的支持。
- 提供了对 redirect:xx 和 forward:xx 前缀的支持。
- 添加了前缀及后缀，并向 View 中加入了必需的属性设置。

2. 页面跳转

当通过 viewName 解析到对应的 View 后，就可以进一步地处理跳转逻辑了。

```
    public void render(Map<String, ?> model, HttpServletRequest request, HttpServletResponse 
response) throws Exception {
            if (logger.isTraceEnabled()) {
                logger.trace("Rendering view with name '" + this.beanName + "' with model " + model +
                        " and static attributes " + this.staticAttributes);
            }

            Map<String, Object> mergedModel = createMergedOutputModel(model, request, 
response);
            prepareResponse(request, response);
            renderMergedOutputModel(mergedModel, request, response);
    }
```

在引导示例中，我们了解到对于 ModelView 的使用，可以将一些属性直接放入其中，然后在页面上直接通过 JSTL 语法或者原始的 request 获取。这是一个很方便也很神奇的功能，但是

11.4 DispatcherServlet 的逻辑处理

实现却并不复杂,无非是把我们将要用到的属性放入 request 中,以便在其他地方可以直接调用,而解析这些属性的工作就是在 createMergedOutputModel 函数中完成的。

```java
protected Map<String, Object> createMergedOutputModel(Map<String, ?> model, HttpServletRequest request,
        HttpServletResponse response) {
    @SuppressWarnings("unchecked")
    Map<String, Object> pathVars = this.exposePathVariables ?
        (Map<String, Object>) request.getAttribute(View.PATH_VARIABLES) : null;

    // Consolidate static and dynamic model attributes.
    int size = this.staticAttributes.size();
    size += (model != null) ? model.size() : 0;
    size += (pathVars != null) ? pathVars.size() : 0;
    Map<String, Object> mergedModel = new HashMap<String, Object>(size);
    mergedModel.putAll(this.staticAttributes);
    if (pathVars != null) {
        mergedModel.putAll(pathVars);
    }
    if (model != null) {
        mergedModel.putAll(model);
    }

    // Expose RequestContext?
    if (this.requestContextAttribute != null) {
        mergedModel.put(this.requestContextAttribute, createRequestContext(request,
            response, mergedModel));
    }

    return mergedModel;
}

//处理页面跳转
protected void renderMergedOutputModel(
        Map<String, Object> model, HttpServletRequest request, HttpServletResponse response) throws Exception {

    // Determine which request handle to expose to the RequestDispatcher.
    HttpServletRequest requestToExpose = getRequestToExpose(request);

    //将 model 中的数据以属性的方式设置到 request 中
    exposeModelAsRequestAttributes(model, requestToExpose);

    // Expose helpers as request attributes, if any.
    exposeHelpers(requestToExpose);

    // Determine the path for the request dispatcher.
    String dispatcherPath = prepareForRendering(requestToExpose, response);

    // Obtain a RequestDispatcher for the target resource (typically a JSP).
    RequestDispatcher rd = getRequestDispatcher(requestToExpose, dispatcherPath);
```

```
                if (rd == null) {
                    throw new ServletException("Could not get RequestDispatcher for [" + getUrl() +
                            "]: Check that the corresponding file exists within your web application archive!");
                }

                // If already included or response already committed, perform include, else forward.
                if (useInclude(requestToExpose, response)) {
                    response.setContentType(getContentType());
                    if (logger.isDebugEnabled()) {
                        logger.debug("Including resource [" + getUrl() + "] in Internal ResourceView '" + getBeanName() + "'");
                    }
                    rd.include(requestToExpose, response);
                }

                else {
                    // Note: The forwarded resource is supposed to determine the content type itself.
                    exposeForwardRequestAttributes(requestToExpose);
                    if (logger.isDebugEnabled()) {
                        logger.debug("Forwarding to resource [" + getUrl() + "] in InternalResourceView '" + getBeanName() + "'");
                    }
                    rd.forward(requestToExpose, response);
                }
    }
```

第 12 章 远程服务

Java 远程方法调用，即 Java RMI（Java Remote Method Invocation），是 Java 编程语言里一种用于实现远程过程调用的应用程序编程接口。它使客户机上运行的程序可以调用远程服务器上的对象。远程方法调用特性使 Java 编程人员能够在网络环境中分布操作。RMI 全部的宗旨就是尽可能地简化远程接口对象的使用。

Java RMI 极大地依赖于接口。在需要创建一个远程对象时，程序员通过传递一个接口来隐藏底层的实现细节。客户端得到的远程对象句柄正好与本地的根代码连接，由后者负责透过网络通信。这样一来，程序员只需关心如何通过自己的接口句柄发送消息。

12.1 RMI

在 Spring 中，同样提供了对 RMI 的支持，使得在 Spring 下的 RMI 开发变得更方便，同样，我们还是通过示例来快速体验 RMI 所提供的功能。

12.1.1 使用示例

以下提供了 Spring 整合 RMI 的使用示例。

1. 建立 RMI 对外接口。

```
public interface HelloRMIService {
    public int getAdd(int a, int b);
}
```

2. 建立接口实现类。

```
public class HelloRMIServiceImpl implements HelloRMIService {

    public int getAdd(int a, int b) {
        return a + b;
    }
}
```

3. 建立服务端配置文件。

```xml
<?xml version="1.0" encoding="UTF-8"?>
<beans xmlns="http://www.Springframework.org/schema/beans"
xmlns:xsi="http://www.w3.org/2001/XMLSchema-instance"
    xsi:schemaLocation="
http://www.Springframework.org/schema/beans
http://www.Springframework.org/schema/beans/Spring-beans-3.0.xsd ">
<!--服务端-->
    <bean id="helloRMIServiceImpl" class="test.remote.HelloRMIServiceImpl" />
    <!-- 将类为一个RMI 服务 -->
    <bean id="myRMI" class="org.Springframework.remoting.RMI.RMIServiceExporter">
        <!-- 服务类 -->
        <property name="service" ref="helloRMIServiceImpl" />
        <!-- 服务名 -->
        <property name="serviceName" value="helloRMI" />
        <!-- 服务接口 -->
        <property name="serviceInterface" value="test.remote.HelloRMIService" />
        <!-- 服务端口 -->
        <property name="registryPort" value="9999" />
        <!-- 其他属性自己查看 org.Springframework.remoting.RMI.RMIServiceExporter 的类,就知道支持的属性了-->
    </bean>
</beans>
```

4. 建立服务端测试。

```java
public class ServerTest {

    public static void main(String[] args) {
        new ClassPathXmlApplicationContext("test/remote/RMIServer.xml");
    }
}
```

到这里,建立 RMI 服务端的步骤已经结束了,服务端发布了一个两数相加的对外接口供其他服务器调用。启动服务端测试类,其他机器或端口便可以通过 RMI 来连接到本机了。

5. 完成了服务端的配置后,还需要在测试端建立测试环境以及测试代码。首先建立测试端配置文件。

```xml
<?xml version="1.0" encoding="UTF-8"?>
<beans xmlns="http://www.Springframework.org/schema/beans"
 xmlns:xsi="http://www.w3.org/2001/XMLSchema-instance"
    xsi:schemaLocation="
http://www.Springframework.org/schema/beans
http://www.Springframework.org/schema/beans/Spring-beans-3.0.xsd ">
    <!--客户端-->
    <bean id="myClient" class="org.Springframework.remoting.RMI.RMIProxyFactoryBean">

        <property name="serviceUrl" value="RMI://127.0.0.1:9999/helloRMI"/>
        <property name="serviceInterface" value="test.remote.HelloRMIService"/>
    </bean>
</beans>
```

6. 编写测试代码。

```java
public class ClientTest {

    public static void main(String[] args) {
```

```
                ApplicationContext context = new ClassPathXmlApplicationContext ("test/remote
/ RMIClient.xml");
                HelloRMIService hms = context.getBean("myClient", HelloRMIService.class);
                System.out.println(hms.getAdd(1, 2));
        }
}
```

通过以上的步骤，实现了测试端的代码调用。你会看到测试端通过 RMI 进行了远程连接，连接到了服务端，并使用对应的实现类 HelloRMIServiceImpl 中提供的方法 getAdd 来计算参数并返回结果，你会看到控制台输出了 3。当然以上的测试用例是使用同一台机器不同的端口来模拟不同机器的 RMI 连接。在企业应用中一般都是使用不同的机器来进行 RMI 服务的发布与访问，你需要将接口打包，并放置在服务端的工程中。

这是一个简单的方法展示，但是却很好地展示了 Spring 中使用 RMI 的流程以及步骤，如果抛弃 Spring 而使用原始的 RMI 发布与连接，则会是一件很麻烦的事情，有兴趣的读者可以查阅相关的资料。在 Spring 中使用 RMI 非常简单，Spring 帮助我们做了大量的工作，这些工作都包括什么呢？接下来我们一起深入分析 Spring 中对 RMI 功能的实现原理。

12.1.2　服务端实现

首先我们从服务端的发布功能开始着手，同样，Spring 中的核心还是配置文件，这是所有功能的基础。在服务端的配置文件中我们可以看到，定义了两个 bean，其中一个是对接口实现类的发布，而另一个则是对 RMI 服务的发布，使用 org.Springframework.remoting.RMI.RMIServiceExporter 类进行封装，其中包括了服务类、服务名、服务接口、服务端口等若干属性，因此我们可以断定，org.Springframework.remoting.RMI.RMIServiceExporter 类应该是发布 RMI 的关键类。我们可以从此类入手进行分析。

根据前面展示的示例，启动 Spring 中的 RMI 服务并没有多余的操作，仅仅是开启 Spring 的环境：new ClassPathXmlApplicationContext("test/remote/RMIServer.xml")，仅此一句。于是，我们分析很可能是 RMIServiceExportern 在初始化的时候做了某些操作完成了端口的发布功能，那么这些操作的入口是在这个类的哪个方法里面呢？

进入这个类，首先分析这个类的层次结构，如图 12-1 所示。

图 12-1　RMIServiceExporter 类层次结构图

根据 Eclipse 提供的功能，我们查看到了 RMIServiceExporter 的层次结构图，那么从这个层次图中我们能得到什么信息呢？

RMIServiceExporter 实现了 Spring 中几个比较敏感的接口：BeanClassLoaderAware、DisposableBean、InitializingBean，其中，DisposableBean 接口保证在实现该接口的 bean 销毁时调用其 destroy 方法，BeanClassLoaderAware 接口保证在实现该接口的 bean 的初始化时调用其 setBeanClassLoader 方法，而 InitializingBean 接口则是保证在实现该接口的 bean 初始化时调用其 afterPropertiesSet 方法，所以我们推断 RMIServiceExporter 的初始化函数入口一定在其 afterPropertiesSet 或者 setBeanClassLoader 方法中。经过查看代码，确认 afterPropertiesSet 为 RMIServiceExporter 功能的初始化入口。

```java
public void afterPropertiesSet() throws RemoteException {
    prepare();
}

public void prepare() throws RemoteException {
    //检查验证 service
    checkService();
    if (this.serviceName == null) {
        throw new IllegalArgumentException("Property 'serviceName' is required");
    }
    //如果用户在配置文件中配置了 clientSocketFactory 或者 serverSocketFactory 的处理
    /*
     * 如果配置中的 clientSocketFactory 同时又实现了 RMIServerSocketFactory 接口那么会忽略
     * 配置中的 serverSocketFactory 而使用 clientSocketFactory 代替
     */
    if (this.clientSocketFactory instanceof RMIServerSocketFactory) {
        this.serverSocketFactory = (RMIServerSocketFactory) this.clientSocketFactory;
    }

    //clientSocketFactory 和 serverSocketFactory 要么同时出现要么都不出现
    if ((this.clientSocketFactory != null && this.serverSocketFactory == null) ||
            (this.clientSocketFactory == null && this.serverSocketFactory != null)) {
        throw new IllegalArgumentException(
                "Both RMIClientSocketFactory and RMIServerSocketFactory or none required");
    }

    /*
     * 如果配置中的 registryClientSocketFactory 同时实现了 RMIServerSocketFactory 接口那么
     * 会忽略配置中的 registryServerSocketFactory 而使用 registryClientSocketFactory 代替
     */
    if (this.registryClientSocketFactory instanceof RMIServerSocketFactory) {
        this.registryServerSocketFactory = (RMIServerSocketFactory) this.registryClientSocketFactory;
    }
    //不允许出现只配置 registryServerSocketFactory 却没有配置 registryClientSocketFactory 的
    //情况出现
    if (this.registryClientSocketFactory == null && this.registryServerSocketFactory != null) {
        throw new IllegalArgumentException(
```

12.1 RMI

```
                        "RMIServerSocketFactory without RMIClientSocketFactory for
registry not supported");
            }

            this.createdRegistry = false;

            //确定 RMI registry
            if (this.registry == null) {
                this.registry = getRegistry(this.registryHost, this.registryPort,
                    this.registryClientSocketFactory, this.registryServerSocketFactory);
                this.createdRegistry = true;
            }

            //初始化以及缓存导出的 Object
            //此时通常情况下是使用 RMIInvocationWrapper 封装的 JDK 代理类,切面为 RemoteInvocation
TraceInterceptor
            this.exportedObject = getObjectToExport();

            if (logger.isInfoEnabled()) {
                logger.info("Binding service '" + this.serviceName + "' to RMI registry:
" + this.registry);
            }

            // Export RMI object.
            if (this.clientSocketFactory != null) {
                /*
                 * 使用由给定的套接字工厂指定的传送方式导出远程对象,以便能够接收传入的调用。
                 * clientSocketFactory:进行远程对象调用的客户端套接字工厂
                 * serverSocketFactory:接收远程调用的服务端套接字工厂
                 */
                UnicastRemoteObject.exportObject(
                    this.exportedObject, this.servicePort, this.clientSocketFactory,
this.serverSocketFactory);
            }
            else {
                //导出 remote object,以使它能接收特定端口的调用
                UnicastRemoteObject.exportObject(this.exportedObject, this.servicePort);
            }

            try {
                if (this.replaceExistingBinding) {
                    this.registry.rebind(this.serviceName, this.exportedObject);
                }
                else {
                    //绑定服务名称到 remote object,外界调用 serviceName 的时候会被 exportedObject 接收
                    this.registry.bind(this.serviceName, this.exportedObject);
                }
            }
            catch (AlreadyBoundException ex) {
                unexportObjectSilently();
                throw new IllegalStateException(
                    "Already an RMI object bound for name '" + this.serviceName + "'
: " + ex.toString());
            }
```

```
        catch (RemoteException ex) {
            unexportObjectSilently();
            throw ex;
        }
}
```

果然，在 afterPropertiesSet 函数中将实现委托给了 prepare，而在 prepare 方法中我们找到了 RMI 服务发布的功能实现，同时，我们也大致清楚了 RMI 服务发布的流程。

1. 验证 service。

此处的 service 对应的是配置中类型为 RMIServiceExporter 的 service 属性，它是实现类，并不是接口。尽管后期会对 RMIServiceExporter 做一系列的封装，但是，无论怎么封装，最终还是会将逻辑引向至 RMIServiceExporter 来处理，所以，在发布之前需要进行验证。

2. 处理用户自定义的 SocketFactory 属性。

在 RMIServiceExporter 中提供了 4 个套接字工厂配置，分别是 clientSocketFactory、serverSocketFactory 和 registryClientSocketFactory、registryServerSocketFactory。那么这两对配置又有什么区别或者说分别是应用在什么样的不同场景呢？

registryClientSocketFactory 与 registryServerSocketFactory 用于主机与 RMI 服务器之间连接的创建，也就是当使用 LocateRegistry.createRegistry(registryPort, clientSocketFactory, serverSocketFactory)方法创建 Registry 实例时会在 RMI 主机使用 serverSocketFactory 创建套接字等待连接，而服务端与 RMI 主机通信时会使用 clientSocketFactory 创建连接套接字。

clientSocketFactory、serverSocketFactory 同样是创建套接字，但是使用的位置不同，clientSocketFactory、serverSocketFactory 用于导出远程对象，serverSocketFactory 用于在服务端建立套接字等待客户端连接，而 clientSocketFactory 用于调用端建立套接字发起连接。

3. 根据配置参数获取 Registry。
4. 构造对外发布的实例。

构建对外发布的实例，当外界通过注册的服务名调用响应的方法时，RMI 服务会将请求引入此类来处理。

5. 发布实例。

在发布 RMI 服务的流程中，有几个步骤可能是我们比较关心的。

1. 获取 registry

对 RMI 稍有了解就会知道，由于底层的封装，获取 Registry 实例是非常简单的，只需要使用一个函数 LocateRegistry.createRegistry(...)创建 Registry 实例就可以了。但是，Spring 中并没有这么做，而是考虑得更多，比如 RMI 注册主机与发布的服务并不在一台机器上，那么需要使用 LocateRegistry.getRegistry(registryHost, registryPort, clientSocketFactory)去远程获取 Registry 实例。

```
protected Registry getRegistry(String registryHost, int registryPort,
            RMIClientSocketFactory clientSocketFactory, RMIServerSocketFactory serverSocketFactory)
                throws RemoteException {
```

```
                if (registryHost != null) {
                    //远程连接测试
                    if (logger.isInfoEnabled()) {
                        logger.info("Looking for RMI registry at port '" + registryPort + "'
of host [" + registryHost + "]");
                    }
                    //如果registryHost不为空则尝试获取对应主机的Registry
                    Registry reg = LocateRegistry.getRegistry(registryHost, registryPort,
clientSocketFactory);
                    testRegistry(reg);
                    return reg;
                }else {
                    //获取本机的Registry
                    return getRegistry(registryPort, clientSocketFactory, serverSocketFactory);
                }
            }
```

如果并不是从另外的服务器上获取Registry连接，那么就需要在本地创建RMI的Registry实例了。当然，这里有一个关键的参数alwaysCreateRegistry，如果此参数配置为true，那么在获取Registry实例时会首先测试是否已经建立了对指定端口的连接，如果已经建立则复用已经创建的实例，否则重新创建。

当然，之前也提到过，创建Registry实例时可以使用自定义的连接工厂，而之前的判断也保证了clientSocketFactory与serverSocketFactory要么同时出现，要么同时不出现，所以这里只对clientSocketFactory是否为空进行了判断。

```
        protected Registry getRegistry(
                int registryPort, RMIClientSocketFactory clientSocketFactory, RMIServerSocketFactory
serverSocketFactory)
                throws RemoteException {

            if (clientSocketFactory != null) {
                if (this.alwaysCreateRegistry) {
                    logger.info("Creating new RMI registry");
                    //使用clientSocketFactory创建Registry
                    return LocateRegistry.createRegistry(registryPort, clientSocketFactory,
serverSocketFactory);
                }
                if (logger.isInfoEnabled()) {
                    logger.info("Looking for RMI registry at port '" + registryPort + "',
using custom socket factory");
                }
                synchronized (LocateRegistry.class) {
                    try {
                        //复用测试
                        Registry reg = LocateRegistry.getRegistry(null, registryPort,
clientSocketFactory);
                        testRegistry(reg);
                        return reg;
                    }
                    catch (RemoteException ex) {
                        logger.debug("RMI registry access threw exception", ex);
```

```
                    logger.info("Could not detect RMI registry - creating new one");
                    return LocateRegistry.createRegistry(registryPort,clientSocketFactory,
serverSocketFactory);
                }
            }
        }else {
            return getRegistry(registryPort);
        }
    }
```

如果创建 Registry 实例时不需要使用自定义的套接字工厂，那么就可以直接使用 LocateRegistry.createRegistry(...)方法来创建了，当然复用的检测还是必要的。

```
protected Registry getRegistry(int registryPort) throws RemoteException {
    if (this.alwaysCreateRegistry) {
        logger.info("Creating new RMI registry");
        return LocateRegistry.createRegistry(registryPort);
    }
    if (logger.isInfoEnabled()) {
        logger.info("Looking for RMI registry at port '" + registryPort + "'");
    }
    synchronized (LocateRegistry.class) {
        try {
            //查看对应当前 registryPort 的 Registry 是否已经创建，如果创建直接使用
            Registry reg = LocateRegistry.getRegistry(registryPort);
            //测试是否可用，如果不可用则抛出异常
            testRegistry(reg);
            return reg;
        }
        catch (RemoteException ex) {
            logger.debug("RMI registry access threw exception", ex);
            logger.info("Could not detect RMI registry - creating new one");
            //根据端口创建 Registry
            return LocateRegistry.createRegistry(registryPort);
        }
    }
}
```

2．初始化将要导出的实体对象

之前有提到过，当请求某个 RMI 服务的时候，RMI 会根据注册的服务名称，将请求引导至远程对象处理类中，这个处理类便是使用 getObjectToExport()进行创建。

```
protected Remote getObjectToExport() {
    //如果配置的 service 属性对应的类实现了 Remote 接口且没有配置 serviceInterface 属性
    if (getService() instanceof Remote &&
            (getServiceInterface() == null || Remote.class.isAssignableFrom (getServiceInterface()))) {
        return (Remote) getService();
    }
    else {
        if (logger.isDebugEnabled()) {
            logger.debug("RMI service [" + getService() + "] is an RMI invoker");
        }
        //对 service 进行封装
```

```
            return new RMIInvocationWrapper(getProxyForService(), this);
    }
}
```

请求处理类的初始化主要处理规则为：如果配置的 service 属性对应的类实现了 Remote 接口且没有配置 serviceInterface 属性，那么直接使用 service 作为处理类；否则，使用 RMIInvocationWrapper 对 service 的代理类和当前类也就是 RMIServiceExporter 进行封装。

经过这样的封装，客户端与服务端便可以达成一致协议，当客户端检测到是 RMIInvocationWrapper 类型 stub 的时候便会直接调用其 invoke 方法，使得调用端与服务端很好地连接在了一起。而 RMIInvocationWrapper 封装了用于处理请求的代理类，在 invoke 中便会使用代理类进行进一步处理。

之前的逻辑已经非常清楚了，当请求 RMI 服务时会由注册表 Registry 实例将请求转向之前注册的处理类去处理，也就是之前封装的 RMIInvocationWrapper，然后由 RMIInvocationWrapper 中的 invoke 方法进行处理，那么为什么不是在 invoke 方法中直接使用 service，而是通过代理再次将 service 封装呢？

这其中的一个关键点是，在创建代理时添加了一个增强拦截器 RemoteInvocationTraceInterceptor，目的是为了对方法调用进行打印跟踪，但是如果直接在 invoke 方法中硬编码这些日志，会使代码看起来很不优雅，而且耦合度很高，使用代理的方式就会解决这样的问题，而且会有很高的可扩展性。

```
protected Object getProxyForService() {
        //验证 service
        checkService();
        //验证 serviceInterface
        checkServiceInterface();

        //使用 JDK 的方式创建代理
        ProxyFactory proxyFactory = new ProxyFactory();
        //添加代理接口
        proxyFactory.addInterface(getServiceInterface());

        if (this.registerTraceInterceptor != null ?
                this.registerTraceInterceptor.booleanValue() : this.interceptors == null) {
            //加入代理的横切面 RemoteInvocationTraceInterceptor 并记录 Exporter 名称
            proxyFactory.addAdvice(new RemoteInvocationTraceInterceptor(getExporterName()));
        }

        if (this.interceptors != null) {
            AdvisorAdapterRegistry adapterRegistry = GlobalAdvisorAdapterRegistry.getInstance();
            for (int i = 0; i < this.interceptors.length; i++) {
                proxyFactory.addAdvisor(adapterRegistry.wrap(this.interceptors[i]));
            }
        }
        //设置要代理的目标类
        proxyFactory.setTarget(getService());
        proxyFactory.setOpaque(true);

        //创建代理
```

```
            return proxyFactory.getProxy(getBeanClassLoader());
}
```

3. RMI 服务激活调用

之前反复提到过，由于在之前 bean 初始化的时候做了服务名称绑定 this.registry.bind
(this.serviceName, this.exportedObject)，其中的 exportedObject 其实是被 RMIInvocationWrapper
进行过封装的，也就是说当其他服务器调用 serviceName 的 RMI 服务时，Java 会为我们封装其
内部操作，而直接会将代码转向 RMIInvocationWrapper 的 invoke 方法中。

```
public Object invoke(RemoteInvocation invocation)
        throws RemoteException, NoSuchMethodException, IllegalAccessException,
InvocationTargetException {

        return this.RMIExporter.invoke(invocation, this.wrappedObject);
}
```

而此时 this.RMIExporter 为之前初始化的 RMIServiceExporter，invocation 为包含着需要激
活的方法参数，而 wrappedObject 则是之前封装的代理类。

```
protected Object invoke(RemoteInvocation invocation, Object targetObject)
        throws NoSuchMethodException, IllegalAccessException, InvocationTargetException {

        return super.invoke(invocation, targetObject);
}

protected Object invoke(RemoteInvocation invocation, Object targetObject)
        throws NoSuchMethodException, IllegalAccessException, InvocationTargetException {

        if (logger.isTraceEnabled()) {
            logger.trace("Executing " + invocation);
        }
        try {
            return getRemoteInvocationExecutor().invoke(invocation, targetObject);
        }
        catch (NoSuchMethodException ex) {
            if (logger.isDebugEnabled()) {
                logger.warn("Could not find target method for " + invocation, ex);
            }
            throw ex;
        }
        catch (IllegalAccessException ex) {
            if (logger.isDebugEnabled()) {
                logger.warn("Could not access target method for " + invocation, ex);
            }
            throw ex;
        }
        catch (InvocationTargetException ex) {
            if (logger.isDebugEnabled()) {
                logger.debug("Target method failed for " + invocation, ex.getTargetException());
            }
            throw ex;
        }
}
```

```
public Object invoke(RemoteInvocation invocation, Object targetObject)
        throws NoSuchMethodException, IllegalAccessException, InvocationTargetException{
    
    Assert.notNull(invocation, "RemoteInvocation must not be null");
    Assert.notNull(targetObject, "Target object must not be null");
    //通过反射方式激活方法
    return invocation.invoke(targetObject);
}

public Object invoke(Object targetObject)
        throws NoSuchMethodException, IllegalAccessException, InvocationTargetException {
    //根据方法名称获取代理中对应的方法
    Method method = targetObject.getClass().getMethod(this.methodName, this.parameterTypes);
    //执行代理中的方法
    return method.invoke(targetObject, this.arguments);
}
```

12.1.3 客户端实现

根据客户端配置文件,锁定入口类为 RMIProxyFactoryBean,同样根据类的层次结构查找入口函数,如图 12-2 所示。

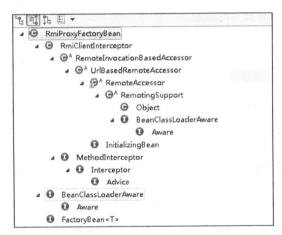

图 12-2 RMIProxyFactoryBean 类的层次结构图

根据层次关系以及之前的分析,我们提取出该类实现的比较重要的接口 InitializingBean、BeanClassLoaderAware 以及 MethodInterceptor。

其中实现了 InitializingBean,则 Spring 会确保在此初始化 bean 时调用 afterPropertiesSet 进行逻辑的初始化。

```
public void afterPropertiesSet() {
    super.afterPropertiesSet();
    if (getServiceInterface() == null) {
```

```
            throw new IllegalArgumentException("Property 'serviceInterface' is required");
        }
        //根据设置的接口创建代理，并使用当前类this作为增强器
        this.serviceProxy = new ProxyFactory(getServiceInterface(), this).getProxy
(getBeanClassLoader());
    }
```

同时，RMIProxyFactoryBean 又实现了 FactoryBean 接口，那么当获取 bean 时并不是直接获取 bean，而是获取该 bean 的 getObject 方法。

```
public Object getObject() {
    return this.serviceProxy;
}
```

这样，我们似乎已近形成了一个大致的轮廓，当获取该 bean 时，首先通过 afterPropertiesSet 创建代理类，并使用当前类作为增强方法，而在调用该 bean 时其实返回的是代理类，既然调用的是代理类，那么又会使用当前 bean 作为增强器进行增强，也就是说会调用 RMIProxyFactoryBean 的父类 RMIClientInterceptor 的 invoke 方法。

我们先从 afterPropertiesSet 中的 super.afterPropertiesSet()方法开始分析。

```
public void afterPropertiesSet() {
    super.afterPropertiesSet();
    prepare();
}
```

继续追踪代码，发现父类的父类，也就是 UrlBasedRemoteAccessor 中的 afterPropertiesSet 方法只完成了对 serviceUrl 属性的验证。

```
public void afterPropertiesSet() {
    if (getServiceUrl() == null) {
        throw new IllegalArgumentException("Property 'serviceUrl' is required");
    }
}
```

所以推断所有的客户端都应该在 prepare 方法中实现，继续查看 prepare()。

1．通过代理拦截并获取 stub

在父类的 afterPropertiesSet 方法中完成了对 serviceUrl 的验证，那么 prepare 函数又完成了什么功能呢？

```
public void prepare() throws RemoteLookupFailureException {
    // Cache RMI stub on initialization?
    //如果配置了lookupStubOnStartup属性便会在启动时寻找stub
    if (this.lookupStubOnStartup) {
        Remote remoteObj = lookupStub();
        if (logger.isDebugEnabled()) {
            if (remoteObj instanceof RMIInvocationHandler) {
                logger.debug("RMI stub [" + getServiceUrl() + "] is an RMI invoker");
            }
            else if (getServiceInterface() != null) {
                boolean isImpl = getServiceInterface().isInstance(remoteObj);
                logger.debug("Using service interface [" + getServiceInterface().
getName() +
                    "] for RMI stub [" + getServiceUrl() + "] - " +
                    (!isImpl ? "not " : "") + "directly implemented");
```

```
            }
        }
        if (this.cacheStub) {
            //将获取的 stub 缓存
            this.cachedStub = remoteObj;
        }
    }
}
```

从上面的代码中,我们了解到了一个很重要的属性 lookupStubOnStartup,如果将此属性设置为 true,那么获取 stub 的工作就会在系统启动时被执行并缓存,从而提高使用时候的响应时间。

获取 stub 是 RMI 应用中的关键步骤,当然你可以使用两种方式进行。

- 使用自定义的套接字工厂。如果使用这种方式,你需要在构造 Registry 实例时将自定义套接字工厂传入,并使用 Registry 中提供的 lookup 方法来获取对应的 stub。
- 直接使用 RMI 提供的标准方法:Naming.lookup(getServiceUrl())。

```
protected Remote lookupStub() throws RemoteLookupFailureException {
    try {
        Remote stub = null;
        if (this.registryClientSocketFactory != null) {

            URL url = new URL(null, getServiceUrl(), new DummyURLStreamHandler());
            String protocol = url.getProtocol();
            //验证传输协议
            if (protocol != null && !"RMI".equals(protocol)) {
                throw new MalformedURLException("Invalid URL scheme '" + protocol + "'");
            }
            //主机
            String host = url.getHost();
            //端口
            int port = url.getPort();
            //服务名
            String name = url.getPath();
            if (name != null && name.startsWith("/")) {
                name = name.substring(1);
            }
            Registry registry = LocateRegistry.getRegistry(host, port, this.registryClientSocketFactory);
            stub = registry.lookup(name);
        }
        else {
            // Can proceed with standard RMI lookup API...
            stub = Naming.lookup(getServiceUrl());
        }
        if (logger.isDebugEnabled()) {
            logger.debug("Located RMI stub with URL [" + getServiceUrl() + "]");
        }
        return stub;
    }
    catch (MalformedURLException ex) {
        throw new RemoteLookupFailureException("Service URL [" + getServiceUrl() + "] is invalid", ex);
```

```
            }
            catch (NotBoundException ex) {
                throw new RemoteLookupFailureException(
                        "Could not find RMI service [" + getServiceUrl() + "] in RMI registry", ex);
            }
            catch (RemoteException ex) {
                throw new RemoteLookupFailureException("Lookup of RMI stub failed", ex);
            }
        }
```

为了使用 registryClientSocketFactory，代码量比使用 RMI 标准获取 stub 方法多出了很多，那么 registryClientSocketFactory 到底是做什么用的呢？

与之前服务端的套接字工厂类似，这里的 registryClientSocketFactory 用来连接 RMI 服务器，用户通过实现 RMIClientSocketFactory 接口来控制用于连接的 socket 的各种参数。

2. 增强器进行远程连接

之前分析了类型为 RMIProxyFactoryBean 的 bean 的初始化中完成的逻辑操作。在初始化时，创建了代理并将本身作为增强器加入了代理中（RMIProxyFactoryBean 间接实现了 MethodInterceptor）。那么这样一来，当在客户端调用代理的接口中的某个方法时，就会首先执行 RMIProxyFactoryBean 中的 invoke 方法进行增强。

```
        public Object invoke(MethodInvocation invocation) throws Throwable {
                    //获取的服务器中对应的注册的 remote 对象，通过序列化传输
                    Remote stub = getStub();
                    try {
                        return doInvoke(invocation, stub);
                    }
                    catch (RemoteConnectFailureException ex) {
                        return handleRemoteConnectFailure(invocation, ex);
                    }
                    catch (RemoteException ex) {
                        if (isConnectFailure(ex)) {
                            return handleRemoteConnectFailure(invocation, ex);
                        }
                        else {
                            throw ex;
                        }
                    }
        }
```

众所周知，当客户端使用接口进行方法调用时是通过 RMI 获取 stub 的，然后再通过 stub 中封装的信息进行服务器的调用，这个 stub 就是在构建服务器时发布的对象，那么，客户端调用时最关键的一步也是进行 stub 的获取了。

```
        protected Remote getStub() throws RemoteLookupFailureException {
            if (!this.cacheStub || (this.lookupStubOnStartup && !this.refreshStubOnConnectFailure)) {
                    //如果有缓存直接使用缓存
                return (this.cachedStub != null ? this.cachedStub : lookupStub());
            }
            else {
```

```
            synchronized (this.stubMonitor) {
                if (this.cachedStub == null) {
                    //获取 stub
                    this.cachedStub = lookupStub();
                }
                return this.cachedStub;
            }
        }
    }
```

当获取到 stub 后便可以进行远程方法的调用了。Spring 中对于远程方法的调用其实是分两种情况考虑的。

- 获取的 stub 是 RMIInvocationHandler 类型的，从服务端获取的 stub 是 RMIInvocationHandler，就意味着服务端也同样使用了 Spring 去构建，那么自然会使用 Spring 中作的约定，进行客户端调用处理。Spring 中的处理方式被委托给了 doInvoke 方法。
- 当获取的 stub 不是 RMIInvocationHandler 类型，那么服务端构建 RMI 服务可能是通过普通的方法或者借助于 Spring 外的第三方插件，那么处理方式自然会按照 RMI 中普通的处理方式进行，而这种普通的处理方式无非是反射。因为在 invocation 中包含了所需要调用的方法的各种信息，包括方法名称以及参数等，而调用的实体正是 stub，那么通过反射方法完全可以激活 stub 中的远程调用。

```
protected Object doInvoke(MethodInvocation invocation, Remote stub) throws Throwable {
    //stub 从服务器传回且经过 Spring 的封装
    if (stub instanceof RMIInvocationHandler) {
        try {
            return doInvoke(invocation, (RMIInvocationHandler) stub);
        }
        catch (RemoteException ex) {
            throw RMIClientInterceptorUtils.convertRMIAccessException(
                invocation.getMethod(), ex, isConnectFailure(ex), getServiceUrl());
        }
        catch (InvocationTargetException ex) {
            Throwable exToThrow = ex.getTargetException();
            RemoteInvocationUtils.fillInClientStackTraceIfPossible(exToThrow);
            throw exToThrow;
        }
        catch (Throwable ex) {
            throw new RemoteInvocationFailureException("Invocation of method [" + invocation.getMethod() +
                "] failed in RMI service [" + getServiceUrl() + "]", ex);
        }
    }
    else {
        try {
            //直接使用反射方法继续激活
            return RMIClientInterceptorUtils.invokeRemoteMethod(invocation, stub);
        }
        catch (InvocationTargetException ex) {
            Throwable targetEx = ex.getTargetException();
            if (targetEx instanceof RemoteException) {
                RemoteException rex = (RemoteException) targetEx;
                throw RMIClientInterceptorUtils.convertRMIAccessException(
```

```
                                    invocation.getMethod(), rex, isConnectFailure(rex),
getServiceUrl());
                    }
                    else {
                        throw targetEx;
                    }
                }
            }
        }
    }
```

之前反复提到了 Spring 中的客户端处理 RMI 的方式。其实，在分析服务端发布 RMI 的方式时，我们已经了解到，Spring 将 RMI 的导出 Object 封装成了 RMIInvocationHandler 类型进行发布，那么当客户端获取 stub 的时候是包含了远程连接信息代理类的 RMIInvocationHandler，也就是说当调用 RMIInvocationHandler 中的方法时会使用 RMI 中提供的代理进行远程连接，而此时，Spring 中要做的就是将代码引向 RMIInvocationHandler 接口的 invoke 方法的调用。

```
    protected Object doInvoke(MethodInvocation methodInvocation, RMIInvocationHandler invocationHandler)
            throws RemoteException, NoSuchMethodException, IllegalAccessException, InvocationTargetException {

        if (AopUtils.isToStringMethod(methodInvocation.getMethod())) {
            return "RMI invoker proxy for service URL [" + getServiceUrl() + "]";
        }
        //将 methodInvocation 中的方法名及参数等信息重新封装到 RemoteInvocation，并通过远程代理方法直接调用
        return invocationHandler.invoke(createRemoteInvocation(methodInvocation));
    }
```

12.2 HttpInvoker

Spring 开发小组意识到在 RMI 服务和基于 HTTP 的服务（如 Hessian 和 Burlap）之间的空白。一方面，RMI 使用 Java 标准的对象序列化，但很难穿越防火墙；另一方面，Hessian/Burlap 能很好地穿过防火墙工作，但使用自己私有的一套对象序列化机制。

就这样，Spring 的 HttpInvoker 应运而生。HttpInvoker 是一个新的远程调用模型，作为 Spring 框架的一部分，来执行基于 HTTP 的远程调用（让防火墙高兴的事），并使用 Java 的序列化机制（这是让程序员高兴的事）。

我们首先看看 HttpInvoker 的使用示例。HttpInvoker 是基于 HTTP 的远程调用，同时也是使用 Spring 中提供的 web 服务作为基础，所以我们的测试需要首先搭建 Web 工程。

12.2.1 使用示例

1. 创建对外接口

```
public interface HttpInvokerTestI {
    public String getTestPo(String desp);
}
```

2. 创建接口实现类

```java
public class HttpInvokertestImpl implements HttpInvokerTestI {

    @Override
    public String getTestPo(String desp) {
        return "getTestPo " + desp;
    }
}
```

3. 创建服务端配置文件 applicationContext-server.xml

```xml
<?xml version="1.0" encoding="UTF-8"?>
<beans xmlns="http://www.Springframework.org/schema/beans"
    xmlns:xsi="http://www.w3.org/2001/XMLSchema-instance"
    xsi:schemaLocation="
http://www.Springframework.org/schema/beans
http://www.Springframework.org/schema/beans/Spring-beans-3.0.xsd ">

    <bean name="httpinvokertest" class="test.HttpInvokertestImpl" />
</beans>
```

4. 在 WEB-INF 下创建 remote-servlet.xml

```xml
<?xml version="1.0" encoding="UTF-8"?>
<beans xmlns="http://www.Springframework.org/schema/beans"
    xmlns:xsi="http://www.w3.org/2001/XMLSchema-instance"
    xsi:schemaLocation="
http://www.Springframework.org/schema/beans
http://www.Springframework.org/schema/beans/Spring-beans-3.0.xsd ">

    <bean name="/hit"
        class="org.Springframework.remoting.httpinvoker.HttpInvokerServiceExporter">

        <property name="service" ref="httpinvokertest" />
        <property name="serviceInterface" value="test.HttpInvokerTestI" />
    </bean>

</beans>
```

至此，服务端的 httpInvoker 服务已经搭建完了，启动 Web 工程后就可以使用我们搭建的 HttpInvoker 服务了。以上代码实现了将远程传入的字符串参数处理加入 "getTestPo" 前缀的功能。服务端搭建完基于 Web 服务的 HttpInvoker 后，客户端还需要使用一定的配置才能进行远程调用。

5. 创建测试端配置 client.xml

```xml
<?xml version="1.0" encoding="UTF-8"?>
<beans xmlns="http://www.Springframework.org/schema/beans"
    xmlns:xsi="http://www.w3.org/2001/XMLSchema-instance"
    xsi:schemaLocation="
http://www.Springframework.org/schema/beans
http://www.Springframework.org/schema/beans/Spring-beans-3.0.xsd ">

    <bean id="remoteService"
        class="org.Springframework.remoting.httpinvoker.HttpInvokerProxyFactoryBean">
```

```
            <property name="serviceUrl" value="http://localhost:8080/httpinvokertest/
remoting/hit" />
            <property name="serviceInterface" value="test.HttpInvokerTestI" />
    </bean>
</beans>
```

6．创建测试类

```
public class Test {
    public static void main(String[] args) {
        ApplicationContext context = new ClassPathXmlApplicationContext ("classpath
: client.xml");
        HttpInvokerTestI httpInvokerTestI = (HttpInvokerTestI) context.getBean
("remoteService");
        System.out.println(httpInvokerTestI.getTestPo("dddd"));
    }
}
```

运行测试类，你会看到打印结果：

```
getTestPo dddd
```

dddd 是我们传入的参数，而 getTestPo 则是在服务端添加的字符串。当然，上面的服务搭建与测试过程中都是在一台机器上进行的，如果需要在不同机器上进行测试，还需要读者对服务端的相关接口打成 JAR 包并加入到客户端的服务器上。

12.2.2 服务端实现

对于 Spring 中 HttpInvoker 服务的实现，我们还是首先从服务端进行分析。

根据 remote-servlet.xml 中的配置，我们分析入口类应该为 org.Springframework.remoting.httpinvoker.HttpInvokerServiceExporter，那么同样，根据这个类分析其入口函数，如图 12-3 所示。

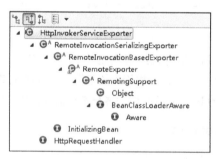

图 12-3　HttpInvokerServiceExporter 类的层次结构图

通过层次关系我们看到 HttpInvokerService Exporter 类实现了 InitializingBean 接口以及 HttpRequestHandler 接口。分析 RMI 服务时我们已经了解到了，当某个 bean 继承自 InitializingBean 接口的时候，Spring 会确保这个 bean 在初始化时调用其 afterPropertiesSet 方法，而对于 HttpRequestHandler 接口，因为我们在配置中已经将此接口配置成 Web 服务，那么当有相应请

12.2 HttpInvoker

求的时候，Spring 的 Web 服务就会将程序引导至 HttpRequestHandler 的 handleRequest 方法中。首先，我们从 afterPropertiesSet 方法开始分析，看看在 bean 的初始化过程中做了哪些逻辑。

1．创建代理

```
public void afterPropertiesSet() {
        prepare();
}

public void prepare() {
      this.proxy = getProxyForService();
}

protected Object getProxyForService() {
        //验证 service
        checkService();
        //验证 serviceInterface
        checkServiceInterface();

        //使用 JDK 的方式创建代理
        ProxyFactory proxyFactory = new ProxyFactory();
        //添加代理接口
        proxyFactory.addInterface(getServiceInterface());

        if (this.registerTraceInterceptor != null ?
              this.registerTraceInterceptor.booleanValue() : this.interceptors == null) {
            //加入代理的横切面 RemoteInvocationTraceInterceptor 并记录 Exporter 名称
            proxyFactory.addAdvice(new RemoteInvocationTraceInterceptor (getExporterName()));
        }

        if (this.interceptors != null) {
            AdvisorAdapterRegistry adapterRegistry = GlobalAdvisorAdapterRegistry.getInstance();
            for (int i = 0; i < this.interceptors.length; i++) {
                proxyFactory.addAdvisor(adapterRegistry.wrap(this.interceptors[i]));
            }
        }
        //设置要代理的目标类
        proxyFactory.setTarget(getService());
        proxyFactory.setOpaque(true);

        //创建代理
        return proxyFactory.getProxy(getBeanClassLoader());
}
```

通过将上面 3 个方法串联，可以看到，初始化过程中实现的逻辑主要是创建了一个代理，代理中封装了对于特定请求的处理方法以及接口等信息，而这个代理的最关键目的是加入了 RemoteInvocationTraceInterceptor 增强器，当然创建代理还有些其他好处，比如代码优雅、方便扩展等。RemoteInvocationTraceInterceptor 中的增强主要是对增强的目标方法进行一些相关信息的日志打印，并没有在此基础上进行任何功能性的增强。那么这个代理究竟是在什么时候使用的呢？暂时留下悬念，我们接下来分析当有 Web 请求时 HttpRequestHandler 的 handleRequest

方法的处理。

2. 处理来自客户端的request

当有 Web 请求时，根据配置中的规则会把路径匹配的访问直接引入对应的 HttpRequest Handler 中。本例中的 Web 请求与普通的 Web 请求是有些区别的，因为此处的请求包含着 HttpInvoker 的处理过程。

```
public void handleRequest(HttpServletRequest request, HttpServletResponse response)
        throws ServletException, IOException {
    try {
        //从request中读取序列化对象
        RemoteInvocation invocation = readRemoteInvocation(request);
        //执行调用
        RemoteInvocationResult result = invokeAndCreateResult(invocation, getProxy());
        //将结果的序列化对象写入输出流
        writeRemoteInvocationResult(request, response, result);
    }
    catch (ClassNotFoundException ex) {
        throw new NestedServletException("Class not found during deserialization", ex);
    }
}
```

在handlerRequest函数中，我们很清楚地看到了HttpInvoker处理的大致框架，HttpInvoker服务简单点说就是将请求的方法，也就是RemoteInvocation对象，从客户端序列化并通过Web请求出入服务端，服务端在对传过来的序列化对象进行反序列化还原RemoteInvocation实例，然后通过实例中的相关信息进行相关方法的调用，并将执行结果再次的返回给客户端。从handleRequest函数中我们也可以清晰地看到程序执行的框架结构。

1. 从request中读取序列化对象。

主要是从 HttpServletRequest 提取相关的信息，也就是提取 HttpServletRequest 中的 RemoteInvocation 对象的序列化信息以及反序列化的过程。

```
protected RemoteInvocation readRemoteInvocation(HttpServletRequest request)
        throws IOException, ClassNotFoundException {

    return readRemoteInvocation(request, request.getInputStream());
}

protected RemoteInvocation readRemoteInvocation(HttpServletRequest request, InputStream is)
        throws IOException, ClassNotFoundException {
    //创建对象输入流
    ObjectInputStream ois = createObjectInputStream(decorateInputStream(request, is));
    try {
        //从输入流中读取序列化对象
        return doReadRemoteInvocation(ois);
    }
    finally {
        ois.close();
    }
}

protected RemoteInvocation doReadRemoteInvocation(ObjectInputStream ois)
```

12.2 HttpInvoker

```
            throws IOException, ClassNotFoundException {

        Object obj = ois.readObject();
        if (!(obj instanceof RemoteInvocation)) {
            throw new RemoteException("Deserialized object needs to be assignable to type [" +
                    RemoteInvocation.class.getName() + "]: " + obj);
        }
        return (RemoteInvocation) obj;
    }
```

对于序列化提取与转换过程其实并没有太多需要解释的东西，这里完全是按照标准的方式进行操作，包括创建 ObjectInputStream 以及从 ObjectInputStream 中提取对象实例。

2. 执行调用。

根据反序列化方式得到的 RemoteInvocation 对象中的信息，进行方法调用。注意，此时调用的实体并不是服务接口或者服务类，而是之前在初始化时候构造的封装了服务接口以及服务类的代理。

完成了 RemoteInvocation 实例的提取，也就意味着可以通过 RemoteInvocation 实例中提供的信息进行方法调用了。

```
protected RemoteInvocationResult invokeAndCreateResult(RemoteInvocation invocation, Object targetObject) {
        try {
            //激活代理类中对应 invocation 中的方法
            Object value = invoke(invocation, targetObject);
            //封装结果以便于序列化
            return new RemoteInvocationResult(value);
        }
        catch (Throwable ex) {
            return new RemoteInvocationResult(ex);
        }
    }
```

这段函数有两点需要说明的地方。

- 对应方法的激活也就是 invoke 方法的调用，虽然经过层层环绕，但是最终还是实现了一个我们熟知的调用 invocation.invoke(targetObject)，也就是执行 RemoteInvocation 类中的 invoke 方法，大致的逻辑还是通过 RemoteInvocation 中对应的方法信息在 targetObject 上去执行，此方法在分析 RMI 功能的时候已经分析过，不再赘述。但是在对于当前方法的 targetObject 参数，此 targetObject 是代理类，调用代理类的时候需要考虑增强方法的调用，这是读者需要注意的地方。
- 对于返回结果需要使用 RemoteInvocationResult 进行封装，之所以需要通过使用 RemoteInvocationResult 类进行封装，是因为无法保证对于所有操作的返回结果都继承 Serializable 接口，也就是说无法保证所有返回结果都可以直接进行序列化，那么，就必须使用 RemoteInvocationResult 类进行统一封装。

3. 将结果的序列化对象写入输出流。

同样这里也包括结果的序列化过程。

```
    protected void writeRemoteInvocationResult(
            HttpServletRequest request, HttpServletResponse response, RemoteInvocation
Result result) throws IOException {

        response.setContentType(getContentType());
        writeRemoteInvocationResult(request, response, result, response.getOutputStream());
    }
    protected void writeRemoteInvocationResult(
            HttpServletRequest request, HttpServletResponse response, RemoteInvocation
Result result, OutputStream os)
            throws IOException {
        //获取输入流
        ObjectOutputStream oos = createObjectOutputStream(decorateOutputStream(request,
response, os));
        try {
            //将序列化对象写入输入流
            doWriteRemoteInvocationResult(result, oos);
        }
        finally {
            oos.close();
        }
    }
    protected void doWriteRemoteInvocationResult(RemoteInvocationResult result, ObjectOutput
Stream oos)throws IOException {

        oos.writeObject(result);
    }
```

12.2.3 客户端实现

分析了服务端的解析以及处理过程后，我们接下来分析客户端的调用过程，在服务端调用的分析中我们反复提到需要从 HttpServletRequest 中提取从客户端传来的 RemoteInvocation 实例，然后进行相应解析。所以，在客户端，一个比较重要的任务就是构建 RemoteInvocation 实例，并传送到服务端。根据配置文件中的信息，我们还是首先锁定 HttpInvokerProxyFactoryBean 类，并查看其层次结构，如图 12-4 所示。

图 12-4　HttpInvokerProxyFactoryBean 类的层次结构图

12.2 HttpInvoker

从层次结构中我们看到，HttpInvokerProxy FactoryBean 类同样实现了 InitializingBean 接口。同时，又实现了 FactoryBean 以及 MethodInterceptor。这已经是老生常谈的问题了，实现这几个接口以及这几个接口在 Spring 中会有什么作用就不再赘述了，我们还是根据实现的 InitializingBean 接口分析初始化过程中的逻辑。

```java
public void afterPropertiesSet() {
        super.afterPropertiesSet();
        if (getServiceInterface() == null) {
            throw new IllegalArgumentException("Property 'serviceInterface' is required");
        }
        //创建代理并使用当前方法为拦截器增强
        this.serviceProxy = new ProxyFactory(getServiceInterface(), this).getProxy(getBeanClassLoader());
    }
```

在 afterPropertiesSet 中主要创建了一个代理，该代理封装了配置的服务接口，并使用当前类也就是 HttpInvokerProxyFactoryBean 作为增强。因为 HttpInvokerProxyFactoryBean 实现了 MethodInterceptor 方法，所以可以作为增强拦截器。

同样，又由于 HttpInvokerProxyFactoryBean 实现了 FactoryBean 接口，所以通过 Spring 中普通方式调用该 bean 时调用的并不是该 bean 本身，而是此类中 getObject 方法返回的实例，也就是实例化过程中所创建的代理。

```java
public Object getObject() {
    return this.serviceProxy;
}
```

那么，综合之前的使用示例，我们再次回顾一下，HttpInvokerProxyFactoryBean 类型 bean 在初始化过程中创建了封装服务接口的代理，并使用自身作为增强拦截器，然后又因为实现了 FactoryBean 接口，所以获取 Bean 的时候返回的其实是创建的代理。那么，汇总上面的逻辑，当调用如下代码时，其实是调用代理类中的服务方法，而在调用代理类中的服务方法时又会使用代理类中加入的增强器进行增强。

```java
ApplicationContext context = new ClassPathXmlApplicationContext("classpath:client.xml");
HttpInvokerTestI httpInvokerTestI = (HttpInvokerTestI) context.getBean("remoteService");
System.out.println(httpInvokerTestI.getTestPo("dddd"));
```

这时，所有的逻辑分析其实已经被转向了对于增强器也就是 HttpInvokerProxyFactoryBean 类本身的 invoke 方法的分析。

在分析 invoke 方法之前，其实我们已经猜出了该方法所提供的主要功能就是将调用信息封装在 RemoteInvocation 中，发送给服务端并等待返回结果。

```java
public Object invoke(MethodInvocation methodInvocation) throws Throwable {
        if (AopUtils.isToStringMethod(methodInvocation.getMethod())) {
            return "HTTP invoker proxy for service URL [" + getServiceUrl() + "]";
        }
        //将要调用的方法封装为 RemoteInvocation
        RemoteInvocation invocation = createRemoteInvocation(methodInvocation);
        RemoteInvocationResult result = null;
        try {
            //远程执行方法
```

```
            result = executeRequest(invocation, methodInvocation);
        }
        catch (Throwable ex) {
            throw convertHttpInvokerAccessException(ex);
        }
        try {
            //提取结果
            return recreateRemoteInvocationResult(result);
        }
        catch (Throwable ex) {
            if (result.hasInvocationTargetException()) {
                throw ex;
            }
            else {
                throw new RemoteInvocationFailureException("Invocation of method [" +
methodInvocation.getMethod() +
                        "] failed in HTTP invoker remote service at [" + getServiceUrl()
+ "]", ex);
            }
        }
    }
```

函数主要有 3 个步骤。

1. 构建 RemoteInvocation 实例。

因为是代理中增强方法的调用，调用的方法及参数信息会在代理中封装至 MethodInvocation 实例中，并在增强器中进行传递，也就意味着当程序进入 invoke 方法时其实是已经包含了调用的接口的相关信息的，那么，首先要做的就是将 MethodInvocation 中的信息提取并构建 RemoteInvocation 实例。

2. 远程执行方法。

3. 提取结果。

考虑到序列化的问题，在 Spring 中约定使用 HttpInvoker 方式进行远程方法调用时，结果使用 RemoteInvocationResult 进行封装，那么在提取结果后还需要从封装的结果中提取对应的结果。

而在这 3 个步骤中最为关键的就是远程方法的执行。执行远程调用的首要步骤就是将调用方法的实例写入输出流中。

```
    protected RemoteInvocationResult executeRequest(
            RemoteInvocation invocation, MethodInvocation originalInvocation) throws Exception {
        return executeRequest(invocation);
    }
    protected RemoteInvocationResult executeRequest(RemoteInvocation invocation) throws Exception {
        return getHttpInvokerRequestExecutor().executeRequest(this, invocation);
    }

    public final RemoteInvocationResult executeRequest(
            HttpInvokerClientConfiguration config, RemoteInvocation invocation) throws
Exception {
        //获取输出流
```

12.2 HttpInvoker

```
            ByteArrayOutputStream baos = getByteArrayOutputStream(invocation);
            if (logger.isDebugEnabled()) {
                logger.debug("Sending HTTP invoker request for service at [" + config.get
ServiceUrl() +
                    "], with size " + baos.size());
            }
            return doExecuteRequest(config, baos);
        }
```

在 doExecuteRequest 方法中真正实现了对远程方法的构造与通信,与远程方法的连接功能实现中,Spring 引入了第三方 JAR:HttpClient。HttpClient 是 Apache Jakarta Common 下的子项目,可以用来提供高效的、最新的、功能丰富的支持 HTTP 协议的客户端编程工具包,并且它支持 HTTP 协议最新的版本和建议。对 HttpClient 的具体使用方法有兴趣的读者可以参考更多的资料和文档。

```
    protected RemoteInvocationResult doExecuteRequest(
            HttpInvokerClientConfiguration config, ByteArrayOutputStream baos)
            throws IOException, ClassNotFoundException {
        //创建 HttpPost
        HttpPost postMethod = createHttpPost(config);
        //设置含有方法的输出流到 post 中
        setRequestBody(config, postMethod, baos);
        try {
            //执行方法并等待结果响应
            HttpResponse response = executeHttpPost(config, getHttpClient(), postMethod);
            //验证
            validateResponse(config, response);
            //提取返回的输入流
            InputStream responseBody = getResponseBody(config, response);
            //从输入流中提取结果
            return readRemoteInvocationResult(responseBody, config.getCodebaseUrl());
        }
        finally {
            if (releaseConnectionMethod != null){
                ReflectionUtils.invokeMethod(releaseConnectionMethod, postMethod);
            }
        }
    }
```

接下来我们逐步分析客户端实现的逻辑。

1. 创建 HttpPost

由于对于服务端方法的调用是通过 Post 方式进行的,那么首先要做的就是构建 HttpPost,构建 HttpPost 过程中可以设置一些必要的参数。

```
    protected PostMethod createPostMethod(HttpInvokerClientConfiguration config) throws
IOException {
            //设置需要访问的 url
            PostMethod postMethod = new PostMethod(config.getServiceUrl());
            LocaleContext locale = LocaleContextHolder.getLocaleContext();
            if (locale != null) {
```

```
            //加入Accept-Language属性
            postMethod.addRequestHeader(HTTP_HEADER_ACCEPT_LANGUAGE, StringUtils.
toLanguageTag(locale.getLocale()));
        }
        if (isAcceptGzipEncoding()) {
            //加入Accept-Encoding属性
            postMethod.addRequestHeader(HTTP_HEADER_ACCEPT_ENCODING, ENCODING_GZIP);
        }
        return postMethod;
    }
```

2. 设置RequestBody

构建好PostMethod实例后便可以将存储RemoteInvocation实例的序列化对象的输出流设置进去，当然这里需要注意的是传入的ContentType类型，一定要传入application/x-java-serialized-object以保证服务端解析时会按照序列化对象的解析方式进行解析。

```
protected void setRequestBody(
        HttpInvokerClientConfiguration config, PostMethod postMethod, ByteArrayOutputStream baos)
            throws IOException {
    //将序列化流加入到postMethod中并声明ContentType类型为application/x-java-serialized-object
    postMethod.setRequestEntity(new ByteArrayRequestEntity(baos.toByteArray(),
getContentType()));
}
```

3. 执行远程方法

通过HttpClient所提供的方法来直接执行远程方法。

```
protected void executePostMethod(
        HttpInvokerClientConfiguration config, HttpClient httpClient, PostMethod
postMethod) throws IOException {
    httpClient.executeMethod(postMethod);
}
```

4. 远程相应验证

对于HTTP调用的响应码处理，大于300则是非正常调用的响应码。

```
protected void validateResponse(HttpInvokerClientConfiguration config, PostMethod postMethod)
        throws IOException {

    if (postMethod.getStatusCode() >= 300) {
        throw new HttpException(
            "Did not receive successful HTTP response: status code = " +
postMethod.getStatusCode() +
            ", status message = [" + postMethod.getStatusText() + "]");
    }
}
```

5. 提取响应信息

从服务器返回的输入流可能是经过压缩的，不同的方式采用不同的办法进行提前。

```
    protected InputStream getResponseBody(HttpInvokerClientConfiguration config, PostMethod
postMethod)
            throws IOException {

        if (isGzipResponse(postMethod)) {
            return new GZIPInputStream(postMethod.getResponseBodyAsStream());
        }
        else {
            return postMethod.getResponseBodyAsStream();
        }
    }
```

6. 提取返回结果

提取结果的流程主要是从输入流中提取响应的序列化信息。

```
    protected RemoteInvocationResult readRemoteInvocationResult(InputStream is, String
codebaseUrl)
            throws IOException, ClassNotFoundException {

        ObjectInputStream ois = createObjectInputStream(decorateInputStream(is),
codebaseUrl);
        try {
            return doReadRemoteInvocationResult(ois);
        }
        finally {
            ois.close();
        }
    }
```

第 13 章 Spring 消息

Java 消息服务（Java Message Service，JMS）应用程序接口是一个 Java 平台中关于面向消息中间件（MOM）的 API，用于在两个应用程序之间或分布式系统中发送消息，并进行异步通信。Java 消息服务是一个与具体平台无关的 API，绝大多数 MOM 提供商都对 JMS 提供支持。

Java 消息服务的规范包括两种消息模式，点对点和发布者/订阅者。许多提供商支持这一通用框架。因此，程序员可以在他们的分布式软件中实现面向消息的操作，这些操作将具有不同面向消息中间件产品的可移植性。

Java 消息服务支持同步和异步的消息处理，在某些场景下，异步消息是必要的，而且比同步消息操作更加便利。

Java 消息服务支持面向事件的方法接收消息，事件驱动的程序设计现在被广泛认为是一种富有成效的程序设计范例，程序员们对其都相当熟悉。

在应用系统开发时，Java 消息服务可以推迟选择面对消息中间件产品，也可以在不同的面对消息中间件之间进行切换。

本章以 Java 消息服务的开源实现产品 ActiveMQ 为例来进行 Spring 整合消息服务功能的实现分析。

13.1 JMS 的独立使用

尽管大多数的 Java 消息服务的使用都会跟 Spring 相结合，但是，我们还是非常有必要了解消息的独立使用方法，这对于我们理解消息的实现原理以及后续与 Spring 整合实现的分析都非常重要。当然在消息服务的使用前，需要先开启消息服务器，如果是 Windows 系统，则可以直接双击 ActiveMQ 安装目录下 bin 目录中的 activemq.bat 文件来启动消息服务器。

消息服务的使用除了要开启消息服务器外，还需要构建消息的发送端与接收端，发送端主要用来将包含业务逻辑的消息发送至消息服务器，而消息接收端则用于将服务器中的消息提取并进行相应的处理。

1. 发送端实现

发送端主要用于发送消息到消息服务器，以下为发送消息测试，尝试发送 3 条消息到消息服务器，消息的内容为"大家好这是个测试"。

```java
public class Sender {
    public static void main(String[] args) throws Exception {
        ConnectionFactory connectionFactory = new ActiveMQConnectionFactory();

        Connection connection = connectionFactory.createConnection();
        //connection.start();

        Session session = connection.createSession(Boolean.TRUE,
            Session.AUTO_ACKNOWLEDGE);
        Destination destination = session.createQueue("my-queue");

        MessageProducer producer = session.createProducer(destination);
        for (int i = 0; i < 3; i++) {
            TextMessage message = session.createTextMessage("大家好这是个测试");
            Thread.sleep(1000);
            // 通过消息生产者发出消息
            producer.send(message);
        }
        session.commit();
        session.close();
        connection.close();
    }
}
```

上面的函数实现很容易让我们联想到数据库的实现，在函数开始时需要一系列冗余的但又必不可少的用于连接的代码，而其中真正用于发送消息的代码其实很简单。

2. 接收端实现

接收端主要用于连接消息服务器并接收服务器上的消息。

```java
public class Receiver {
    public static void main(String[] args) throws Exception {
        ConnectionFactory connectionFactory = new ActiveMQConnectionFactory();

        Connection connection = connectionFactory.createConnection();
        connection.start();

        final Session session = connection.createSession(Boolean.TRUE, Session.AUTO_ACKNOWLEDGE);

        Destination destination = session.createQueue("my-queue");

        MessageConsumer consumer = session.createConsumer(destination);

        int i=0;
        while(i<3) {
            i++;
            TextMessage message = (TextMessage) consumer.receive();
            session.commit();
```

```
            //TODO something....
            System.out.println("收到消息: " + message.getText());
        }

        session.close();
        connection.close();
    }
}
```

程序测试的顺序是,首先开启发送端,然后向服务器发送消息,接着再开启接收端,不出意外,就会接收到发送端发出的消息。

13.2 Spring 整合 ActiveMQ

整个消息的发送与接收过程非常简单,但是其中却掺杂着大量的冗余代码,比如 Connection 的创建与关闭,Session 的创建与关闭等,为了消除这一冗余工作量,Spring 进行了进一步的封装。Spring 下的 ActiveMQ 使用方式如下。

1. Spring 配置文件。

配置文件是 Spring 的核心,Spring 整合消息服务的使用也从文件配置开始。类似于数据库操作,Spring 也将 ActiveMQ 中的操作封装至 JmsTemplate 中,以方便我们统一使用。所以,在 Spring 的核心配置文件中首先要注册 JmsTemplate 类型的 bean。当然,ActiveMQConnection Factory 用于连接消息服务器,是消息服务的基础,也要注册。ActiveMQQueue 则用于指定消息的目的地。

```xml
<beans>
    <bean id="connectionFactory" class="org.apache.activemq.ActiveMQConnectionFactory">
        <property name="brokerURL">
            <value>tcp://localhost:61616</value>
        </property>
    </bean>

    <bean id="jmsTemplate" class="org.Springframework.jms.core.JmsTemplate">
        <property name="connectionFactory">
            <ref bean="connectionFactory" />
        </property>
    </bean>

    <bean id="destination" class="org.apache.activemq.command.ActiveMQQueue">
        <constructor-arg index="0">
            <value>HelloWorldQueue</value>
        </constructor-arg>
    </bean>
</beans>
```

2. 发送端。

有了以上的配置,Spring 就可以根据配置信息简化我们的工作量。Spring 中发送消息到消

13.2 Spring 整合 ActiveMQ

息服务器，省去了冗余的 Connection 以及 Session 等的创建与销毁过程，简化了工作量。

```java
public class HelloWorldSender {
    public static void main(String args[]) throws Exception {
        ApplicationContext context = new ClassPathXmlApplicationContext(
                new String[] { "test/activeMQ/Spring/applicationContext.xml" });

        JmsTemplate jmsTemplate = (JmsTemplate) context.getBean("jmsTemplate");
        Destination destination = (Destination) context.getBean("destination");

        jmsTemplate.send(destination, new MessageCreator() {
            public Message createMessage(Session session) throws JMSException {
                return session.createTextMessage("大家好这个是测试！");
            }
        });
    }
}
```

3. 接收端。

同样，在 Spring 中接收消息也非常方便，Spring 中连接服务器接收消息的示例如下：

```java
public class HelloWorldReciver {

    public static void main(String args[]) throws Exception {
        ApplicationContext context = new ClassPathXmlApplicationContext(
                new String[] { "test/activeMQ/Spring/applicationContext.xml" });
        JmsTemplate jmsTemplate = (JmsTemplate) context.getBean("jmsTemplate");
        Destination destination = (Destination) context.getBean("destination");

        TextMessage msg = (TextMessage) jmsTemplate.receive(destination);
        System.out.println("reviced msg is:" + msg.getText());
    }
}
```

到这里我们已经完成了 Spring 消息的发送与接收操作。但是，如 HelloWorldReciver 中所示的代码，使用 jmsTemplate.receive(destination)方法只能接收一次消息，如果未接收到消息，则会一直等待。当然用户可以通过设置 timeout 属性来控制等待时间，但是一旦接收到消息，本次接收任务就会结束，虽然用户可以通过 while(true)的方式来实现循环监听消息服务器上的消息，但是还有一种更好的解决办法——创建消息监听器。消息监听器的使用方式如下。

1. 创建消息监听器。

用于监听消息，一旦有新消息，Spring 就会将消息引导至消息监听器，以方便用户进行相应的逻辑处理。

```java
public class MyMessageListener implements MessageListener{

    @Override
    public void onMessage(Message arg0) {
        TextMessage msg = (TextMessage) arg0;
        try {
            System.out.println(msg.getText());
        } catch (JMSException e) {
            e.printStackTrace();
        }
```

        }
    }

2. 修改配置文件。

为了使用消息监听器，需要在配置文件中注册消息容器，并将消息监听器注入到容器中。
```
<beans>
    <bean id="connectionFactory" class="org.apache.activemq.ActiveMQConnectionFactory">
        <property name="brokerURL">
            <value>tcp://localhost:61616</value>
        </property>
    </bean>

    <bean id="jmsTemplate" class="org.Springframework.jms.core.JmsTemplate">
        <property name="connectionFactory">
            <ref bean="connectionFactory" />
        </property>
    </bean>

    <bean id="destination" class="org.apache.activemq.command.ActiveMQQueue">
        <constructor-arg index="0">
            <value>HelloWorldQueue</value>
        </constructor-arg>
    </bean>

    <bean id="myTextListener" class="test.activeMQ.Spring.MyMessageListener" />

    <bean id="javaConsumer"
        class="org.Springframework.jms.listener.DefaultMessageListenerContainer">
        <property name="connectionFactory" ref="connectionFactory" />
        <property name="destination" ref="destination" />
        <property name="messageListener" ref="myTextListener" />
    </bean>

</beans>
```

通过以上的修改便可以进行消息监听的功能了，一旦有消息传入至消息服务器，就会被消息监听器监听到，并由 Spring 将消息内容引导至消息监听器的处理函数中，等待用户的进一步逻辑处理。

13.3 源码分析

尽管消息接收可以使用消息监听器的方式替代模板方法，但是在发送阶段模板方法仍然是无法替代的，在 Spring 中必须要使用 JmsTemplate 提供的方法来进行发送操作，可见 JmsTemplate 类的重要性，那么我们对于 Spring 整合消息服务的分析就从 JmsTemplate 开始。

13.3.1 JmsTemplate

在代码与 Spring 整合的实例中，我们看到 Spring 采用了与 JDBC 等一贯的套路，为我们提供了 JmsTemplate 来封装常用操作。查看 JmsTemplate 的类型层级结构图，如图 13-1 所示。

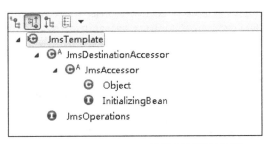

图 13-1　JmsTemplate 的类型层级结构图

首先还是按照一贯的分析套路，提取我们感兴趣的接口 InitializingBean，接口方法实现在 JmsAccessor 类中，如下：

```
public void afterPropertiesSet() {
    if (getConnectionFactory() == null) {
        throw new IllegalArgumentException("Property 'connectionFactory' is required");
    }
}
```

发现函数中只是一个验证的功能，并没有逻辑实现。丢掉这个线索，我们转向实例代码的分析。首先以发送为例，在 Spring 中发送消息可以通过 JmsTemplate 中提供的方法来实现。

```
public void send(final Destination destination, final MessageCreator messageCreator) throws JmsException
```

使用方式如下：

```
jmsTemplate.send(destination, new MessageCreator() {
    public Message createMessage(Session session) throws JMSException {
        return session.createTextMessage("大家好这个是测试！");
    }
});
```

我们就跟着程序流，进入函数 send 查看其源代码：

```
public void send(final Destination destination, final MessageCreator messageCreator) throws JmsException {
    execute(new SessionCallback<Object>() {
        public Object doInJms(Session session) throws JMSException {
            doSend(session, destination, messageCreator);
            return null;
        }
    }, false);
}
```

现在的风格不得不让我们回想起 JdbcTemplate 的类实现风格，它们极为相似，都是提取一个公共的方法作为最底层、最通用的功能实现，然后又通过回调函数的不同来区分个性化的功能。我们首先查看通用代码的抽取实现。

1. 通用代码抽取

根据之前分析 JdbcTemplate 的经验,我们推断,在 execute 中一定是封装了 Connection 以及 Session 的创建操作。

```java
public <T> T execute(SessionCallback<T> action, boolean startConnection) throws JmsException {
    Assert.notNull(action, "Callback object must not be null");
    Connection conToClose = null;
    Session sessionToClose = null;
    try {
        Session sessionToUse = ConnectionFactoryUtils.doGetTransactionalSession(
                getConnectionFactory(), this.transactionalResourceFactory, startConnection);
        if (sessionToUse == null) {
            //创建 connection
            conToClose = createConnection();
            //根据 connection 创建 session
            sessionToClose = createSession(conToClose);
            //是否开启向服务器推送连接信息,只有接收信息时需要,发送时不需要
            if (startConnection) {
                conToClose.start();
            }
            sessionToUse = sessionToClose;
        }
        if (logger.isDebugEnabled()) {
            logger.debug("Executing callback on JMS Session: " + sessionToUse);
        }
        //调用回调函数
        return action.doInJms(sessionToUse);
    }
    catch (JMSException ex) {
        throw convertJmsAccessException(ex);
    }
    finally {
        //关闭 session
        JmsUtils.closeSession(sessionToClose);
        //释放连接
        ConnectionFactoryUtils.releaseConnection(conToClose, getConnectionFactory(), startConnection);
    }
}
```

在单独使用 activeMQ 时,我们知道为了发送一条消息需要做很多工作,它需要很多的辅助代码,而这些代码又都是千篇一律的,没有任何的差异,所以 execute 方法的目的就是帮助我们抽离这些冗余代码,从而使我们更加专注于业务逻辑的实现。从函数中看,这些冗余代码包括创建 Connection、创建 Session,当然也包括关闭 Session 和关闭 Connection。而在准备工作结束后,将调用回调函数将程序引入用户自定义实现的个性化处理。至于如何创建 Session 与 Connection,有兴趣的读者可以进一步研究 Spring 的源码。

2. 发送消息的实现

有了基类辅助的实现，使得 Spring 更加专注于个性的处理，也就是说 Spring 使用 execute 方法封装了冗余代码，而将个性化的代码实现放在了回调函数 doInJms 函数中。在发送消息的功能中通过局部类实现回调函数。

```
new SessionCallback<Object>() {
            public Object doInJms(Session session) throws JMSException {
                doSend(session, destination, messageCreator);
                return null;
            }
}
```

此时的发送逻辑已经完全被转向了 doSend 方法，这样使整个功能实现变得更加清晰。

```
protected void doSend(Session session, Destination destination, MessageCreator messageCreator)
            throws JMSException {

    Assert.notNull(messageCreator, "MessageCreator must not be null");
    MessageProducer producer = createProducer(session, destination);
    try {
        Message message = messageCreator.createMessage(session);
        if (logger.isDebugEnabled()) {
            logger.debug("Sending created message: " + message);
        }
        doSend(producer, message);
        // Check commit - avoid commit call within a JTA transaction.
        if (session.getTransacted() && isSessionLocallyTransacted(session)) {
            // Transacted session created by this template -> commit.
            JmsUtils.commitIfNecessary(session);
        }
    }
    finally {
        JmsUtils.closeMessageProducer(producer);
    }
}
protected void doSend(MessageProducer producer, Message message) throws JMSException {
        if (isExplicitQosEnabled()) {
            producer.send(message, getDeliveryMode(), getPriority(), getTimeToLive());
        }
        else {
            producer.send(message);
        }
}
```

在演示独立使用消息功能的时候，我们大体了解了消息发送的基本套路，虽然这些步骤已经被 Spring 拆得支离破碎，但是我们还是能捕捉到一些影子。发送消息时还是遵循着消息发送的规则，比如根据 Destination 创建 MessageProducer、Message，并使用 MessageProducer 实例来发送消息。

3. 接收消息

我们通常使用 jmsTemplate.receive(destination) 来接收简单的消息，那么 Spring 是如何封装

这个功能的呢？

```java
    public Message receive(Destination destination) throws JmsException {
        return receiveSelected(destination, null);
    }

    public Message receiveSelected(final Destination destination, final String messageSelector)
throws JmsException {
            return execute(new SessionCallback<Message>() {
                public Message doInJms(Session session) throws JMSException {
                    return doReceive(session, destination, messageSelector);
                }
            }, true);
    }

    protected Message doReceive(Session session, Destination destination, String messageSelector)
            throws JMSException {

            return doReceive(session, createConsumer(session, destination, messageSelector));
    }

    protected Message doReceive(Session session, MessageConsumer consumer) throws JMSException {
            try {
                // Use transaction timeout (if available).
                long timeout = getReceiveTimeout();
                JmsResourceHolder resourceHolder =
                        (JmsResourceHolder) TransactionSynchronizationManager.getResource (getConnectionFactory());
                if (resourceHolder != null && resourceHolder.hasTimeout()) {
                    timeout = Math.min(timeout, resourceHolder.getTimeToLiveInMillis());
                }
                Message message = doReceive(consumer, timeout);
                if (session.getTransacted()) {
                    // Commit necessary - but avoid commit call within a JTA transaction.
                    if (isSessionLocallyTransacted(session)) {
                        // Transacted session created by this template -> commit.
                        JmsUtils.commitIfNecessary(session);
                    }
                }
                else if (isClientAcknowledge(session)) {
                    // Manually acknowledge message, if any.
                    if (message != null) {
                        message.acknowledge();
                    }
                }
                return message;
            }
            finally {
                JmsUtils.closeMessageConsumer(consumer);
            }
    }

    private Message doReceive(MessageConsumer consumer, long timeout) throws JMSException {
            if (timeout == RECEIVE_TIMEOUT_NO_WAIT) {
                return consumer.receiveNoWait();
            }
```

```
        else if (timeout > 0) {
            return consumer.receive(timeout);
        }
        else {
            return consumer.receive();
        }
    }
```

实现的套路与发送差不多，同样还是使用 execute 函数来封装冗余的公共操作，而最终的目标还是通过 consumer.receive() 来接收消息，其中的过程就是对于 MessageConsumer 的创建以及一些辅助操作。

13.3.2 监听器容器

消息监听器容器是一个用于查看 JMS 目标等待消息到达的特殊 bean，一旦消息到达它就可以获取到消息，并通过调用 onMessage() 方法将消息传递给一个 MessageListener 实现。Spring 中消息监听器容器的类型如下。

- SimpleMessageListenerContainer：最简单的消息监听器容器，只能处理固定数量的 JMS 会话，且不支持事务。
- DefaultMessageListenerContainer：这个消息监听器容器建立在 SimpleMessageListener Container 容器之上，添加了对事务的支持。
- serversession.ServerSessionMessage.ListenerContainer：这是功能最强大的消息监听器，与 DefaultMessageListenerContainer 相同，除了支持事务，它还允许动态地管理 JMS 会话。

下面以 DefaultMessageListenerContainer 为例进行分析，了解消息监听器容器的实现。在之前消息监听器的使用示例中，我们了解到在使用消息监听器容器时一定要将自定义的消息监听器置入到容器中，这样容器才可以在收到信息时，把消息转向监听器处理。查看 DefaultMessageListenerContainer 层次结构图，如图 13-2 所示。

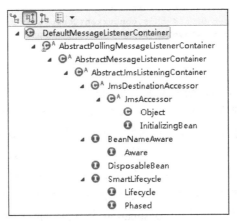

图 13-2 DefaultMessageListenerContainer 层次结构图

同样，我们看到此类实现了 InitializingBean 接口，按照以往的风格我们还是首先查看接口方法 afterPropertiesSet() 中的逻辑，其方法实现在其父类 AbstractJmsListeningContainer 中。

```
public void afterPropertiesSet() {
        //验证 connectionFactory
        super.afterPropertiesSet();
        //验证配置文件
        validateConfiguration();
        //初始化
        initialize();
}
```

监听器容器的初始化只包含了三句代码，其中前两句只用于属性的验证，比如 connectionFacory 或者 destination 等属性是否为空等，而真正用于初始化的操作委托在 initialize() 中执行。

```
public void initialize() throws JmsException {
        try {
                //lifecycleMonitor 用于控制生命周期的同步处理
                synchronized (this.lifecycleMonitor) {
                        this.active = true;
                        this.lifecycleMonitor.notifyAll();
                }
                doInitialize();
        }
        catch (JMSException ex) {
                synchronized (this.sharedConnectionMonitor) {
                        ConnectionFactoryUtils.releaseConnection(this.sharedConnection,
getConnectionFactory(), this.autoStartup);
                        this.sharedConnection = null;
                }
                throw convertJmsAccessException(ex);
        }
}

protected void doInitialize() throws JMSException {
        synchronized (this.lifecycleMonitor) {
                for (int i = 0; i < this.concurrentConsumers; i++) {
                        scheduleNewInvoker();
                }
        }
}
```

这里用到了 concurrentConsumers 属性，网络中对此属性用法的说明如下。

消息监听器允许创建多个 Session 和 MessageConsumer 来接收消息。具体的个数由 concurrentConsumers 属性指定。需要注意的是，应该只是在 Destination 为 Queue 的时候才使用多个 MessageConsumer（Queue 中的一个消息只能被一个 Consumer 接收），虽然使用多个 MessageConsumer 会提高消息处理的性能，但是消息处理的顺序却得不到保证。消息被接收的顺序仍然是消息发送时的顺序，但是由于消息可能会被并发处理，因此消息处理的顺序可能和消息发送的顺序不同。此外，不应该在 Destination 为 Topic 的时候使用多个 MessageConsumer，因为多个 MessageConsumer 会接收到同样的消息。

对于具体的实现逻辑我们只能继续查看源码:

```
private void scheduleNewInvoker() {
        AsyncMessageListenerInvoker invoker = new AsyncMessageListenerInvoker();
        if (rescheduleTaskIfNecessary(invoker)) {
// This should always be true, since we're only calling this when active.
            this.scheduledInvokers.add(invoker);
        }
}

protected final boolean rescheduleTaskIfNecessary(Object task) {
        if (this.running) {
            try {
                doRescheduleTask(task);
            }
            catch (RuntimeException ex) {
                logRejectedTask(task, ex);
                this.pausedTasks.add(task);
            }
            return true;
        }
        else if (this.active) {
            this.pausedTasks.add(task);
            return true;
        }
        else {
            return false;
        }
}

protected void doRescheduleTask(Object task) {
        this.taskExecutor.execute((Runnable) task);
}
```

分析源码得知，根据 concurrentConsumers 数量建立了对应数量的线程，即使读者不了解线程池的使用，根据以上代码至少可以推断出 doRescheduleTask 函数其实是在开启一个线程执行 Runnable。我们反追踪这个传入的参数，可以看到它其实是 AsyncMessageListenerInvoker 类型实例。因此我们可以推断，Spring 根据 concurrentConsumers 数量建立了对应数量的线程，而每个线程都作为一个独立的接收者在循环接收消息。

于是我们把所有的焦点转向 AsyncMessageListenerInvoker 这个类的实现，因为它作为一个 Runnable 角色去执行，所以对这个类的分析从 run 方法开始。

```
public void run() {
            //并发控制
            synchronized (lifecycleMonitor) {
                activeInvokerCount++;
                lifecycleMonitor.notifyAll();
            }
            boolean messageReceived = false;
            try {
                //根据每个任务设置的最大处理消息数量而作不同处理
                //小于 0 默认为无限制，一直接收消息
```

```
                    if (maxMessagesPerTask < 0) {
                        messageReceived = executeOngoingLoop();
                    }
                    else {
                        int messageCount = 0;
                        //消息数量控制，一旦超出数量则停止循环
                        while (isRunning() && messageCount < maxMessagesPerTask) {
                            messageReceived = (invokeListener() || messageReceived);
                            messageCount++;
                        }
                    }
                }
                catch (Throwable ex) {
                    //清理操作，包括关闭session等
                    clearResources();
                    if (!this.lastMessageSucceeded) {
                        // We failed more than once in a row - sleep for recovery interval
                        // even before first recovery attempt.
                        sleepInbetweenRecoveryAttempts();
                    }
                    this.lastMessageSucceeded = false;
                    boolean alreadyRecovered = false;
                    synchronized (recoveryMonitor) {
                        if (this.lastRecoveryMarker == currentRecoveryMarker) {
                            handleListenerSetupFailure(ex, false);
                            recoverAfterListenerSetupFailure();
                            currentRecoveryMarker = new Object();
                        }
                        else {
                            alreadyRecovered = true;
                        }
                    }
                    if (alreadyRecovered) {
                        handleListenerSetupFailure(ex, true);
                    }
                }
                finally {
                    synchronized (lifecycleMonitor) {
                        decreaseActiveInvokerCount();
                        lifecycleMonitor.notifyAll();
                    }
                    if (!messageReceived) {
                        this.idleTaskExecutionCount++;
                    }
                    else {
                        this.idleTaskExecutionCount = 0;
                    }
                    synchronized (lifecycleMonitor) {
                        if (!shouldRescheduleInvoker(this.idleTaskExecutionCount) ||
!reschedule TaskIfNecessary(this)) {
                            // We're shutting down completely.
                            scheduledInvokers.remove(this);
                            if (logger.isDebugEnabled()) {
```

13.3 源码分析

```
                            logger.debug("Lowered scheduled invoker count: " +
scheduledInvokers.size());
                        }
                        lifecycleMonitor.notifyAll();
                        clearResources();
                    }
                    else if (isRunning()) {
                        int nonPausedConsumers = getScheduledConsumerCount() -
getPausedTaskCount();
                        if (nonPausedConsumers < 1) {
                            logger.error("All scheduled consumers have been paused,
probably due to tasks having been rejected. " +
                                "Check your thread pool configuration! Manual
recovery necessary through a start() call.");
                        }
                        else if (nonPausedConsumers < getConcurrentConsumers()) {
                            logger.warn("Number of scheduled consumers has dropped
below concurrentConsumers limit, probably " +
                                "due to tasks having been rejected. Check your
thread pool configuration! Automatic recovery " +
                                "to be triggered by remaining consumers.");
                        }
                    }
                }
            }
        }
```

以上函数主要根据变量 maxMessagesPerTask 的值来分情况处理，当然，函数中还使用了大量的代码处理异常机制的数据维护，但是我相信大家跟我一样更加关注程序的正常流程是如何处理的。

其实核心的处理就是调用 invokeListener 来接收消息并激活消息监听器，但是之所以两种情况分开处理，正是考虑到在无限制循环接收消息的情况下，用户可以通过设置标志位 running 来控制消息接收的暂停与恢复，并维护当前消息监听器的数量。

```
private boolean executeOngoingLoop() throws JMSException {
    boolean messageReceived = false;
    boolean active = true;
    while (active) {
        synchronized (lifecycleMonitor) {
            boolean interrupted = false;
            boolean wasWaiting = false;
            //如果当前任务已经处于激活状态但是却给了暂时终止的命令
            while ((active = isActive()) && !isRunning()) {
                if (interrupted) {
                    throw new IllegalStateException("Thread was interrupted
while waiting for " +
                        "a restart of the listener container, but container
is still stopped");
                }
                if (!wasWaiting) {
                    //如果并非处于等待状态则说明是第一次执行，需要将激活任务数量减少
                    decreaseActiveInvokerCount();
```

```
                    }
                    //开始进入等待状态，等待任务的恢复命令
                    wasWaiting = true;
                    try {
                        //通过 wait 等待，也就是等待 notify 或者 notifyAll
                        lifecycleMonitor.wait();
                    }
                    catch (InterruptedException ex) {
                        // Re-interrupt current thread, to allow other threads to react.
                        Thread.currentThread().interrupt();
                        interrupted = true;
                    }
                }
                if (wasWaiting) {
                    activeInvokerCount++;
                }
                if (scheduledInvokers.size() > maxConcurrentConsumers) {
                    active = false;
                }
            }
            //正常处理流程
            if (active) {
                messageReceived = (invokeListener() || messageReceived);
            }
        }
        return messageReceived;
    }
```

如果按照正常的流程其实是不会进入 while 循环中的，而是直接进入函数 invokeListener() 来接收消息并激活监听器，但是，不可能让循环一直持续下去，我们要考虑到暂停线程或者恢复线程的情况，这时，isRunning() 函数就派上用场了。

isRunning() 用来检测标志位 this.running 状态进而判断是否需要进入 while 循环。由于要维护当前线程激活数量，所以引入了 wasWaiting 变量，用来判断线程是否处于等待状态。如果线程首次进入等待状态，则需要减少线程来激活数量计数器。

当然，还有一个地方需要提一下，就是线程等待不是一味地采用 while 循环来控制，因为如果单纯地采用 while 循环会浪费 CPU 的时钟周期，给资源造成巨大的浪费。这里，Spring 采用全局控制变量 lifecycleMonitor 的 wait() 方法来暂停线程，所以，如果终止线程需要再次恢复的话，除了更改 this.running 标志位外，还需要调用 lifecycleMonitor.notify 或者 lifecycle Monitor.notifyAll 来使线程恢复。

接下来就是消息接收的处理了。

```
    private boolean invokeListener() throws JMSException {
        //初始化资源包括首次创建的时候创建 session 与 consumer
        initResourcesIfNecessary();
        boolean messageReceived = receiveAndExecute(this, this.session, this.consumer);
        //改变标志位，信息成功处理
        this.lastMessageSucceeded = true;
        return messageReceived;
    }
```

```java
protected boolean receiveAndExecute(Object invoker, Session session, MessageConsumer consumer)
        throws JMSException {

    if (this.transactionManager != null) {
        // Execute receive within transaction.
        TransactionStatus status = this.transactionManager.getTransaction(this.transactionDefinition);
        boolean messageReceived;
        try {
            messageReceived = doReceiveAndExecute(invoker, session, consumer, status);
        }
        catch (JMSException ex) {
            rollbackOnException(status, ex);
            throw ex;
        }
        catch (RuntimeException ex) {
            rollbackOnException(status, ex);
            throw ex;
        }
        catch (Error err) {
            rollbackOnException(status, err);
            throw err;
        }
        this.transactionManager.commit(status);
        return messageReceived;
    }

    else {
        // Execute receive outside of transaction.
        return doReceiveAndExecute(invoker, session, consumer, null);
    }
}
```

在介绍消息监听器容器的分类时，已介绍了 DefaultMessageListenerContainer 消息监听器容器建立在 SimpleMessageListenerContainer 容器之上，它添加了对事务的支持，那么此时，事务特性的实现就开始了。如果用户配置了 this.transactionManager，也就是配置了事务，那么，消息的接收会被控制在事务之内，一旦出现任何异常都会被回滚，而回滚操作也会交由事务管理器统一处理，比如 this.transactionManager.rollback(status)。

doReceiveAndExecute 包含了整个消息的接收处理过程，由于掺杂着事务，所以并没有复用模板中的方法。

```java
protected boolean doReceiveAndExecute(
        Object invoker, Session session, MessageConsumer consumer, TransactionStatus status)
        throws JMSException {

    Connection conToClose = null;
    Session sessionToClose = null;
    MessageConsumer consumerToClose = null;
    try {
        Session sessionToUse = session;
        boolean transactional = false;
        if (sessionToUse == null) {
```

```java
            sessionToUse = ConnectionFactoryUtils.doGetTransactionalSession(
                    getConnectionFactory(), this.transactionalResourceFactory, true);
            transactional = (sessionToUse != null);
        }
        if (sessionToUse == null) {
            Connection conToUse;
            if (sharedConnectionEnabled()) {
                conToUse = getSharedConnection();
            }
            else {
                conToUse = createConnection();
                conToClose = conToUse;
                conToUse.start();
            }
            sessionToUse = createSession(conToUse);
            sessionToClose = sessionToUse;
        }
        MessageConsumer consumerToUse = consumer;
        if (consumerToUse == null) {
            consumerToUse = createListenerConsumer(sessionToUse);
            consumerToClose = consumerToUse;
        }
        //接收消息
        Message message = receiveMessage(consumerToUse);
        if (message != null) {
            if (logger.isDebugEnabled()) {
                logger.debug("Received message of type [" + message.getClass() + "] from consumer [" +
                        consumerToUse + "] of " + (transactional ? "transactional " : "") + "session [" +
                        sessionToUse + "]");
            }

            //模板方法，当消息接收且在未处理前给子类机会做相应处理，当期空实现
            messageReceived(invoker, sessionToUse);
            boolean exposeResource = (!transactional && isExposeListenerSession() &&
                    !TransactionSynchronizationManager.hasResource (getConnectionFactory()));
            if (exposeResource) {
                TransactionSynchronizationManager.bindResource(
                        getConnectionFactory(), new LocallyExposedJmsResourceHolder(sessionToUse));
            }
            try {
                //激活监听器
                doExecuteListener(sessionToUse, message);
            }
            catch (Throwable ex) {
                if (status != null) {
                    if (logger.isDebugEnabled()) {
                        logger.debug("Rolling back transaction because of listener exception thrown: " + ex);
                    }
                    status.setRollbackOnly();
```

```
                    }
                    handleListenerException(ex);
                    // Rethrow JMSException to indicate an infrastructure problem
                    // that may have to trigger recovery...
                    if (ex instanceof JMSException) {
                        throw (JMSException) ex;
                    }
                }
                finally {
                    if (exposeResource) {
                        TransactionSynchronizationManager.unbindResource(getConnectionFactory());
                    }
                }
                // Indicate that a message has been received.
                return true;
            }
            else {
                if (logger.isTraceEnabled()) {
                    logger.trace("Consumer [" + consumerToUse + "] of " + (transactional ? "transactional " : "") +
                            "session [" + sessionToUse + "] did not receive a message");
                }
                //接收到空消息的处理
                noMessageReceived(invoker, sessionToUse);
                // Nevertheless call commit, in order to reset the transaction timeout (if any).
                // However, don't do this on Tibco since this may lead to a deadlock there.
                if (shouldCommitAfterNoMessageReceived(sessionToUse)) {
                    commitIfNecessary(sessionToUse, message);
                }
                // Indicate that no message has been received.
                return false;
            }
        }
        finally {
            JmsUtils.closeMessageConsumer(consumerToClose);
            JmsUtils.closeSession(sessionToClose);
            ConnectionFactoryUtils.releaseConnection(conToClose, getConnectionFactory(), true);
        }
    }
```

上面函数代码看似繁杂，但是真正的逻辑并不多，大多是固定的套路，而我们最关心的问题就是监听器的激活处理。

```
    protected void doExecuteListener(Session session, Message message) throws JMSException {
        if (!isAcceptMessagesWhileStopping() && !isRunning()) {
            if (logger.isWarnEnabled()) {
                logger.warn("Rejecting received message because of the listener container " +
                        "having been stopped in the meantime: " + message);
            }
            rollbackIfNecessary(session);
            throw new MessageRejectedWhileStoppingException();
        }
```

```
            try {
                invokeListener(session, message);
            }
            catch (JMSException ex) {
                rollbackOnExceptionIfNecessary(session, ex);
                throw ex;
            }
            catch (RuntimeException ex) {
                rollbackOnExceptionIfNecessary(session, ex);
                throw ex;
            }
            catch (Error err) {
                rollbackOnExceptionIfNecessary(session, err);
                throw err;
            }
            commitIfNecessary(session, message);
        }

    protected void invokeListener(Session session, Message message) throws JMSException {
        Object listener = getMessageListener();
        if (listener instanceof SessionAwareMessageListener) {
            doInvokeListener((SessionAwareMessageListener) listener, session, message);
        }
        else if (listener instanceof MessageListener) {
            doInvokeListener((MessageListener) listener, message);
        }
        else if (listener != null) {
            throw new IllegalArgumentException(
                    "Only MessageListener and SessionAwareMessageListener supported: " + listener);
        }
        else {
            throw new IllegalStateException("No message listener specified - see property 'messageListener'");
        }
    }
    protected void doInvokeListener(MessageListener listener, Message message) throws JMSException {
        listener.onMessage(message);
    }
```

通过层层调用，最终提取监听器并使用 listener.onMessage(message)对其进行了激活，也就是激活了用户自定义的监听器逻辑。这里还有一句重要的代码很容易被忽略掉，那就是 commitIfNecessary(session, message)，它完成的功能是 session.commit()。完成消息服务的事务提交，涉及两个事务，我们常说的 DefaultMessageListenerContainer 增加了事务的支持，是通用的事务，也就是说我们在消息接收过程中如果产生其他操作，比如向数据库中插入数据，一旦出现异常时就需要全部回滚，包括回滚插入数据库中的数据。但是，除了我们常说的事务之外，对于消息本身还有一个事务，当接收一个消息的时候，必须使用事务提交的方式，这是在告诉消息服务器本地已经正常接收消息，消息服务器接收到本地的事务提交后便可以将此消息删除，否则，当前消息会被其他接收者重新接收。

第 3 部分　Spring Boot

第 14 章　Spring Boot 体系原理

第 14 章 Spring Boot 体系原理

Spring Boot 是由 Pivotal 团队提供的全新框架，其设计目的是用来简化新 Spring 应用的初始搭建以及开发过程。该框架使用了特定的方式来进行配置，从而使开发人员不再需要定义样板化的配置。通过这种方式，Spring Boot 将致力于在蓬勃发展的快速应用开发领域（Rapid Application Development）成为领导者。

Spring Boot 特点如下：
- 创建独立的 Spring 应用程序；
- 嵌入的 Tomcat，无须部署 WAR 文件；
- 简化 Maven 配置；
- 自动配置 Spring；
- 提供生产就绪型功能，如指标、健康检查和外部配置；
- 绝对没有代码生成，以及对 XML 没有配置要求。

当然，这样的介绍似乎太过于官方化，好像并没有帮助我们理解 Spring Boot 到底做了什么，我们不妨通过一个小例子来快速了解 Spring Boot。

首先我们搭建一个 maven 工程，pom 如下：

```xml
<?xml version="1.0" encoding="UTF-8"?>
<project xmlns="http://maven.apache.org/POM/4.0.0"
         xmlns:xsi="http://www.w3.org/2001/XMLSchema-instance"
         xsi:schemaLocation="http://maven.apache.org/POM/4.0.0 http://maven.apache.org/xsd/maven-4.0.0.xsd">

    <parent>
        <groupId>org.springframework.boot</groupId>
        <artifactId>spring-boot-starter-parent</artifactId>
        <version>2.0.1.RELEASE</version>
        <relativePath/>
    </parent>

    <modelVersion>4.0.0</modelVersion>
    <groupId>com.springstudy</groupId>
```

```xml
<artifactId>study-web</artifactId>
<version>1.0-SNAPSHOT</version>

<dependencies>
    <dependency>
        <groupId>org.springframework.boot</groupId>
        <artifactId>spring-boot-starter-web</artifactId>
    </dependency>
</dependencies>

<build>
    <plugins>
        <plugin>
            <groupId>org.springframework.boot</groupId>
            <artifactId>spring-boot-maven-plugin</artifactId>
        </plugin>
    </plugins>
</build>


</project>
```

然后我们建立一个 controller 类：

```java
package com.springstudy.controller;
import com.spring.study.module.HelloService;
import org.springframework.beans.factory.annotation.Autowired;
import org.springframework.web.bind.annotation.RequestMapping;
import org.springframework.web.bind.annotation.RestController;

@RestController
public class TestController {
    @RequestMapping("/")
    String home() {
        return "helloworld";
    }
}
```

最后我们再加入启动整个项目的 main 函数：

```java
package com.springstudy;

import org.springframework.boot.SpringApplication;
import org.springframework.boot.autoconfigure.SpringBootApplication;

@SpringBootApplication
public class SpringBootDemo1Application {
    public static void main(String[] args) {
        SpringApplication.run(SpringBootDemo1Application.class, args);
    }
}
```

以上就是我们要准备的示例的所有内容，最后我们尝试启动 main 函数并在浏览器中输入 localhost:8080，发现浏览器显示如图 14-1 所示的界面。

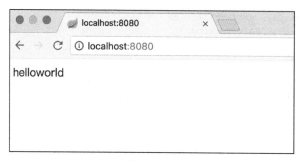

图 14-1　第一个 Spring Boot

这一切都似乎完全超乎了我们的预料，按照之前的经验，如果要构建这样一套 MVC 体系，似乎是非常麻烦的，至少要引入一大堆的 pom 依赖，同时，最为神奇的是整个过程中我们似乎根本没有启动过 Tomcat，但是当我们运行函数的时候 Tomcat 居然自动起来了，而且还能通过浏览器访问，这一切都是怎么回事呢？这里留下悬念，我们稍后探索。

当然，如果你认为 Spring Boot 仅仅是封装了 Tomcat 那就大错特错了，一个流行的框架一定是一个理念的创新，它绝对不是一个简简单单的封装就能搞的定的。

在我们正式进入 Spring Boot 的原理探索之前，首先我们还是尝试去下载及安装其源码。

14.1　Spring Boot 源码安装

同样，Spring Boot 通过 Github 维护，我们打开 Github 官网，输入 spring-boot，如图 14-2 所示。

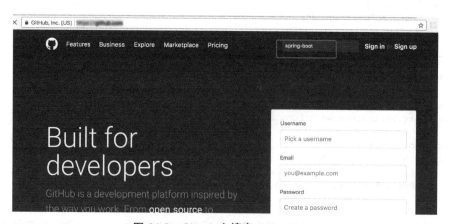

图 14-2　Github 上搜索 spring-boot

之后的源码下载（界面见图 14-3）和安装的流程与第 1 章中介绍的 Spring 源码下载和安装流程相同，在这里就不再赘述。

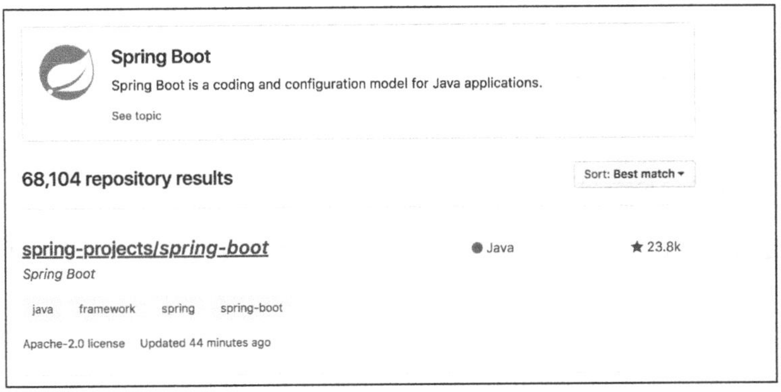

図 14-3　Github 上搜索 Spring Boot

14.2　第一个 Starter

在我看来，Spring Boot 之所以流行，是因为 spring starter 模式的提出。spring starter 的出现，可以让模块开发更加独立化，相互间依赖更加松散以及可以更加方便地集成。从前言中介绍的例子来看，正是由于在 pom 文件中引入了下述代码：

```xml
<dependency>
        <groupId>org.springframework.boot</groupId>
        <artifactId>spring-boot-starter-web</artifactId>
</dependency>
```

这就让 Spring 自动化地帮我们做了非常多的事情，当然现阶段我们去分析 spring-boot-starter-web 的实现原理似乎跨越还有些大，我们可以从一个更为简单的例子来看。

同样，为了不跟之前的 Web 工程混淆，我们另建一个 maven 工程，pom 如下：

```xml
<?xml version="1.0" encoding="UTF-8"?>
<project xmlns="http://maven.apache.org/POM/4.0.0"
        xmlns:xsi="http://www.w3.org/2001/XMLSchema-instance"
        xsi:schemaLocation="http://maven.apache.org/POM/4.0.0 http://maven.apache.org/xsd/maven-4.0.0.xsd">

        <modelVersion>4.0.0</modelVersion>
        <groupId>com.springstudy</groupId>
        <artifactId>study-client-starter</artifactId>
        <version>1.0-SNAPSHOT</version>
        <dependencies>
            <dependency>
                <groupId>org.springframework.boot</groupId>
                <artifactId>spring-boot-autoconfigure</artifactId>
            </dependency>
        </dependencies>
        <dependencyManagement>
            <dependencies>
                <dependency>
                    <!-- Import dependency management from Spring Boot -->
                    <groupId>org.springframework.boot</groupId>
```

```xml
            <artifactId>spring-boot-dependencies</artifactId>
            <version>2.0.1.RELEASE</version>
            <type>pom</type>
            <scope>import</scope>
        </dependency>
    </dependencies>
</dependencyManagement>
</project>
```

然后我们定义一个接口，可以认为它是当前独立业务开发模块对外暴露的可以直接调用的接口，如下：

```java
package com.spring.study.module;
public interface HelloService {
    public String sayHello();
}
```

我们对这个接口做一个简单的实现，返回 hello 字符串：

```java
package com.spring.study.module;
import org.springframework.stereotype.Component;

@Component
public class HelloServiceImpl implements HelloService {
    public String sayHello(){
        return "hello!! ";
    }

}
```

以上实现为了尽量屏蔽 Spring Boot 基础理论以外的东西，把演示设计得尽量简单，如果是真实的业务，这个接口以及接口实现可能会非常复杂，甚至还会间接依赖于非常多的其他的 bean。它基本上就是一个独立的业务模块，当然这个模块并不是自己部署，而是运行在依赖它的主函数中。如果我们开发到这种程度，想要主函数感知的话也不是不可以，但是至少要让主工程知道当前业务的 bean 路径并加入 scan 列表中，否则在 Spring 启动的过程中没有办法把 client 中所有的 bean 载入 Spring 容器，逻辑也就没法生效了，但是，随着业务的增长，模块也会越来越多、越来越分散，大量的配置在主函数中维护，这会造成主函数非常臃肿及冲突严重，而且根据职责划分原则，以上的例子中主模块只关心自己是否使用外部依赖的模块以及对应的接口就好了，再去让主模块感知对应的路径等细节信息显然是不合适的。于是乎，在 Spring Boot 出来之前我们会尝试把 Scan 等配置项写入 XML 里面，然后让主函数直接引用配置项，这样，主函数知道的事情就进一步减少了，但是还有没有更好的解决方式呢，或者，还有没有更好的办法能让主函数做更少的事情呢？Spring Boot 做到了这一点，继续追加代码，添加自动配置项：

```java
package com.spring.study;

import org.springframework.context.annotation.ComponentScan;
import org.springframework.context.annotation.Configuration;

@Configuration
@ComponentScan({"com.spring.study.module"})
public class HelloServiceAutoConfiguration    {

}
```

我们发现，在 HelloServiceAutoConfiguration 类中并没有逻辑实现，它存在的目的仅仅是通过注解进行配置的声明，我们可以在 ComponentScan 中加入这个模块的容器扫描路径。

当然，如果仅仅是到此，Starter 还是没有开发完成，还需要最后一步，那就是声明这个配置文件的路径，在 Spring 的跟路径下建立 **META-INF/spring.factories** 文件，并声明配置项路径（见图 14-4）：

```
org.springframework.boot.autoconfigure.EnableAutoConfiguration=\
com.spring.study.HelloServiceAutoConfiguration
```

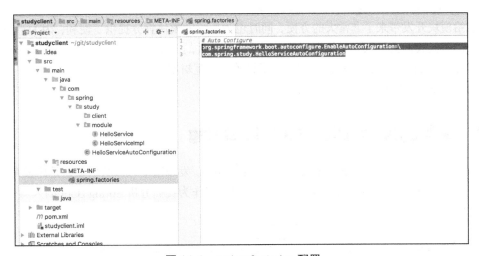

图 14-4　spring.factories 配置

到此，一个标准的 Starter 就开发完成了，它有什么用或者说它怎么使用呢？我们来看一下它的使用方式。

修改前言中的 Web 工程，加入依赖：

```
<dependency>
    <groupId>com.springstudy</groupId>
    <artifactId>study-client-starter</artifactId>
    <version>1.0-SNAPSHOT</version>
</dependency>
```

同时，更改 Controller 逻辑，将模块的逻辑引入，如下所示：

```
package com.springstudy.controller;

import com.spring.study.module.HelloService;
import org.springframework.beans.factory.annotation.Autowired;
import org.springframework.web.bind.annotation.RequestMapping;
import org.springframework.web.bind.annotation.RestController;

@RestController
public class TestController {

    @Autowired
    private HelloService helloService;
```

```
@RequestMapping("/")
String home() {
    return helloService.sayHello();
}
}
```

读者会发现，我们刚才开发的 Starter 对于使用者来说非常的方便，除了在 pom 中引入依赖，什么都不做就可以直接使用模块内部的接口注入：

```
@Autowired
    private HelloService helloService;
```

这给模块开发带来了非常大的方便，同时也为后续的模块拆分提供了便利，因为当业务逐渐复杂的时候我们会引入大量的中间件，而这些中间件的配置、依赖、以及初始化是非常麻烦的，现在有了 Starter 模式，它帮我们做到了只关注于逻辑本身。

那么，Spring Boot 是如何做到的呢？

14.3　探索 SpringApplication 启动 Spring

我们找到主函数入口 SpringBootDemo1Application，发现这个入口的启动还是比较奇怪的，这也是 Spring Boot 启动的必要做法，那么，这也可以作为我们分析 Spring Boot 的入口：

```
@SpringBootApplication
public class SpringBootDemo1Application {
    public static void main(String[] args) {
        SpringApplication.run(SpringBootDemo1Application.class, args);
    }
}
```

当顺着 SpringApplication.run 方法进入的时候我们找到了 SpringApplication 的一个看似核心逻辑的方法：

```
public ConfigurableApplicationContext run(String... args) {
    StopWatch stopWatch = new StopWatch();
    stopWatch.start();
    ConfigurableApplicationContext context = null;
    FailureAnalyzers analyzers = null;
    configureHeadlessProperty();
    SpringApplicationRunListeners listeners = getRunListeners(args);
    listeners.starting();
    try {
        ApplicationArguments applicationArguments = new DefaultApplicationArguments(
                args);
        ConfigurableEnvironment environment = prepareEnvironment(listeners,
                applicationArguments);
        Banner printedBanner = printBanner(environment);
        context = createApplicationContext();
        analyzers = new FailureAnalyzers(context);
        prepareContext(context, environment, listeners, applicationArguments,
                printedBanner);
        refreshContext(context);
        afterRefresh(context, applicationArguments);
```

```
         listeners.finished(context, null);
         stopWatch.stop();
         if (this.logStartupInfo) {
            new StartupInfoLogger(this.mainApplicationClass)
                  .logStarted(getApplicationLog(), stopWatch);
         }
         return context;
      }
      catch (Throwable ex) {
         handleRunFailure(context, listeners, analyzers, ex);
         throw new IllegalStateException(ex);
      }
   }
```

在这里，我们发现了几个关键字眼：

```
context = createApplicationContext();
refreshContext(context);
      afterRefresh(context, applicationArguments);
```

如果读者看过之前的内容，就会知道，我们曾经在第 5 章介绍过 Spring 完整的初始化方案，其中就最为核心的就是 SpringContext 的创建、初始化、刷新等。那么我们可以直接进入查看其中的逻辑，同时，Spring 作为一个全球都在使用的框架，会有非常多的需要考虑的问题，我们在阅读源码的过程中只需要关系核心的主流程，了解其工作原理，并在阅读的过程中感受它的代码风格以及设计理念就好了，如果真的追求理解每一行代码真的是非常耗时的一件事情，毕竟我们阅读源码的目的大多数是成长而不是真的要去维护 Spring。

14.3.1　SpringContext 创建

```
protected ConfigurableApplicationContext createApplicationContext() {
   Class<?> contextClass = this.applicationContextClass;
   if (contextClass == null) {
      try {
         contextClass = Class.forName(this.webEnvironment
               ? DEFAULT_WEB_CONTEXT_CLASS : DEFAULT_CONTEXT_CLASS);
      }
      catch (ClassNotFoundException ex) {
         throw new IllegalStateException(
               "Unable create a default ApplicationContext, "
                     + "please specify an ApplicationContextClass",
               ex);
      }
   }
   return (ConfigurableApplicationContext) BeanUtils.instantiate(contextClass);
}
```

这个函数似乎没有什么特别之处，无非就是实例化一个 ApplicationContext，因为 ApplicationContext 是 Spring 存在的基础。而对应的 SpringContext 候选类如下：

```
public static final String DEFAULT_WEB_CONTEXT_CLASS = "org.springframework."
      + "boot.context.embedded.AnnotationConfigEmbeddedWebApplicationContext";

private static final String[] WEB_ENVIRONMENT_CLASSES = { "javax.servlet.Servlet",
      "org.springframework.web.context.ConfigurableWebApplicationContext" };
```

这里有个关键的判断，this.webEnvironment，如果读者没有看过代码很容易会忽略，但是这里将成为在前言中提到的 Spring 如何自动化启动 Tomcat 的关键，我们将会在后续章节详细介绍。

14.3.2 bean 的加载

继续返回追踪 prepareContext：

```
private void prepareContext(ConfigurableApplicationContext context,
        ConfigurableEnvironment environment, SpringApplicationRunListeners listeners,
        ApplicationArguments applicationArguments, Banner printedBanner) {
    context.setEnvironment(environment);
    postProcessApplicationContext(context);
    applyInitializers(context);
    listeners.contextPrepared(context);
    if (this.logStartupInfo) {
        logStartupInfo(context.getParent() == null);
        logStartupProfileInfo(context);
    }

    // Add boot specific singleton beans
    context.getBeanFactory().registerSingleton("springApplicationArguments",
            applicationArguments);
    if (printedBanner != null) {
        context.getBeanFactory().registerSingleton("springBootBanner", printedBanner);
    }

    // Load the sources
    Set<Object> sources = getSources();
    Assert.notEmpty(sources, "Sources must not be empty");
    load(context, sources.toArray(new Object[sources.size()]));
    listeners.contextLoaded(context);
}
```

这里面的 load 函数是我们比较感兴趣的，代码如下：

```
protected void load(ApplicationContext context, Object[] sources) {
    if (logger.isDebugEnabled()) {
        logger.debug(
                "Loading source " + StringUtils.arrayToCommaDelimitedString(sources));
    }
    BeanDefinitionLoader loader = createBeanDefinitionLoader(
            getBeanDefinitionRegistry(context), sources);
    if (this.beanNameGenerator != null) {
        loader.setBeanNameGenerator(this.beanNameGenerator);
    }
    if (this.resourceLoader != null) {
        loader.setResourceLoader(this.resourceLoader);
    }
    if (this.environment != null) {
        loader.setEnvironment(this.environment);
    }
    loader.load();
}
```

相信当读者看到 BeanDefinitionLoader 这个类的时候基本上就已经知道后续的逻辑了，bean 的加载作为本书中最核心的部分早在第 1 章就已经开始分析了。

14.3.3 Spring 扩展属性的加载

```
protected void refresh(ApplicationContext applicationContext) {
    Assert.isInstanceOf(AbstractApplicationContext.class, applicationContext);
    ((AbstractApplicationContext) applicationContext).refresh();
}
```

对于 Spring 的扩展属性加载则更为简单，因为这些都是 Spring 本身原有的东西，Spring Boot 仅仅是使用 refresh 激活下而已，如果读者想回顾 refresh 的详细逻辑，可以回到第 5 章进一步查看。

14.3.4 总结

分析下来，Spring Boot 的启动并不是我们想象的那么神秘，按照约定大于配置的原则，内置了 Spring 原有的启动类，并在启动的时候启动及刷新，仅此而已。

```
org.springframework.context.annotation.AnnotationConfigApplicationContext
```

14.4 Starter 自动化配置原理

我们已经知道了 Spring Boot 如何启动 Spring，但是目前为止我们并没有揭开 Spring Boot 的面纱，究竟 Starter 是如何生效的呢？这些逻辑现在看来只能体现在注解 SpringBootApplication 本身了。

继续追查代码，看一看 SpringBootApplication 注解内容：

```
@Target(ElementType.TYPE)
@Retention(RetentionPolicy.RUNTIME)
@Documented
@Inherited
@SpringBootConfiguration
@EnableAutoConfiguration
@ComponentScan(excludeFilters = {
        @Filter(type = FilterType.CUSTOM, classes = TypeExcludeFilter.class),
        @Filter(type = FilterType.CUSTOM, classes = AutoConfigurationExcludeFilter.class) })
public @interface SpringBootApplication {

    ... xxx 内容忽略...
}
```

这其中我们更关注 SpringBootApplication 上的注解内容，因为注解具有传递性，EnableAutoConfiguration 是个非常特别的注解，它是 Spring Boot 的全局开关，如果把这个注解去掉，则一切 Starter 都会失效，这就是约定大于配置的潜规则，那么，Spring Boot 的核心很可能就藏在这个注解里面：

```
@SuppressWarnings("deprecation")
@Target(ElementType.TYPE)
@Retention(RetentionPolicy.RUNTIME)
```

```
@Documented
@Inherited
@AutoConfigurationPackage
@Import(EnableAutoConfigurationImportSelector.class)
public @interface EnableAutoConfiguration {
    xxx 内容忽略…

}
```

EnableAutoConfigurationImportSelector 作为 Starter 自动化导入的关键选项终于浮现出来，那么 Spring 是怎么识别并让这个注解起作用的呢？我们看到这个类中只有一个方法，那么只要看一看到底是哪个方法调用了它，就可以顺藤摸瓜找到最终的调用点。

```
public class EnableAutoConfigurationImportSelector extends AutoConfigurationImportSelector {

    @Override
    protected boolean isEnabled(AnnotationMetadata metadata) {
        if (getClass().equals(EnableAutoConfigurationImportSelector.class)) {
            return getEnvironment().getProperty(
                EnableAutoConfiguration.ENABLED_OVERRIDE_PROPERTY, Boolean.class, true);
        }
        return true;
    }

}
```

14.4.1 spring.factories 的加载

顺着思路反向查找，看一看究竟是谁在哪里调用了 isEnabled 函数，强大的编译器很容器帮我们定位到了 AutoConfigurationImportSelector 类的方法：

```
@Override
public String[] selectImports(AnnotationMetadata annotationMetadata) {
    if (!isEnabled(annotationMetadata)) {
        return NO_IMPORTS;
    }
    try {
        AutoConfigurationMetadata autoConfigurationMetadata = AutoConfigurationMetadataLoader
                .loadMetadata(this.beanClassLoader);
        AnnotationAttributes attributes = getAttributes(annotationMetadata);
        List<String> configurations = getCandidateConfigurations(annotationMetadata,
                attributes);
        configurations = removeDuplicates(configurations);
        configurations = sort(configurations, autoConfigurationMetadata);
        Set<String> exclusions = getExclusions(annotationMetadata, attributes);
        checkExcludedClasses(configurations, exclusions);
        configurations.removeAll(exclusions);
        configurations = filter(configurations, autoConfigurationMetadata);
        fireAutoConfigurationImportEvents(configurations, exclusions);
        return configurations.toArray(new String[configurations.size()]);
    }
    catch (IOException ex) {
        throw new IllegalStateException(ex);
    }
}
```

它是一个非常核心的函数,可以帮我们解释很多问题。在上面的函数中,有一个是我们比较关注的 getCandidateConfigurations 函数:

```
protected List<String> getCandidateConfigurations(AnnotationMetadata metadata,
        AnnotationAttributes attributes) {
    List<String> configurations = SpringFactoriesLoader.loadFactoryNames(
            getSpringFactoriesLoaderFactoryClass(), getBeanClassLoader());
    Assert.notEmpty(configurations,
            "No auto configuration classes found in META-INF/spring.factories. If you "
                + "are using a custom packaging, make sure that file is correct.");
    return configurations;
}
```

从上面的函数中我们看到了 META-INF/spring.factories,在我们之前演示的环节,按照约定大于配置的原则,Starter 如果要生效则必须要在 META-INF 文件下下建立 spring.factories 文件,并把相关的配置类声明在里面,虽然这仅仅是一个报错异常提示,但是其实我们已经可以推断出来这一定就是这个逻辑的处理之处,继续进入 SpringFactoriesLoader 类:

```
public static List<String> loadFactoryNames(Class<?> factoryClass, ClassLoader classLoader) {
    String factoryClassName = factoryClass.getName();
    try {
        Enumeration<URL> urls = (classLoader != null ? classLoader.getResources
(FACTORIES_RESOURCE_LOCATION) :
                ClassLoader.getSystemResources(FACTORIES_RESOURCE_LOCATION));
        List<String> result = new ArrayList<String>();
        while (urls.hasMoreElements()) {
            URL url = urls.nextElement();
            Properties properties = PropertiesLoaderUtils.loadProperties(new UrlResource(url));
            String factoryClassNames = properties.getProperty(factoryClassName);
            result.addAll(Arrays.asList(StringUtils.commaDelimitedListToStringArray
(factoryClassNames)));
        }
        return result;
    }
    catch (IOException ex) {
        throw new IllegalArgumentException("Unable to load [" + factoryClass.getName() +
            "] factories from location [" + FACTORIES_RESOURCE_LOCATION + "]", ex);
    }
}
```

而上面函数中对 FACTORIES_RESOURCE_LOCATION 的定义为:

```
public static final String FACTORIES_RESOURCE_LOCATION = "META-INF/spring.factories";
```

至此,我们终于明白了为什么 Starter 的生效必须要依赖于配置 META-INF/spring.factories 文件,因为在启动过程中有一个硬编码的逻辑就是会扫描各个包中的对应文件,并把配置捞取出来,但是,捞取出来后又是怎么跟 Spring 整合的呢?或者说 AutoConfigurationImportSelector.selectImports 方法后把加载的类又委托给谁继续处理了呢?

14.4.2 factories 调用时序图

META-INF/spring.factories 中的配置文件是如何与 Spring 整合的呢? 其路径还是比较深的,这里就不大段地放代码了,作者通过一个图去理清它的逻辑。

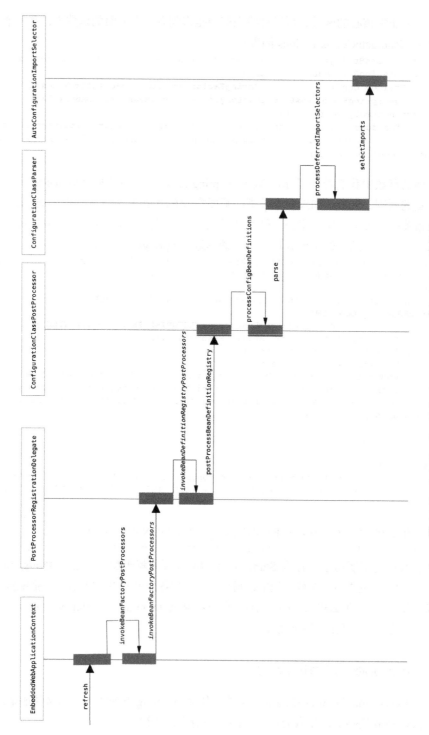

图 14-5 AutoConfigurationImportSelector 与 Spring 的整合

图 14-5 中梳理了从 EmbeddedWebApplicationContext 到 AutoConfigurationImportSelector 的调用链路，当然这个链路还有非常多的额外分支被忽略。不过至少从上图中我们可以很清晰地看到 AutoConfigurationImportSelector 与 Spring 的整合过程，在这个调用链中最核心的就是 Spring Boot 使用了 Spring 提供的 BeanDefinitionRegistryPostProcessor 扩展点并实现了 ConfigurationClassPostProcessor 类，从而实现了 spring 之上的一系列逻辑扩展，让我们看一下 ConfigurationClassPostProcessor 的继承关系，如图 14-6 所示：

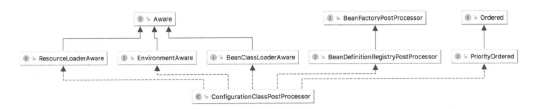

图 14-6 AutoConfigurationImportSelector 继承关系

当然 Spring 还提供了非常多不同阶段的扩展点，读者可以通过前几章的内容获取详细的扩展点以及实现原理。

14.4.3 配置类的解析

截止到目前我们知道了 Starter 为什么要求默认将自身入口配置写在 META-INF 文件下的 spring.factories 文件中，以及 AutoConfigurationImportSelector 的上下文调用链路，但是通过 AutoConfigurationImportSelector. selectImports 方法返回后的配置类又是如何进一步处理的呢？对照图 14-5 我们抽出 ConfigurationClassParser 的 processDeferredImportSelectors 方法代码查看：

```java
private void processDeferredImportSelectors() {
    List<DeferredImportSelectorHolder> deferredImports = this.deferredImportSelectors;
    this.deferredImportSelectors = null;
    Collections.sort(deferredImports, DEFERRED_IMPORT_COMPARATOR);

    for (DeferredImportSelectorHolder deferredImport : deferredImports) {
        ConfigurationClass configClass = deferredImport.getConfigurationClass();
        try {
            String[] imports = deferredImport.getImportSelector().selectImports(configClass.getMetadata());
            processImports(configClass, asSourceClass(configClass), asSourceClasses(imports), false);
        }
        catch (BeanDefinitionStoreException ex) {
            throw ex;
        }
        catch (Throwable ex) {
            throw new BeanDefinitionStoreException(
                "Failed to process import candidates for configuration class [" +
                configClass.getMetadata().getClassName() + "]", ex);
```

        }
      }
    }
其中

```
String[] imports =
  deferredImport.getImportSelector().selectImports(configClass.getMetadata());
```

的 imports 对应的就是我们定义的配置文件中配置的类,通过断点调试,如图 14-7 所示。

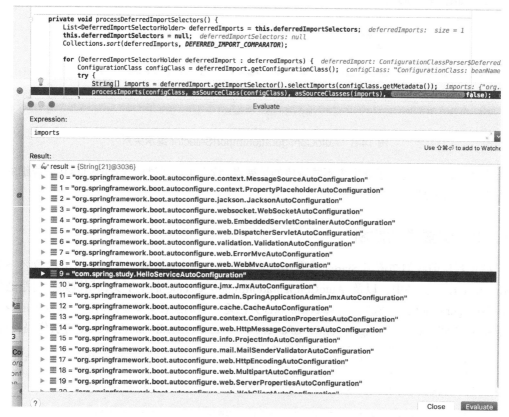

图 14-7　imports 返回值查看

也就是说在 Spring 启动的时候会扫描所有 JAR 中的 spring.factories 定义的类,而这些对于用户来说如果不是通过调试信息可能根本就感知不到。

那么也就是说,下面运行代码就是配置文件的处理逻辑:

```
processImports(configClass,asSourceClass(configClass),asSourceClasses(imports), false);
```

这个逻辑其实还是非常复杂的,其内部包含了各种分支的处理,我们不妨先通过时序图从全局的角度了解一下它的处理全貌。

14.4 Starter 自动化配置原理

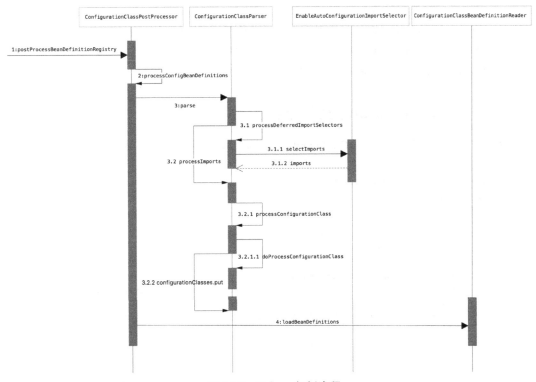

图 14-8 Parser 解析流程

图 14-8 是在图 14-5 基础上更细粒度的突出解析过程的时序图，从时序图中我们可以大致看到 Spring 的全局处理流程。

- ConfigurationClassPostProcessor 作为 Spring 扩展点是 Spring Boot 一系列功能的基础入口。
- ConfigurationClassParser 作为解析职责的基本处理类，涵盖了各种解析处理的逻辑，如@Import、@Bean、@ImportResource 、@PropertySource、@ComponentScan 等注解都是在这个注解类中完成的，而这个类对外开放的函数入口就是 parse 方法。对应时序图中的步骤 3。
- 在完成步骤 3 后，所有解析的结果已经通过 3.2.2 步骤放在了 parse 的 configurationClasses 属性中，这时候对这个属性进行统一的 spring bean 硬编码注册，注册逻辑统一委托给 ConfigurationClassBeanDefinitionReader，对外的接口是 loadBeanDefinitions，对应步骤 4。
- 当然，在 parse 中的处理是最复杂的，parse 中首先会处理自己本身能扫描到的 bean 注册逻辑，然后才会处理 spring.factories 定义的配置。处理 spring.factories 定义的配置首先就是要加载配置类，这个时候 EnableAutoConfigurationImportSelector 提供的 selectImports 就被派上用场了，它返回的配置类需要进行进一步解析，因为这些配置类中可能对应不同的类型，如@Import、@Bean、@ImportResource、@PropertySource、@ComponentScan，而这些类型又有不同的处理逻辑，例如 ComponentScan，我们就能猜到这里面除了解析外一定还会有递归解析的处理逻辑，因为很有可能通过 ComponentScan 又扫描出了另一个 ComponentScan 配置。

14.4.4　Componentscan 的切入点

这里重点讲解一下 doProcessConfigurationClass 函数，我们熟悉的很多注解逻辑的实现都在这里：

```
protected final SourceClass doProcessConfigurationClass(ConfigurationClass configClass,
SourceClass sourceClass)
        throws IOException {

    // Recursively process any member (nested) classes first
    processMemberClasses(configClass, sourceClass);

    // Process any @PropertySource annotations
    for (AnnotationAttributes propertySource : AnnotationConfigUtils.attributesForRepeatable(
            sourceClass.getMetadata(), PropertySources.class,
            org.springframework.context.annotation.PropertySource.class)) {
        if (this.environment instanceof ConfigurableEnvironment) {
            processPropertySource(propertySource);
        }
        else {
            logger.warn("Ignoring @PropertySource annotation on [" + sourceClass.getMetadata().getClassName() +
                    "]. Reason: Environment must implement ConfigurableEnvironment");
        }
    }

    // Process any @ComponentScan annotations
    Set<AnnotationAttributes> componentScans = AnnotationConfigUtils.attributesForRepeatable(
            sourceClass.getMetadata(), ComponentScans.class, ComponentScan.class);
    if (!componentScans.isEmpty() &&
            !this.conditionEvaluator.shouldSkip(sourceClass.getMetadata(), ConfigurationPhase.REGISTER_BEAN)) {
        for (AnnotationAttributes componentScan : componentScans) {
            // The config class is annotated with @ComponentScan -> perform the scan immediately
            Set<BeanDefinitionHolder> scannedBeanDefinitions =
                    this.componentScanParser.parse(componentScan, sourceClass.getMetadata().getClassName());
            // Check the set of scanned definitions for any further config classes and parse recursively if needed
            for (BeanDefinitionHolder holder : scannedBeanDefinitions) {
                //对扫描出来的类进行过滤
                if (ConfigurationClassUtils.checkConfigurationClassCandidate(
                        holder.getBeanDefinition(), this.metadataReaderFactory)) {
                    //将所有扫描出来的类委托到 parse 方法中递归处理
                    parse(holder.getBeanDefinition().getBeanClassName(), holder.getBeanName());
                }
```

14.4 Starter 自动化配置原理

```
      }
    }
  }

  // Process any @Import annotations
  processImports(configClass, sourceClass, getImports(sourceClass), true);

  // Process any @ImportResource annotations
  if (sourceClass.getMetadata().isAnnotated(ImportResource.class.getName())) {
    AnnotationAttributes importResource =
        AnnotationConfigUtils.attributesFor(sourceClass.getMetadata(), ImportResource.class);
    String[] resources = importResource.getStringArray("locations");
    Class<? extends BeanDefinitionReader> readerClass = importResource.getClass("reader");
    for (String resource : resources) {
      String resolvedResource = this.environment.resolveRequiredPlaceholders(resource);
      configClass.addImportedResource(resolvedResource, readerClass);
    }
  }

  // Process individual @Bean methods
  Set<MethodMetadata> beanMethods = retrieveBeanMethodMetadata(sourceClass);
  for (MethodMetadata methodMetadata : beanMethods) {
    configClass.addBeanMethod(new BeanMethod(methodMetadata, configClass));
  }

  // Process default methods on interfaces
  processInterfaces(configClass, sourceClass);

  // Process superclass, if any
  if (sourceClass.getMetadata().hasSuperClass()) {
    String superclass = sourceClass.getMetadata().getSuperClassName();
    if (!superclass.startsWith("java") && !this.knownSuperclasses.containsKey(superclass)) {
      this.knownSuperclasses.put(superclass, configClass);
      // Superclass found, return its annotation metadata and recurse
      return sourceClass.getSuperClass();
    }
  }

  // No superclass -> processing is complete
  return null;
}
```

而以上函数中传递过来的参数 ConfigurationClass configClass 就是 spring.factories 中定义的配置类，这里我们重点关注一下 ComponentScan 注解的实现逻辑，首先通过代码

```
Set<AnnotationAttributes> componentScans = AnnotationConfigUtils.attributesForRepeatable(
    sourceClass.getMetadata(), ComponentScans.class, ComponentScan.class);
```

获取对应的注解配置信息，也就是对应的@ComponentScan({"com.spring.study.module"}) 中最主要的扫描路径信息，然后委托给 ComponentScanAnnotationParser 的 parse 进一步扫描：

```
parse(holder.getBeanDefinition().getBeanClassName(), holder.getBeanName());
```

当然，顺着思路继续跟进 parse 方法，这里面还会有一些额外的处理分支，我们顺着主流程一层一层跟进，直到进入一个核心解析类 ComponentScanAnnotationParser 的函数中。

```java
public Set<BeanDefinitionHolder> parse(AnnotationAttributes componentScan, final String declaringClass) {
    Assert.state(this.environment != null, "Environment must not be null");
    Assert.state(this.resourceLoader != null, "ResourceLoader must not be null");

    ClassPathBeanDefinitionScanner scanner = new ClassPathBeanDefinitionScanner(this.registry,
            componentScan.getBoolean("useDefaultFilters"), this.environment, this.resourceLoader);

    Class<? extends BeanNameGenerator> generatorClass = componentScan.getClass("nameGenerator");
    boolean useInheritedGenerator = (BeanNameGenerator.class == generatorClass);
    scanner.setBeanNameGenerator(useInheritedGenerator ? this.beanNameGenerator :
            BeanUtils.instantiateClass(generatorClass));

    // scopedProxy 属性构造
    ScopedProxyMode scopedProxyMode = componentScan.getEnum("scopedProxy");
    if (scopedProxyMode != ScopedProxyMode.DEFAULT) {
        scanner.setScopedProxyMode(scopedProxyMode);
    }
    else {
        Class<? extends ScopeMetadataResolver> resolverClass = componentScan.getClass("scopeResolver");
        scanner.setScopeMetadataResolver(BeanUtils.instantiateClass(resolverClass));
    }

    // resourcePattern 属性构造
    scanner.setResourcePattern(componentScan.getString("resourcePattern"));

    // includeFilters 设置
    for (AnnotationAttributes filter : componentScan.getAnnotationArray("includeFilters")) {
        for (TypeFilter typeFilter : typeFiltersFor(filter)) {
            scanner.addIncludeFilter(typeFilter);
        }
    }

    // excludeFilters 属性设置
    for (AnnotationAttributes filter : componentScan.getAnnotationArray("excludeFilters")) {
        for (TypeFilter typeFilter : typeFiltersFor(filter)) {
            scanner.addExcludeFilter(typeFilter);
        }
    }

    boolean lazyInit = componentScan.getBoolean("lazyInit");
    if (lazyInit) {
        scanner.getBeanDefinitionDefaults().setLazyInit(true);
    }

    //basePackages 设置
    Set<String> basePackages = new LinkedHashSet<String>();
    String[] basePackagesArray = componentScan.getStringArray("basePackages");
    for (String pkg : basePackagesArray) {
        String[] tokenized = StringUtils.tokenizeToStringArray(this.environment.resolvePlaceholders(pkg),
                ConfigurableApplicationContext.CONFIG_LOCATION_DELIMITERS);
```

```
        basePackages.addAll(Arrays.asList(tokenized));
    }
    for (Class<?> clazz : componentScan.getClassArray("basePackageClasses")) {
        basePackages.add(ClassUtils.getPackageName(clazz));
    }
    if (basePackages.isEmpty()) {
        basePackages.add(ClassUtils.getPackageName(declaringClass));
    }

    scanner.addExcludeFilter(new AbstractTypeHierarchyTraversingFilter(false, false) {
        @Override
        protected boolean matchClassName(String className) {
            return declaringClass.equals(className);
        }
    });
    return scanner.doScan(StringUtils.toStringArray(basePackages));
}
```

而上面提到的最为核心的解析工具类 ClassPathBeanDefinitionScanner 就是 Spring 原生的解析类，这是 Spring 核心解析类，它通过字节码扫描的方式，效率要比通常我们用的反正机制效率要高很多，如果读者在日常工作中有扫描路径下类的需求，哪怕脱离了 Spring 环境也可以直接使用这个工具类。不知道读者是否还清楚，在介绍整合 Mybatis 那一章中 Mybatis 的动态扫描就是封装了类似的类，读者不妨回过头再去回顾下那一章的内容。

14.5 Conditional 机制实现

14.5.1 Conditional 使用

Spring 提供了一个更通用的基于条件的 bean 的创建——使用@Conditional 注解。@Conditional 根据满足的某一个特定条件创建一个特定的 bean。比方说，当某一个 JAR 包在一个类路径下的时候，会自动配置一个或多个 bean；或者只有某个 bean 被创建后才会创建另外一个 bean。总的来说，就是根据特定条件来控制 bean 的创建行为，这样我们可以利用这个特性进行一些自动的配置。当然，Conditional 注解有非常多的使用方式，我们仅仅通过 ConditionalOnProperty 来深入探讨它的运行机制。我们通过一个示例来详细了解。

更改示例中的内容，在配置类中增加 ConditionalOnProperty 注解：

```
@Configuration
@ComponentScan({"com.spring.study.module"})
@ConditionalOnProperty(prefix = "study", name = "enable", havingValue = "true")
public class HelloServiceAutoConfiguration   {

}
```

上面的声明想要表达的逻辑是如果配置属性中显示的声明 study.enable=true，则当前的整套体系才生效，我们可以进行验证，如图 14-9 所示。

```
严重:

***************************
APPLICATION FAILED TO START
***************************

Description:

Field helloService in com.springstudy.controller.TestController required a bean of type 'com.spring.study.module.HelloService' that could not be found.

Action:

Consider defining a bean of type 'com.spring.study.module.HelloService' in your configuration.

Process finished with exit code 1
```

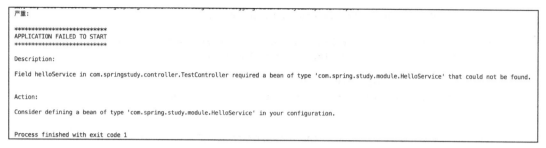

图 14-9　studyweb 启动失败

发现启动失败，而报错信息则是注入时找不到对应的 bean，这说明 Starter 中的 bean 并未生效。当我们加入 study.enabled=true 配置后，如图 14-10 所示。

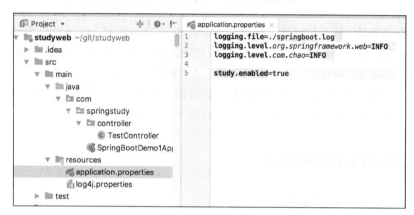

图 14-10　studyweb 加入配置

继续启动，发现启动成功，如图 14-11 所示。

```
四月 22, 2018 10:23:02 上午 org.springframework.boot.web.servlet.FilterRegistrationBean configure
信息: Mapping filter: 'requestContextFilter' to: [/*]
四月 22, 2018 10:23:03 上午 org.springframework.web.servlet.mvc.method.annotation.RequestMappingHandlerAdapter initControllerAdviceC
信息: Looking for @ControllerAdvice: org.springframework.boot.context.embedded.AnnotationConfigEmbeddedWebApplicationContext@262b2c8
四月 22, 2018 10:23:03 上午 org.springframework.web.servlet.mvc.method.annotation.RequestMappingHandlerMapping register
信息: Mapped "{[/]}" onto java.lang.String com.springstudy.controller.TestController.home()
四月 22, 2018 10:23:03 上午 org.springframework.web.servlet.mvc.method.annotation.RequestMappingHandlerMapping register
信息: Mapped "{[/error]}" onto public org.springframework.http.ResponseEntity<java.util.Map<java.lang.String, java.lang.Object>> org
四月 22, 2018 10:23:03 上午 org.springframework.web.servlet.mvc.method.annotation.RequestMappingHandlerMapping register
信息: Mapped "{[/error],produces=[text/html]}" onto public org.springframework.web.servlet.ModelAndView org.springframework.boot.aut
四月 22, 2018 10:23:03 上午 org.springframework.web.servlet.handler.SimpleUrlHandlerMapping registerHandler
信息: Mapped URL path [/webjars/**] onto handler of type [class org.springframework.web.servlet.resource.ResourceHttpRequestHandler]
四月 22, 2018 10:23:03 上午 org.springframework.web.servlet.handler.SimpleUrlHandlerMapping registerHandler
信息: Mapped URL path [/**] onto handler of type [class org.springframework.web.servlet.resource.ResourceHttpRequestHandler]
四月 22, 2018 10:23:03 上午 org.springframework.web.servlet.handler.SimpleUrlHandlerMapping registerHandler
信息: Mapped URL path [/**/favicon.ico] onto handler of type [class org.springframework.web.servlet.resource.ResourceHttpRequestHand
四月 22, 2018 10:23:03 上午 org.springframework.jmx.export.annotation.AnnotationMBeanExporter afterSingletonsInstantiated
信息: Registering beans for JMX exposure on startup
四月 22, 2018 10:23:03 上午 org.springframework.boot.context.embedded.tomcat.TomcatEmbeddedServletContainer start
信息: Tomcat started on port(s): 8080 (http)
四月 22, 2018 10:23:03 上午 com.springstudy.SpringBootDemo1Application logStarted
信息: Started SpringBootDemo1Application in 1.717 seconds (JVM running for 2.117)
```

图 14-11　studyweb 启动成功

14.5.2 Conditional 原理

好，了解了 ConditionalOnProperty 的使用后我们继续深入探索它的内部实现机制。继续按照之前的思路，如果想反推 ConditionalOnProperty 的实现机制，那么在代码中必然会存在 ConditionalOnProperty.class 的调用，于是我们搜索 ConditionalOnProperty.class，如图 14-12 所示。

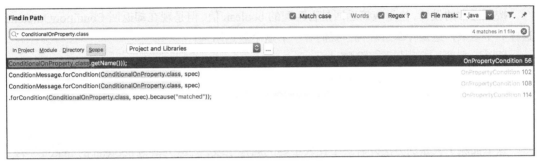

图 14-12　ConditionalOnProperty.class 搜索结果

发现所有的调用都出现在一个类 OnPropertyCondition 中，于是进入这个类，如图 14-13 所示，好在其中仅仅有一个 public 方法，这会大大减少我们的分析范围。

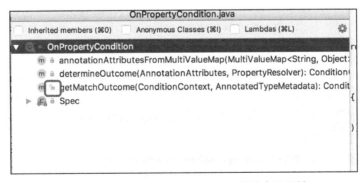

图 14-13　ConditionalOnProperty 类的方法属性

OnPropertyCondition 类的 getMatchOutcome 方法如下：

```
public ConditionOutcome getMatchOutcome(ConditionContext context,
        AnnotatedTypeMetadata metadata) {
    List<AnnotationAttributes> allAnnotationAttributes = annotationAttributesFromMultiValueMap(
            metadata.getAllAnnotationAttributes(
                    ConditionalOnProperty.class.getName()));
    List<ConditionMessage> noMatch = new ArrayList<ConditionMessage>();
    List<ConditionMessage> match = new ArrayList<ConditionMessage>();
```

```
    for (AnnotationAttributes annotationAttributes : allAnnotationAttributes) {
        ConditionOutcome outcome = determineOutcome(annotationAttributes,
            context.getEnvironment());
        (outcome.isMatch() ? match : noMatch).add(outcome.getConditionMessage());
    }
    if (!noMatch.isEmpty()) {
        return ConditionOutcome.noMatch(ConditionMessage.of(noMatch));
    }
    return ConditionOutcome.match(ConditionMessage.of(match));
}
```

按照通常的设计,这里应该返回是否匹配的 boolean 值,但是现在却返回 ConditionOutcome 这样一个对象,这是什么道理呢?我们看一下这个数据结构:

```
public class ConditionOutcome {

    private final boolean match;
    private final ConditionMessage message;

    xxx xxx …
}
```

这里面除了大量的方法外有一个比较重要的属性字段,就是类型为 boolean 的 match 字段,根据直觉,大致可以断定这个属性很重要,再来看

`ConditionOutcome.noMatch(ConditionMessage.of(noMatch))`

对应的构造逻辑:

```
public static ConditionOutcome noMatch(ConditionMessage message) {
    return new ConditionOutcome(false, message);
}
```

以及

`ConditionOutcome.match(ConditionMessage.of(match));`

对应的构造逻辑:

```
public static ConditionOutcome match(ConditionMessage message) {
    return new ConditionOutcome(true, message);
}
```

差别仅仅是这个属性的初始化值,那么根据这个信息可以断定,getMatchOutcome 方法中 noMatch 这个属性的逻辑一定是整个逻辑的核心。

我们重新再去分析 getMatchOutcome 函数中的逻辑:

```
List<AnnotationAttributes> allAnnotationAttributes =
    annotationAttributesFromMultiValueMap(
        metadata.getAllAnnotationAttributes(ConditionalOnProperty.class.getName())
    );
```

这句代码是要扫描出 ConditionalOnProperty 的注解信息,例如我们刚才配置的

`@ConditionalOnProperty(prefix = "study", name = "enabled", havingValue = "true")`

我们通过 Debug 进一步确认:

通过上面的断点信息,我们可以看到 name 对应的 enabled 属性已经被读取,如图 14-14 所示。那么,现在核心的验证逻辑就应该在 ConditionOutcome outcome = determineOutcome (annotationAttributes, context.getEnvironment())中了。顺着函数继续进行下一步探索:

14.5 Conditional 机制实现

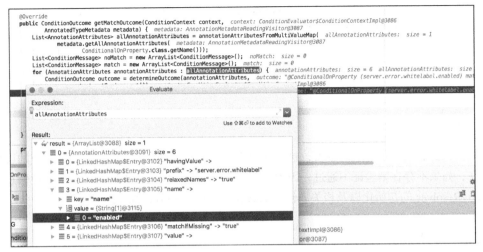

图 14-14 ConditionalOnProperty 配置获取

```
private ConditionOutcome determineOutcome(AnnotationAttributes annotationAttributes,
      PropertyResolver resolver) {
   Spec spec = new Spec(annotationAttributes);
   List<String> missingProperties = new ArrayList<String>();
   List<String> nonMatchingProperties = new ArrayList<String>();
   spec.collectProperties(resolver, missingProperties, nonMatchingProperties);
   if (!missingProperties.isEmpty()) {
      return ConditionOutcome.noMatch(
            ConditionMessage.forCondition(ConditionalOnProperty.class, spec)
                  .didNotFind("property", "properties")
                  .items(Style.QUOTE, missingProperties));
   }
   if (!nonMatchingProperties.isEmpty()) {
      return ConditionOutcome.noMatch(
            ConditionMessage.forCondition(ConditionalOnProperty.class, spec)
                  .found("different value in property",
                        "different value in properties")
                  .items(Style.QUOTE, nonMatchingProperties));
   }
   return ConditionOutcome.match(ConditionMessage
         .forCondition(ConditionalOnProperty.class, spec).because("matched"));
}
```

这个逻辑表明，不匹配有两种情况：missingProperties 对应属性缺失的情况；nonMatchingProperties 对应不匹配的情况。而这两个属性的初始化都在 spec.collectProperties(resolver, missingProperties, nonMatchingProperties)中，于是进入这个函数：

```
private void collectProperties(PropertyResolver resolver, List<String> missing,
      List<String> nonMatching) {
   if (this.relaxedNames) {
      resolver = new RelaxedPropertyResolver(resolver, this.prefix);
   }
   for (String name : this.names) {
      String key = (this.relaxedNames ? name : this.prefix + name);
      if (resolver.containsProperty(key)) {
```

```
            if (!isMatch(resolver.getProperty(key), this.havingValue)) {
                nonMatching.add(name);
            }
        }
        else {
            if (!this.matchIfMissing) {
                missing.add(name);
            }
        }
    }
}
```

终于，我们找到了对应的逻辑，这个函数尝试使用 PropertyResolver 来验证对应的属性是否存在，如果不存在则验证不通过，因为 PropertyResolver 中包含了所有的配置属性信息。而 PropertyResolver 的初始化以及相关属性的加载我们会在下一节详细介绍。

14.5.3 调用切入点

那么现在的问题是，OnPropertyCondition.getMatchOutcome 方法是谁去调用的呢？或者说这个类是如何与 Spring 整合在一起的呢？它又是怎么样影响 bean 的加载逻辑的呢？我们再从全局的角度来梳理下 Conditional 的实现逻辑。读者可以继续看图 14-8 中的 bean 的 parse 解析链路，在 3.2.1 processConfigurationClass 步骤中的主要逻辑是要对即将解析的注解做预处理，如图 14-15 所示。

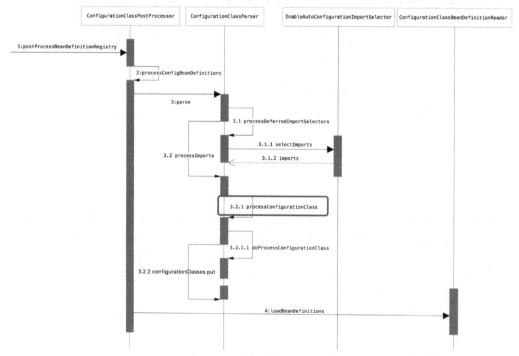

图 14-15　processConfigurationClass 位置

14.5 Conditional 机制实现

图 14-15 很清晰地展示了 Spring 整个配置类解析及加载的全过程，那么通过分析代码定位到原来整个判断逻辑的切入点就是在 processConfigurationClass 中，代码如下：

```
protected void processConfigurationClass(ConfigurationClass configClass) throws IOException {
    if (this.conditionEvaluator.shouldSkip(configClass.getMetadata(), ConfigurationPhase.PARSE_CONFIGURATION)) {
        return;
    }

    ConfigurationClass existingClass = this.configurationClasses.get(configClass);
    if (existingClass != null) {
        if (configClass.isImported()) {
            if (existingClass.isImported()) {
                existingClass.mergeImportedBy(configClass);
            }
            // Otherwise ignore new imported config class; existing non-imported class overrides it.
            return;
        }
        else {
            // Explicit bean definition found, probably replacing an import.
            // Let's remove the old one and go with the new one.
            this.configurationClasses.remove(configClass);
            for (Iterator<ConfigurationClass> it = this.knownSuperclasses.values().iterator(); it.hasNext();) {
                if (configClass.equals(it.next())) {
                    it.remove();
                }
            }
        }
    }

    // Recursively process the configuration class and its superclass hierarchy.
    SourceClass sourceClass = asSourceClass(configClass);
    do {
        sourceClass = doProcessConfigurationClass(configClass, sourceClass);
    }
    while (sourceClass != null);

    this.configurationClasses.put(configClass, configClass);
}
```

代码的第一行就是整个 Conditional 逻辑生效的切入点，如果验证不通过则会直接忽略掉后面的解析逻辑，那么这个类的属性以及 componentScan 之类的配置也自然不会得到解析了。这个方法会拉取所有的 condition 属性，onConditionProperty 就是这里拉取的：

```
public boolean shouldSkip(AnnotatedTypeMetadata metadata, ConfigurationPhase phase) {
    if (metadata == null || !metadata.isAnnotated(Conditional.class.getName())) {
        return false;
    }

    if (phase == null) {
        if (metadata instanceof AnnotationMetadata &&
                ConfigurationClassUtils.isConfigurationCandidate((AnnotationMetadata) metadata)) {
            return shouldSkip(metadata, ConfigurationPhase.PARSE_CONFIGURATION);
        }
        return shouldSkip(metadata, ConfigurationPhase.REGISTER_BEAN);
```

```
        }

        List<Condition> conditions = new ArrayList<Condition>();
        for (String[] conditionClasses : getConditionClasses(metadata)) {
            for (String conditionClass : conditionClasses) {
                Condition condition = getCondition(conditionClass, this.context.getClassLoader());
                conditions.add(condition);
            }
        }

        AnnotationAwareOrderComparator.sort(conditions);

        for (Condition condition : conditions) {
            ConfigurationPhase requiredPhase = null;
            if (condition instanceof ConfigurationCondition) {
                requiredPhase = ((ConfigurationCondition) condition).getConfigurationPhase();
            }
            if (requiredPhase == null || requiredPhase == phase) {
                if (!condition.matches(this.context, metadata)) {
                    return true;
                }
            }
        }

        return false;
    }
```

这段方法里面有几个关键的地方。

- condition 的获取。

通过代码 getConditionClasses(metadata)调用，因为代码走到这里已经是对某一个特定类的解析，metadata 中包含了完整的配置类信息，只要通过 metadata.getAllAnnotationAttributes(Conditional.class.getName(), true)即可获取，所以这一步的逻辑并不复杂。

- condition 的运行匹配。

通过代码 condition.matches(this.context, metadata)调用，因为我们的配置为@ConditionalOnProperty(prefix = "study", name = "enabled", havingValue = "true")

所以此时 condition 对应的运行态类为 OnPropertyCondition，这样就跟 14.5.2 中讲的内容结合起来了。

14.6 属性自动化配置实现

14.6.1 示例

通过 14.5.2 节的介绍，我们了解到对于下述注解，Spring Boot 会读取配置拼装成 study.enabled 并作为 key，然后尝试使用 PropertyResolver 来验证对应的属性是否存在，如果不存在则验证不通过，自然也就不会继续后面的解析流程，因为 PropertyResolver 中包含了所有的配置属性信息。

14.6 属性自动化配置实现

```
@ConditionalOnProperty(prefix = "study", name = "enabled", havingValue = "true")
```

那么，PropertyResolver 又是如何被初始化的呢？同样，这一功能并不仅仅供 Spring 内部使用，在现在的 Spring 中我们也可以通过 Value 注解直接将属性赋值给类的变量。这两个问题都涉及 Spring 的属性处理逻辑。我们在研究它的属性处理逻辑前先体验一下通过 Value 注解注入属性的样例。

```
@Component
public class HelloServiceImpl implements HelloService {

    @Value("${study.testStr}")
    private String testStr;

    public String sayHello(){
        return "hello!! "+testStr;
    }

    public String getTestStr() {
        return testStr;
    }

    public HelloServiceImpl setTestStr(String testStr) {
        this.testStr = testStr;
        return this;
    }

}
```

在 studyweb 中的 application.properties 加入 study.testStr=this is testStr，如图 14-16 所示。

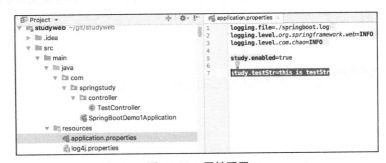

图 14-16 属性配置

运行后显示了我们配置的结果的属性，证明属性生效，结果如图 14-17 所示。

图 14-17 属性生效验证

14.6.2 原理

同样，要探索它的实现原理，按照之前的思路，我们首先定位关键字然后反推代码逻辑。我们通过搜索 Value.class 进行反推，如图 14-18 所示。

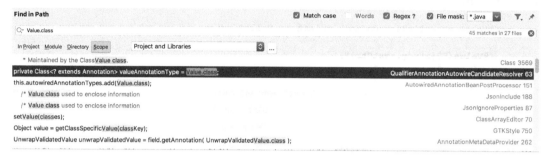

图 14-18 Value.class 搜索结果

找到了一个看起来像是调用点的地方，进入 QualifierAnnotationAutowireCandidateResolver 这个类查看代码：

```
private Class<? extends Annotation> valueAnnotationType = Value.class;
```

这是一个属性定义，那么进一步查看使用这个属性的地方：

```
protected Object findValue(Annotation[] annotationsToSearch) {
    AnnotationAttributes attr = AnnotatedElementUtils.getMergedAnnotationAttributes(
            AnnotatedElementUtils.forAnnotations(annotationsToSearch), this.valueAnnotationType);
    if (attr != null) {
        return extractValue(attr);
    }
    return null;
}
```

然后我们设置断点来看一看系统在启动的时候是否在此停留，进而验证我们的判断，如图 14-19 所示。

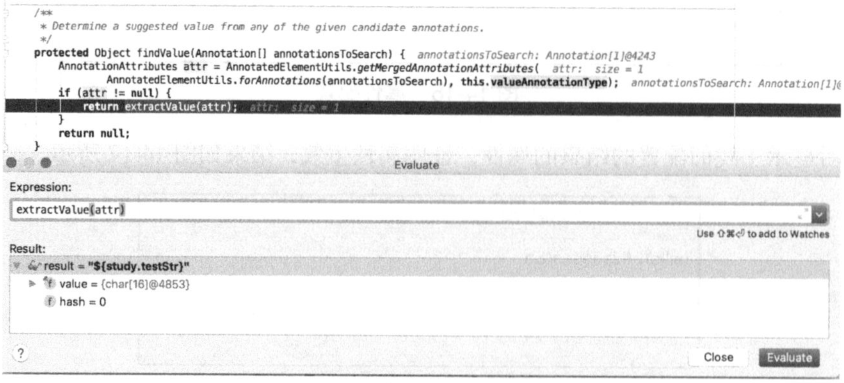

图 14-19 Value 注解处理逻辑验证

14.6 属性自动化配置实现

果然，尝试运行代码后程序在断点处停住，而尝试 evaluate 断点处的方法后能看到返回的就是我们在@Value("${study.testStr}")中配置的值。因为属性注解已经找到，所以获取注解中的属性就比较简单了：

```
protected Object extractValue(AnnotationAttributes attr) {
    Object value = attr.get(AnnotationUtils.VALUE);
    if (value == null) {
        throw new IllegalStateException("Value annotation must have a value attribute");
    }
    return value;
}
```

现在要解决两个疑问。
- 表达式对应的值是在哪里被替换的？
- 表达式替换后的值又是如何与原有的 bean 整合的？

带着这两个疑问，我们顺着调用栈继续找线索，发现当获取到 Value 的表达式属性后程序进入了 DefaultListableBeanFactory 类的 resolveEmbeddedValue 方法，并且在尝试 evaluate 后发现返回的值正是属性替换后的值，如图 14-20 所示。

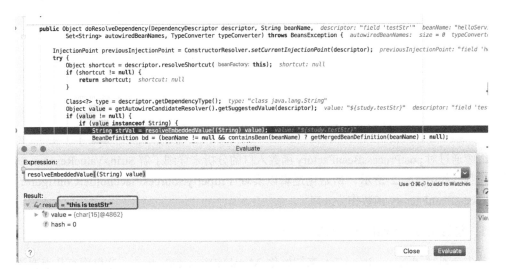

图 14-20　表达式 evaluate 结果

那么现在问题就比较清晰了，替换的逻辑一定是在 resolveEmbeddedValue 方法中：

```
@Override
public String resolveEmbeddedValue(String value) {
    if (value == null) {
        return null;
    }
    String result = value;
    for (StringValueResolver resolver : this.embeddedValueResolvers) {
        result = resolver.resolveStringValue(result);
```

```
            if (result == null) {
                return null;
            }
        }
        return result;
    }
```

通过代码逻辑我们看到，对于属性的解析已经委托给了 StringValueResolver 对应的实现类，接下来我们就要分析一下这个 StringValueResolver 是如何初始化的。

StringValueResolver 功能实现依赖 Spring 的切入点是 PropertySourcesPlaceholderConfigurer，我们看一下它的依赖结构。如图 14-21 所示，它的关键是实现了 BeanFactoryPostProcessor 接口，从而利用实现对外扩展函数 postProcessBeanFactory 来进行对 Spring 的扩展。

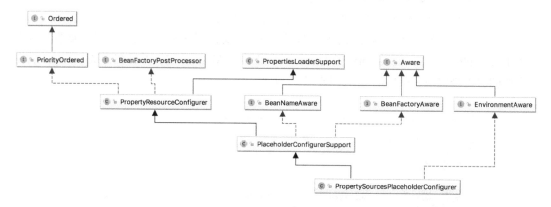

图 14-21 PropertySourcesPlaceholderConfigurer 结构

继续通过对 postProcessBeanFactory 函数入口的分析来详细了解 StringValueResolver 初始化的全过程。如图 14-22 所示，初始化逻辑以实现 PropertySourcesPlaceholderConfigurer 类的 postProcessBeanFactory 函数作为入口。

1. 初始化 MutablePropertySources

首先会通过 this.environment 来初始化 MutablePropertySources。这里面有几点要说明，environment 是 Spring 属性加载的基础，里面包含了 Spring 已经加载的各个属性，而之所以使用 MutablePropertySources 封装，是因为 MutablePropertySources 还能实现单独加载自定义的额外属性的功能。

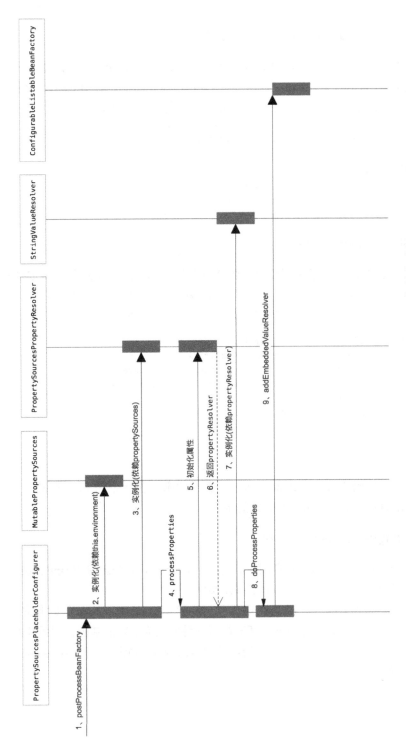

图 14-22 StringValueResolver 初始化过程

2. 初始化 PropertySourcesPropertyResolver

使用 PropertySourcesPropertyResolver 对 MutablePropertySources 的操作进行进一步封装，使得操作多个文件属性对外部不感知。当然 PropertySourcesPropertyResolver 还提供一个重要的功能就是对变量的解析，例如，它的初始化过程会包含这样的设置：

```
propertyResolver.setPlaceholderPrefix(this.placeholderPrefix);
propertyResolver.setPlaceholderSuffix(this.placeholderSuffix);
propertyResolver.setValueSeparator(this.valueSeparator);
```

而对应的变量定义如下：

```
public static final String DEFAULT_PLACEHOLDER_PREFIX = "${";
public static final String DEFAULT_PLACEHOLDER_SUFFIX = "}";
public static final String DEFAULT_VALUE_SEPARATOR = ":";
```

3. StringValueResolver 初始化

StringValueResolver 存在的目的主要是对解析逻辑的进一步封装，例如通过变量 ignoreUnresolvablePlaceholders 来控制是否对变量做解析，它的初始化代码如下：

```
StringValueResolver valueResolver = new StringValueResolver() {
    @Override
    public String resolveStringValue(String strVal) {
        String resolved = (ignoreUnresolvablePlaceholders ?
                propertyResolver.resolvePlaceholders(strVal) :
                propertyResolver.resolveRequiredPlaceholders(strVal));
        if (trimValues) {
            resolved = resolved.trim();
        }
        return (resolved.equals(nullValue) ? null : resolved);
    }
};
```

在上面的代码中 resolvePlaceholders 表示如果变量无法解析则忽略，resolveRequiredPlaceholders 表示如果变量无法解析则抛异常。

4. StringValueResolver 注册

最后将 StringValueResolver 实例注册到单例 ConfigurableListableBeanFactory 中，也就是在真正解析变量时使用的 StringValueResolver 实例。

这里面有一个关键点，就是在初始化 MutablePropertySources 的时候依赖的一个变量 environment。Environment 是 Spring 所有配置文件转换为 KV 的基础，而后续的一系列操作都是在 environment 基础上做的进一步封装，那么我们就再来探索一下 environment 的实现路径，如图 14-23 所示。

14.6 属性自动化配置实现

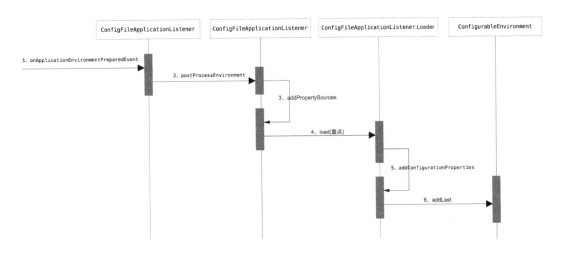

图 14-23 environment 初始化过程

由图 14-23 可知，environment 初始化过程并不是之前通用的在 PostProcessor 类型的扩展口上做扩展，而是通过 ConfigFileApplicationListener 监听机制完成。当然这里面要重点提到步骤 4 load 方法，它是整个流程的核心点：

```java
public void load() {
    this.propertiesLoader = new PropertySourcesLoader();
    this.activatedProfiles = false;
    this.profiles = Collections.asLifoQueue(new LinkedList<Profile>());
    this.processedProfiles = new LinkedList<Profile>();

    // 通过profile标记不同的环境，可以通过设置spring.profiles.active和spring.profiles.default。
    //如果设置了active,default便失去了作用。如果两个都没有设//置，那么带有profiles的bean都不会生成。
    Set<Profile> initialActiveProfiles = initializeActiveProfiles();
    this.profiles.addAll(getUnprocessedActiveProfiles(initialActiveProfiles));
    if (this.profiles.isEmpty()) {
        for (String defaultProfileName : this.environment.getDefaultProfiles()) {
            Profile defaultProfile = new Profile(defaultProfileName, true);
            if (!this.profiles.contains(defaultProfile)) {
                this.profiles.add(defaultProfile);
            }
        }
    }

    //支持不添加任何profile注解的bean的加载
    this.profiles.add(null);

    while (!this.profiles.isEmpty()) {
        Profile profile = this.profiles.poll();

//Spring Boot 默认从4个位置查找application.properties文件就是从**getSearchLocations()方法返回**：
//1. 当前目录下的/config目录
```

```
//2. 当前目录
//3. 类路径下的/config目录
//4. 类路径根目录
        for (String location : getSearchLocations()) {
            if (!location.endsWith("/")) {
                // location is a filename already, so don't search for more
                // filenames
                load(location, null, profile);
            }
            else {
// 如果没有配置则默认从application.properties中加载，约定大于配置
                for (String name : getSearchNames()) {
                    load(location, name, profile);
                }
            }
        }
        this.processedProfiles.add(profile);
    }

    addConfigurationProperties(this.propertiesLoader.getPropertySources());
}
```

这里面涉及我们经常使用的 profile 机制的实现，profile 机制是 Spring 提供的一个用来标明当前运行环境的注解。我们在正常开发的过程中经常遇到这样的问题，开发环境是一套环境，QA 测试是一套环境，线上部署又是一套环境。从开发到测试再到部署，会对程序中的配置修改多次，尤其是从 QA 到上线这个环节，经过 QA 测试的也不敢保证改了哪个配置之后能不能在线上运行。

为了解决上面的问题，我们一般会使用一种方法——配置文件，然后通过不同的环境读取不同的配置文件，从而在不同的场景中运行我们的程序。

Spring 中的 profile 机制的作用就体现在这里。在 Spring 使用 DI 来依赖注入的时候，能够根据当前制定的运行环境来注入相应的 bean。最常见的就是使用不同的环境对应不同的数据源。

这个机制的实现就是在 load(location, name, profile)这段代码中控制，这里只会加载当前设置 profile 对应的配置文件。

14.7 Tomcat 启动

截止到目前，我们已经完成了对 Spring Boot 基本功能的分析，包括 Spring Boot 的启动、属性自动化配置、conditional 实现以及 starter 运行模式原理。那么，在之前的理论基础上再来分析 Spring Boot 是如何集成 Tomcat 会更为简单。

分析 Tomcat 嵌入原理首先要找到扩展入口，我们可以从启动信息开始，如图 14-24 所示。

14.7 Tomcat 启动

```
onto public org.springframework.http.ResponseEntity<java.util.Map<java.lang.String, java.lang.
 org.springframework.web.servlet.mvc.method.annotation.RequestMappingHandlerMapping register
oduces=[text/html]}" onto public org.springframework.web.servlet.ModelAndView org.springframew
 org.springframework.web.servlet.handler.SimpleUrlHandlerMapping registerHandler
jars/**] onto handler of type [class org.springframework.web.servlet.resource.ResourceHttpReq
 org.springframework.web.servlet.handler.SimpleUrlHandlerMapping registerHandler
 onto handler of type [class org.springframework.web.servlet.resource.ResourceHttpRequestHand
 org.springframework.web.servlet.handler.SimpleUrlHandlerMapping registerHandler
/favicon.ico] onto handler of type [class org.springframework.web.servlet.resource.ResourceHtt
 org.springframework.jmx.export.annotation.AnnotationMBeanExporter afterSingletonsInstantiate
 JMX exposure on startup
 org.springframework.boot.context.embedded.tomcat.TomcatEmbeddedServletContainer start
rt(s): 8080 (http)
```

图 14-24 Tomcat 启动信息

当然，为了整个说明的连贯性我们还是从入口处讲起。在讲解 14.3.1 springContext 创建的时候我们曾经提到过一段代码：

```java
protected ConfigurableApplicationContext createApplicationContext() {
    Class<?> contextClass = this.applicationContextClass;
    if (contextClass == null) {
        try {
            contextClass = Class.forName(this.webEnvironment
                ? DEFAULT_WEB_CONTEXT_CLASS : DEFAULT_CONTEXT_CLASS);
        }
        catch (ClassNotFoundException ex) {
            throw new IllegalStateException(
                "Unable create a default ApplicationContext, "
                    + "please specify an ApplicationContextClass",
                ex);
        }
    }
    return (ConfigurableApplicationContext) BeanUtils.instantiate(contextClass);
}
```

其中，下述代码就是默认配置：

```java
public static final String DEFAULT_WEB_CONTEXT_CLASS = "org.springframework."
    + "boot.context.embedded.AnnotationConfigEmbeddedWebApplicationContext";
```

这也是 Web 扩展的关键。在第 5 章我们曾经花了很大的篇幅讲解了 AbstractApplicationContext 的一个函数 refresh()，它是 springcontext 扩展的关键，再次来回顾一下：

```java
public void refresh() throws BeansException, IllegalStateException {
        synchronized (this.startupShutdownMonitor) {
            //准备刷新的上下文环境
            prepareRefresh();

            // Tell the subclass to refresh the internal bean factory.
            //初始化 BeanFactory，并进行 XML 文件读取
            ConfigurableListableBeanFactory beanFactory = obtainFreshBeanFactory();

            // Prepare the bean factory for use in this context.
            //对 BeanFactory 进行各种功能填充
            prepareBeanFactory(beanFactory);

            try {
```

```java
// Allows post-processing of the bean factory in context subclasses.
// 子类覆盖方法做额外的处理
postProcessBeanFactory(beanFactory);

//激活各种 BeanFactory 处理器
invokeBeanFactoryPostProcessors(beanFactory);

// 注册拦截 Bean 创建的 Bean 处理器,这里只是注册,真正的调用是在 getBean 时候
registerBeanPostProcessors(beanFactory);

// 为上下文初始化 Message 源,即不同语言的消息体 ,国际化处理
initMessageSource();

// Initialize event multicaster for this context.
// 初始化应用消息广播器,并放入 "applicationEventMulticaster" bean 中
initApplicationEventMulticaster();

// Initialize other special beans in specific context subclasses.
// 留给子类来初始化其他的 Bean
onRefresh();

// Check for listener beans and register them.
// 在所有注册的 bean 中查找 Listener bean,注册到消息广播器中
registerListeners();

// Instantiate all remaining (non-lazy-init) singletons.
// 初始化剩下的单实例(非惰性的)
finishBeanFactoryInitialization(beanFactory);

// Last step: publish corresponding event.
// 完成刷新过程,通知生命周期处理器 lifecycleProcessor 刷新过程,同时发出
//ContextRefreshEvent 通知别人
finishRefresh();
        }

        catch (BeansException ex) {
            // Destroy already created singletons to avoid dangling resources.
            destroyBeans();

            // Reset 'active' flag.
            cancelRefresh(ex);

            // Propagate exception to caller.
            throw ex;
        }
    }
}
```

而刚才说的 AnnotationConfigEmbeddedWebApplicationContext 正是扩展了这个类,我们看一下 AnnotationConfigEmbeddedWebApplicationContext 类的层次结构,如图 14-25 所示。

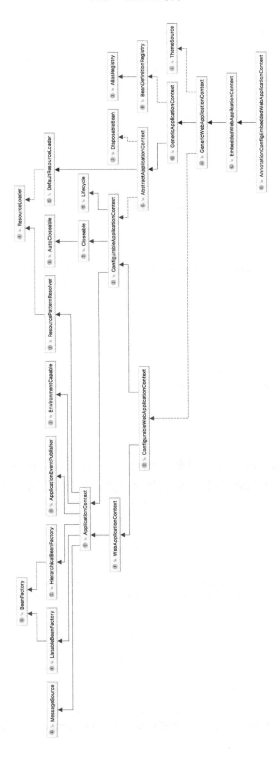

图 14-25 AnnotationConfigEmbeddedWebApplicationContext 类的层次结构

而 EmbeddedWebApplicationContext 类对于 Tomcat 嵌入的一个关键点就是 onRefresh()函数的重写：

```
@Override
protected void onRefresh() {
    super.onRefresh();
    try {
        createEmbeddedServletContainer();
    }
    catch (Throwable ex) {
        throw new ApplicationContextException("Unable to start embedded container",
            ex);
    }
}
```

createEmbeddedServletContainer()函数对应代码：

```
private void createEmbeddedServletContainer() {
    EmbeddedServletContainer localContainer = this.embeddedServletContainer;
    ServletContext localServletContext = getServletContext();
    if (localContainer == null && localServletContext == null) {
        EmbeddedServletContainerFactory containerFactory = getEmbeddedServletContainerFactory();
        this.embeddedServletContainer = containerFactory
            .getEmbeddedServletContainer(getSelfInitializer());
    }
    else if (localServletContext != null) {
        try {
            getSelfInitializer().onStartup(localServletContext);
        }
        catch (ServletException ex) {
            throw new ApplicationContextException("Cannot initialize servlet context",
                ex);
        }
    }
    initPropertySources();
}
```

EmbeddedServletContainerFactory 是服务器启动的上层抽象，无论是 Tomcat 还是 Jetty 都要通过这个类实现对 Spring 服务器的注册。现在我们通过断点来看看它的返回结果，如图 14-26 所示。

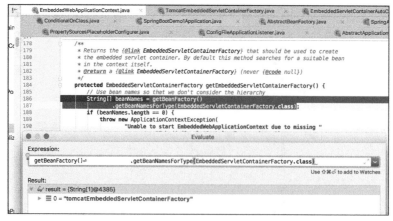

图 14-26　getEmbeddedServletContainerFactory()返回结果

正如我们所料，它返回的就是 Tomcat 对应的子类实现。于是我们找到 TomcatEmbeddedServletContainerFactory 来查看它的实现逻辑，但是却发现这个类既没有打一些 Spring 注册的注解也没有配置在任何配置文件中，那么它是如何注入到 Spring 容器中的呢？

带着这个疑问，我们搜索代码，看一看是否会有其他地方对这个类进行了硬编码的注册，如图 14-27 所示。

图 14-27　TomcatEmbeddedServletContainerFactory 搜索结果

果然，发现 EmbeddedServletContainerAutoConfiguration 这个类进行了调用，这是 Spring 自动化整合各种服务器注册的非常关键的入口类：

```
@AutoConfigureOrder(Ordered.HIGHEST_PRECEDENCE)
@Configuration
@ConditionalOnWebApplication
@Import(BeanPostProcessorsRegistrar.class)
public class EmbeddedServletContainerAutoConfiguration {

    /**
     * Nested configuration if Tomcat is being used.
     */
    @Configuration
    @ConditionalOnClass({ Servlet.class, Tomcat.class })
    @ConditionalOnMissingBean(value = EmbeddedServletContainerFactory.class, search = SearchStrategy.CURRENT)
    public static class EmbeddedTomcat {

        @Bean
        public TomcatEmbeddedServletContainerFactory tomcatEmbeddedServletContainerFactory() {
            return new TomcatEmbeddedServletContainerFactory();
        }

    }
```

```java
    /**
     * Nested configuration if Jetty is being used.
     */
    @Configuration
    @ConditionalOnClass({ Servlet.class, Server.class, Loader.class,
            WebAppContext.class })
    @ConditionalOnMissingBean(value = EmbeddedServletContainerFactory.class, search =
SearchStrategy.CURRENT)
    public static class EmbeddedJetty {

        @Bean
        public JettyEmbeddedServletContainerFactory jettyEmbeddedServletContainerFactory() {
            return new JettyEmbeddedServletContainerFactory();
        }

    }

    /**
     * Nested configuration if Undertow is being used.
     */
    @Configuration
    @ConditionalOnClass({ Servlet.class, Undertow.class, SslClientAuthMode.class })
    @ConditionalOnMissingBean(value = EmbeddedServletContainerFactory.class, search =
SearchStrategy.CURRENT)
    public static class EmbeddedUndertow {

        @Bean
        public UndertowEmbeddedServletContainerFactory undertowEmbeddedServletContainer
Factory() {
            return new UndertowEmbeddedServletContainerFactory();
        }

    }
    xxx xxx 忽略 xxx xxx
    }

}
```

这个类中包含了 Tomcat、Jetty、Undertow 3 种类型的服务器自动注册逻辑,而选择条件则是通过@ConditionalOnClass 注解控制。我们之前讲解过 ConditionalOnProperty 注解的实现逻辑,而@ConditionalOnClass 实现逻辑与之类似,对应的类在 classpath 目录下存在时,才会去解析对应的配置文件。这也就解释了之所以 Spring 默认会启动 Tomcat 正是由于在启动的类目录下存在 Servlet.class、Tomcat.class,而这个依赖是由 Spring 自己在 spring-boot-starter-web 中默认引入,如图 14-28 所示。

按照代码逻辑,如果我们默认的服务器不希望使用 Tomcat 而是希望使用 Jetty,那么我们只需要将 Tomcat 对应的 jar 从 spring-boot-starter-web 中排除掉,然后加入 Jetty 依赖即可。

14.7 Tomcat 启动

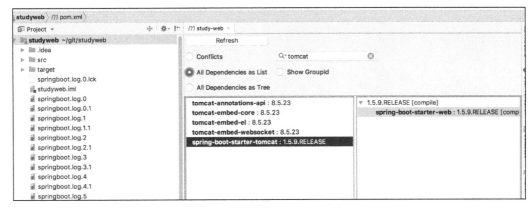

图 14-28 Tomcat 反向依赖

TomcatEmbeddedServletContainerFactory 类的 getEmbeddedServletContainer()实现类如下，Tomcat 会在 getTomcatEmbeddedServletContainer(tomcat)代码中异步启动。

```
@Override
public EmbeddedServletContainer getEmbeddedServletContainer(
        ServletContextInitializer... initializers) {
    Tomcat tomcat = new Tomcat();
    File baseDir = (this.baseDirectory != null ? this.baseDirectory
            : createTempDir("tomcat"));
    tomcat.setBaseDir(baseDir.getAbsolutePath());
    Connector connector = new Connector(this.protocol);
    tomcat.getService().addConnector(connector);
    customizeConnector(connector);
    tomcat.setConnector(connector);
    tomcat.getHost().setAutoDeploy(false);
    configureEngine(tomcat.getEngine());
    for (Connector additionalConnector : this.additionalTomcatConnectors) {
        tomcat.getService().addConnector(additionalConnector);
    }
    prepareContext(tomcat.getHost(), initializers);
//异步启动 Tomcat
    return getTomcatEmbeddedServletContainer(tomcat);
}
```